Continental Rifted Margins 2

SCIENCES

Geoscience, Field Director – Yves Lagabrielle

Dynamics of the Continental Lithosphere,
Subject Head – Sylvie Leroy

Continental Rifted Margins 2

Case Examples

Coordinated by
Gwenn Peron-Pinvidic

iSTE

WILEY

First published 2022 in Great Britain and the United States by ISTE Ltd and John Wiley & Sons, Inc.

ISTE Ltd
27-37 St George's Road
London SW19 4EU
UK

www.iste.co.uk

John Wiley & Sons, Inc.
111 River Street
Hoboken, NJ 07030
USA

www.wiley.com

Library of Congress Control Number: 2021950811

British Library Cataloguing-in-Publication Data
A CIP record for this book is available from the British Library
ISBN 978-1-78945-062-0

ERC code:
PE10 Earth System Science
PE10_5 Geology, tectonics, volcanology
PE10_7 Physics of earth's interior, seismology, volcanology
PE10_11 Geochemistry, crystal chemistry, isotope geochemistry, thermodynamics
PE10_12 Sedimentology, soil science, palaeontology, earth evolution

Contents

Introduction ix

Gwenn PERON-PINVIDIC

Chapter 1. The West Iberian Margin: Past and Current Research Concepts and Future Challenges 1

Gaël LYMER and Tim J. RESTON

1.1. Introduction: the West Iberian Margin, the "classic" magma-poor margin? . . 2
1.2. Structures of the West Iberian Margin 6
1.2.1. The Deep Galicia Margin (DGM) 9
1.2.2. The Southern Iberia Abyssal Plain (SIAP) 12
1.2.3. Conjugated West Iberian and Newfoundland Margins 13
1.3. Synrift tectono-stratigraphy and age and evolution of extension across the West Iberian Margin 13
1.4. Current models of development of the West Iberian Margin 18
1.5. Remaining questions at the West Iberian Margin and other magma-poor margins 21
1.6. Conclusions and outlook 25
1.7. References 26

Chapter 2. Tectonic Evolution of the Mid-Norwegian Rifted Margin 35

Per Terje OSMUNDSEN and Gwenn PERON-PINVIDIC

2.1. Introduction 36
2.2. Large-scale structure 39
2.2.1. The proximal margin and the proximal breakaway complex 40
2.2.2. The necking domain and the necking breakaway complexes 45
2.2.3. The distal domain and the distal breakaway complex 48
2.2.4. The outer domain 51
2.3. Syn-rift basin configuration and evolution 51

2.4. Breakup 57
2.5. Post-rift evolution 58
2.6. Future challenges 59
2.7. Acknowledgments 60
2.8. References 60

Chapter 3. South and Equatorial Atlantic Margins 69

Daniel ASLANIAN and Maryline MOULIN

3.1. Introduction 69
3.2. Geodynamics and understanding passive margins genesis 76
3.3. Crustal structures in the South Atlantic and equatorial margins 79
3.4. Conclusion 84
3.5. References 87

Chapter 4. The Continental Margins of the Gulf of Aden 93

Sylvie LEROY and Chloé NONN

4.1. Introduction 93
4.2. Geological and geodynamic context and heritage 94
4.3. Continental rifting timing, initiation and depositional environments 96
4.4. Structure of the conjugate margins 98
4.5. Ocean-continent transition (OCT) and the birth of an oceanic ridge 102
4.6. Opening of the Gulf of Aden 103
4.7. References 104

Chapter 5. The South China Sea – Rifting and the Evolution of a Marginal Basin 107

Manuel PUBELLIER, Sung-Ping CHANG and Matthias DELESCLUSE

5.1. Introduction 108
5.2. Geodynamic setting and particularities 109
5.3. Propagators and implications on the rifting and the timing of breakup 113
5.3.1. Extreme thinning and detachment faulting away from narrow ocean–continent transition zones: far field and deep processes versus structural inheritance 115
5.4. A wide rift with boudinage crust resulting from contrasted crustal rheologies 116
5.4.1. Hyperextended continental crust 116
5.4.2. Reactivation of the Mesozoic Granite Belt 118
5.5. Impact of the subduction and collision of the neighboring basin 118
5.6. Post rifting evolution; uplift of margins and delayed subsidence of the basin 121

5.7. Conclusion 122
5.8. Acknowledgments 123
5.9. References 124

Chapter 6. Arctic Continental Margins 133

Carmen GAINA

6.1. Introduction 133
6.2. Passive continental margins in the Arctic 134
6.2.1. The Canada Basin continental margins 134
6.2.2. The Eurasia Basin continental margins 138
6.3. Discussion and concluding remarks 141
6.4. References 145

Chapter 7. Alpine Tethys Margins 149

Gianreto MANATSCHAL and Pauline CHENIN

7.1. Background and historical overview 149
7.1.1. Preamble 149
7.1.2. Geological and paleogeographical settings 150
7.1.3. Historical overview 154
7.2. From dispersed outcrops to the 2D architecture of the Alpine Tethys rift system 157
7.2.1. Datasets and building blocks (BB) 157
7.2.2. Building a spatial and temporal template for the Alpine Tethys rifting . . 160
7.2.3. What is missing? 164
7.3. Evolution of the Alpine Tethys rift system 165
7.4. Continuous deformation vs. succession of distinct deformation phases. 168
7.5. Three-dimensional aspect of the Alpine Tethys rift system 169
7.6. The Alpine Tethys remnants in the Alps: a laboratory to investigate and calibrate rift systems 171
7.6.1. The role of rift-inheritance in building collisional orogens 171
7.6.2. Calibration of rifting processes 172
7.6.3. The benefit of data-based scientific debate 172
7.7. References 173

Chapter 8. A Fossil Magma-rich Rifted Margin in the Scandinavian Caledonides 185

Johannes JAKOB, Hans Jørgen KJØLL and Torgeir Bjørge ANDERSEN

8.1. Introduction 186
8.2. Opening of the Iapetus Ocean and formation of the Central Iapetus Magmatic Province 187

8.2.1. The Seiland Igneous Province 190
8.2.2. The Scandinavian Dyke Complex 191
8.3. Discussion 193
8.4. Conclusion 194
8.5. Acknowledgments 195
8.6. References 195

Chapter 9. Cretaceous Rifting at the Pyrenean Iberia–Eurasia Border: Crustal Thinning under a Pre-rift Cover – Application to Passive Margins 203

Yves LAGABRIELLE

9.1. The North Pyrenean area within the Pyrenees. 203
9.2. Mechanisms of crustal thinning deduced from the study of the NPZ 206
9.3. Crustal thinning at the Iberian–Eurasian plate boundary: smooth-slope basins and the legacy of Triassic paleogeography 209
9.3.1. Cretaceous basins along the diffuse Iberian–Eurasian border. 209
9.3.2. General characteristics of "smooth-slope" basins – consideration of continental crust thinning modes 211
9.3.3. The opposition between Pyrenean- and Alpine-type rifts: role of Triassic palaeogeography on the modes of crustal thinning 215
9.4. Conclusion: from the Pyrenean basins to the formation mechanisms of passive continental margins. 219
9.5. Acknowledgments 222
9.6. References 222

List of Authors 235

Index. 237

Summary of Volume 1 239

Introduction

Gwenn PERON-PINVIDIC[1,2]
[1] *Geological Survey of Norway (NGU), Trondheim, Norway*
[2] *Department of Geoscience and Petroleum, Norwegian University of Science and Technology, Trondheim, Norway*

Volume 2 of this book presents concise reviews of several examples that represent today's most discussed sites for rifted margin investigations. These cases span a wide range of rifted margin categories (magma-poor vs. magma-rich, sediment-starved vs. sediment-rich, wide vs. narrow, divergent vs. oblique), with both onshore fossil margins (the Pyrenees, Alps and Scandinavian Caledonides) and offshore modern cases (the Arctic, Iberia, Mid-Norway, the South Atlantic, Gulf of Aden and the South China Sea). The summaries provided here are only short descriptions; if the reader is interested in learning more, we direct you to the lists of key recommended scientific contributions, presented in the "Further reading" sections within the chapters.

Figure I.1 shows where the cases discussed in Volume 2 are located, with their chapter number for reference:

1) Iberia;

2) Mid-Norway;

3) South and Equatorial Atlantic;

For a color version of all the figures in this chapter, see www.iste.co.uk/peronpinvidic/margins2.zip.

Continental Rifted Margins 2,
coordinated by Gwenn PERON-PINVIDIC.

4) Gulf of Aden;

5) South China Sea;

6) Arctic;

7) Alpine Tethys;

8) Scandinavian Caledonides;

9) Pyrenees.

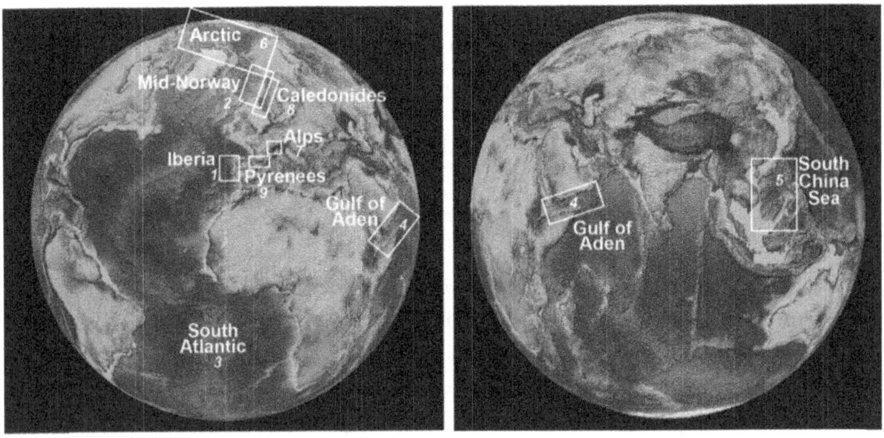

Figure I.1. *Location of the case examples described in the nine chapters of Volume 2*

1

The West Iberian Margin: Past and Current Research Concepts and Future Challenges

Gaël LYMER[1] and Tim J. RESTON[2]

[1] *Fault Analysis Group/iCRAG (Irish Centre for Research in Applied Geosciences), School of Earth Sciences, University College Dublin, Ireland*

[2] *School of Geography, Earth and Environmental Science, University of Birmingham, UK*

The West Iberian Margin has historically been at the forefront of fundamental research on rifting and breakup. Decades of drilling, sampling and geophysical campaigns, including a recent high-resolution 3D seismic volume, have placed the West Iberian Margin among the best-documented continental margins worldwide. Data from the West Iberian Margin allowed initial development of the concepts of hyper-extension, detachment faulting, "undercrusting" by serpentinites and exhumed mantle rocks. Above all, they have revolutionized our modern scientific concepts of rifting by allowing us to demonstrate that the rifting process is diachronous across margins and involves the migration and/or the focusing of extension towards the future oceanic spreading center during rift evolution. These concepts resulted in the early theories about the processes of oceanic basin development and still contribute to defining the latest models of continental breakup. Three major models of continental breakup have been defined from observations at the West Iberian Margin:

For a color version of all the figures in this chapter, see www.iste.co.uk/peronpinvidic/margins2.zip.

Continental Rifted Margins 2,
coordinated by Gwenn PERON-PINVIDIC.

depth-dependent stretching (DDS), cross-cutting polyphase faulting and sequential faulting. The fundamental disparities between these models – in terms of the timing of faulting, the number of faulting phases and rheologies – demonstrate that our knowledge of rifting and breakup remains fundamentally incomplete, as long as the timing of geological events at rifted margins remains undefined.

In this chapter, we summarize the historical investigation of the West Iberian Margin and the current geological knowledge on the features formed during their rift evolution and breakup. We present an overview of their structures and litho-stratigraphy, and the latest ideas for their evolution, highlighting the key remaining questions, how they might be addressed and why answering these questions would represent a paradigm shift in our knowledge of the concepts of development of rifted margins worldwide.

1.1. Introduction: the West Iberian Margin, the "classic" magma-poor margin?

The West Iberian Margin (WIM) is the rifted margin marking the western edge of the Iberian Peninsula (Figure 1.1). It includes three main segments: the Galicia Margin (GM) in the north, the South Iberia Abyssal Plain (SIAP) and the Tagus Abyssal Plain (TAP) in the south. From land to sea, the GM comprises a narrow continental shelf, bounded to the west by the Galicia Interior Basin (GIB), the Galicia Bank (BG) and the Deep Galicia Margin (DGM), where the smooth seafloor is locally marked by local ridges (Figure 1.1b and 1.1c). To the south, the continental shelf includes intra-continental basins (Porto and Lusitania basins – PB and LB), bounded to the west by the SIAP showing a smooth and relatively flat seafloor. Finally, the southern segment of the WIM is marked by the Estramadura Spur (ES) and the Tore Seamounts (TS) that form the northern boundary of the TAP. The segmentation of the margin has been related to the south to north propagation of the North Atlantic rifting and crustal breakup (e.g. Brune *et al.* 2014; Srivastava *et al.* 1990; Malod and Mauffret 1990; Tucholke *et al.* 2007; Brune *et al.* 2014), favored by orthogonal fractures (AF, NFZ, TF on Figure 1.1), either considered to be inherited from Late Hercynian fabric (e.g. Boillot and Malod 1988; Manatschal *et al.* 2015) or to have developed during a Late Triassic–Early Jurassic phase of rifting (e.g. Vegas *et al.* 2016).

While the thinned continental crust is still relatively thick at the GIB/GB (≥10 km, although locally ~8 km at the center of the GIB, e.g. Reston 2005; Peron-Pinvidic *et al.* 2013; Druet *et al.* 2018) and at the PB/LB (Figure 1.2), the continental crust at the DGM, SIAP and TAP has been hyper-thinned to less than

5–10 km during rifting (e.g. Pérez-Gussinyé and Reston 2001; Lymer *et al.* 2019). Where hyper-thinned, the crust exhibits arrays of tilted fault blocks below a thin sedimentary cover (Figures 1.3 and 1.4), but the observed thinning greatly outstrips the amount of extension inferred from fault geometries (Ziegler 1983; Sibuet 1992; Davis and Kusznir 2004, pp. 92–136; Reston *et al.* 2007; Reston 2009; Figure 1.5). This problem, known as extensional discrepancy (Reston *et al.* 2007; Reston 2009), has been explained by distinct models involving different fault geometries and timing of fault activity to describe the rift evolution of the margin (see section 1.4). Beneath the hyper-thinned crust (Figures 1.3 and 1.4), the mantle with reduced seismic velocity (Figure 1.2) has been interpreted, but not yet proven, as being partially serpentinized, "undercrusting" the hyper-extended domain (Boillot *et al.* 1989; Whitmarsh *et al.* 2001; Bayrakci *et al.* 2016; Davy *et al.* 2016). This subcrustal layer is thought to form when the entire crust becomes brittle as a result of the ingress of seawater from above, through the thinned continental crust (Pérez-Gussinyé and Reston 2001; Bayrakci *et al.* 2016; Prada *et al.* 2017), and seems to continue beyond the distal edge of the crust (Figure 1.1, green drill sites, and Figure 1.2) as an expanse of partially serpentinized mantle, locally exhumed to the seafloor during the final stages of rifting (e.g. Boillot *et al.* 1987; Krawczyk *et al.* 1996; see Peridotite Ridges in Figure 1.1). The boundary between the hyper-extended crust and the underlying serpentinized mantle corresponds at the DGM to a set of bright reflections forming a major detachment surface known as the S reflector (Figure 1.3; e.g., Krawczyk *et al.* 1996; Reston *et al.* 2007; Schuba *et al.* 2018; Lymer *et al.* 2019). West of the exhumed mantle domain, the location of the transition to the oceanic domain remains debated (e.g. Sibuet *et al.* 2007; Welford *et al.* 2010; Peron-Pinvidic *et al.* 2013). Oceanic crust is commonly identified based on the presence of seafloor-spreading magnetic anomalies (e.g. Eagles *et al.* 2015), with the use of the oldest isochrons to define the approximate landward limit of the oceanic domain, but magnetic anomalies have also been observed within exhumed mantle or hyper-extended continental crust (Whitmarsh and Miles 1995; Funck *et al.* 2003). Two magnetic anomalies of debated nature are observed at the deep WIM (Figure 1.1; Tucholke *et al.* 2007): the M3 magnetic anomaly, within the exhumed mantle domain, and the M0 magnetic anomaly, generally interpreted as corresponding to the first identified oceanic crust (Srivastava *et al.* 1990). After its Early Cretaceous breakup, the WIM underwent a period of relative tectonic quiescence in the Late Cretaceous, until the progressive development of a compressive field in the northern margin, related to the Alpine orogeny during the Cenozoic (Thinon *et al.* 2001; Tugend *et al.* 2014). Compression led to the reactivation and inversion of structures formed during the rifting, as well as the development of new compressional structures, including thrusts, reverse faults and folds (e.g. Murillas *et al.* 1990; Druet *et al.* 2018).

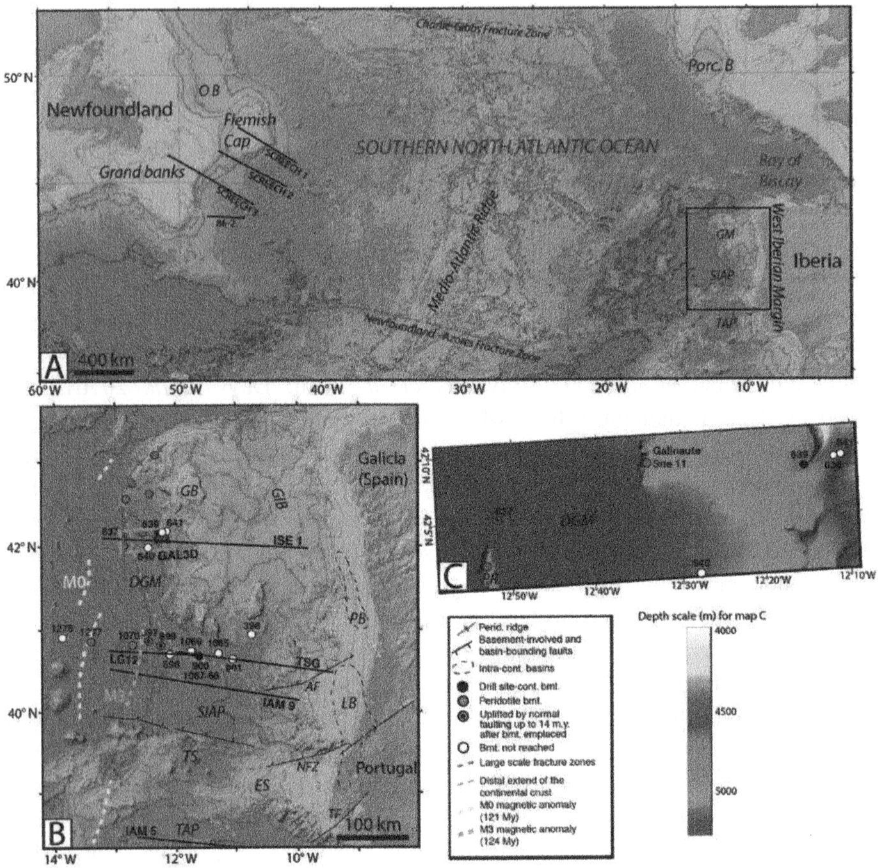

Figure 1.1. *Location maps of the Western Iberian Margin. a) Bathymetric and topographic map of the Southern North Atlantic Ocean and surrounding continental margins. Black rectangle indicates the location of the map of the Western Iberian Margin shown in b); and b) Bathymetric and topographic map of the Western Iberian Margin*

CONTINUATION OF CAPTION FOR FIGURE 1.1.– *Black and orange dashed lines are from Tucholke et al. (2007) and respectively show basins within continental crust containing uppermost Triassic to Lower Jurassic evaporite deposits and the oceanward extent of the continental crust. Peridotite ridges are marked by red lines (Tucholke et al. 2007; Druet et al. 2018). Green and yellow dashed lines respectively show the locations of magnetic anomalies M3 (124 Myr) and M0 (121 Myr) from Miles et al. (2012). DSDP and ODP drill sites (Boillot et al. 1988; Whitmarsh et al. 1998, Tucholke et al. 2007) are shown and coded according to whether they reached apparent continental basement (black circle), peridotite basement (green circle), or no basement (white circle); Sites with good evidence that peridotite basement was faulted and uplifted 3–14 Myr after it was emplaced have black dots at circle centers. Pink circles show the locations of dredge sampling (Boillot et al. 1988). Blue rectangle indicates the location the 68.5 km x 20 km Galicia 3D volume and of the map of the seafloor shown in c); c) Bathymetric map of the DGM generated within the Galicia 3D volume. The color bar shows the depth scale in meters. Black lines indicate the locations of the sections shown in Figure 1.2 across the Western Iberian and Newfoundland margins (Datasources: SCREECH1, Funck et al. 2003; Hopper et al. 2004; SCREECH2, Van Avendonk et al. 2006; Shillington et al. 2006; IAM9, Dean et al. 2000; Pickup et al. 1996). AF: Aveiro Fault; DGM: Deep Galicia Margin; ES: Estramadura Spur; GB: Galicia Bank; GM: Galicia Margin; GIB: Galicia Interior Basin; LB: Lusitanian Basin; NF: Nazare Fault; OB: Orphan Basin; PR: Peridotite Ridge; Porc. B: Porcupine Basin; PB: Porto Basin; SIAP: South Iberian Abyssal Plain; TAP: Tagus Abyssal Plain; TF: Tagus Fault; TS: Tore Seamounts. Bathymetric data for a) and b) are shown in meters and are from https://www.ngdc.noaa.gov/.*

Crustal hyper-thinning, extensional discrepancy, crustal embrittlement, development of detachment faulting, mantle serpentinization and exhumation are all defining characteristics of individual magma-poor margins observed at other margins worldwide (e.g. Reston 2009; Turner and Wilson 2009; Autin *et al.* 2010; Zalán *et al.* 2011; Ball *et al.* 2013; Gillard *et al.* 2015; Osmundsen *et al.* 2016) that were first observed and described at either the WIM or at the neighboring Biscay margin. de Charpal *et al.* (1978) recognized tilted fault blocks above the S detachment, and similar detachment faults at the crust-mantle boundary (CMB) were subsequently identified at the SIAP (see section 1.2.2, "H" detachment, Figure 1.4; Hoffmann and Reston 1992; Krawczyk *et al.* 1996). Exhumed mantle was first recovered, by dredging then by drilling, at the DGM (Boillot *et al.* 1980; Boillot and Winterer 1988), and first shown to be an expanse ~100 km wide at the SIAP (Pickup *et al.* 1996). Le Pichon and Barbier (1987) and subsequently Sibuet (1992) pointed out extensional discrepancy in the Biscay and Galicia margins (Figure 1.5). Serpentine undercrust was proposed (Boillot *et al.* 1989) and modeled for the first time within the Galicia Margin (Pérez-Gussinyé and Reston 2001). It was also there where seismic velocities were first of sufficient resolution to both map out the degree of serpentinization and relate the causal fluid ingress to slip along the overlying crustal faults (Bayrakci *et al.* 2016). Furthermore, the most current models of continental breakup (see section 1.4) have been designed from observation at the Galicia Margin, namely: crustal DDS (Davis and Kusznir 2004, pp. 92–136), polyphase faulting (Reston *et al.* 2007) and sequential faulting (Ranero and Pérez-Gussinyé 2010).

The WIM is therefore considered a "classic" example of a magma-poor margin. In this chapter, we present an overview of the structure of the WIM and summarize the main tectono-stratigraphic models proposed to explain its evolution. We then highlight the key remaining questions concerning the geodynamics of that margin, and also other magma-poor margins, thus exploring why the WIM has been and still remains at the forefront of research into continental breakup.

1.2. Structures of the West Iberian Margin

The WIM is sediment-starved and magma-poor (Boillot and Winterer 1988), allowing for optimal geophysical imaging of the margin's structures, and have thus been studied since the early stages of deep offshore scientific exploration.

First geophysical and drilling data: early reflection data across the GM were the Galice-Portugal profiles, collected in 1975 and 1980 by the Institut Français du Pétrole (Mauffret and Montadert 1987). These data were used as the basis for the first dedicated drilling leg on the margin, ODP Leg 103 (Boillot and Winterer 1988). In the late 1980s, these data were supplemented by the Lusigal profiles, which also covered the SIAP (Krawczyk *et al.* 1996). The widely spaced Iberian Atlantic Margin (IAM) profiles, acquired in 1993, included the IAM-9 across the SIAP, and IAM-5 across the TAP (Figures 1.1 and 1.2) and wide-angle/refraction profiles. In 1997, the Iberia Seismic Experiment collected reflection profiles on both the GM and the SIAP during the UK's Discovery 215 cruise (Whitmarsh *et al.* 2000), which also collected wide-angle data using a fleet of ocean bottom seismometers (OBSs). The Discovery and Lusigal reflection profiles supported two ODP legs on the SIAP margin (149 and 173 – Sawyer *et al.* 1994; Whitmarsh *et al.* 1998; Whitmarsh *et al.* 2000).

The Galicia 3D volume: enhancing the data available at the GM, a 3D volume of seismic reflection data was acquired in 2013 and provides spectacular new observations of the 3D structure of the DGM (Figures 1.1 and 1.3). These 3D data provide exceptional images of the edge of the hyper-extended crust down to 14 s TWT around the ODP Sites 638, 639 and 641, as well as images of the exhumed mantle around Site 637, at a resolution level never achieved before along this margin. Wide-angle data within the 3D volume were recorded by 72 OBS and hydrophones along a 2D profile (Davy *et al.* 2018), which were subsequently distributed into four irregular lines of 18 instruments for 3D inversion (Bayrakci *et al.* 2016). Together with the 2D data and the ODP drilling results, the 3D volume provides one of the best existing databases at rifted margins worldwide.

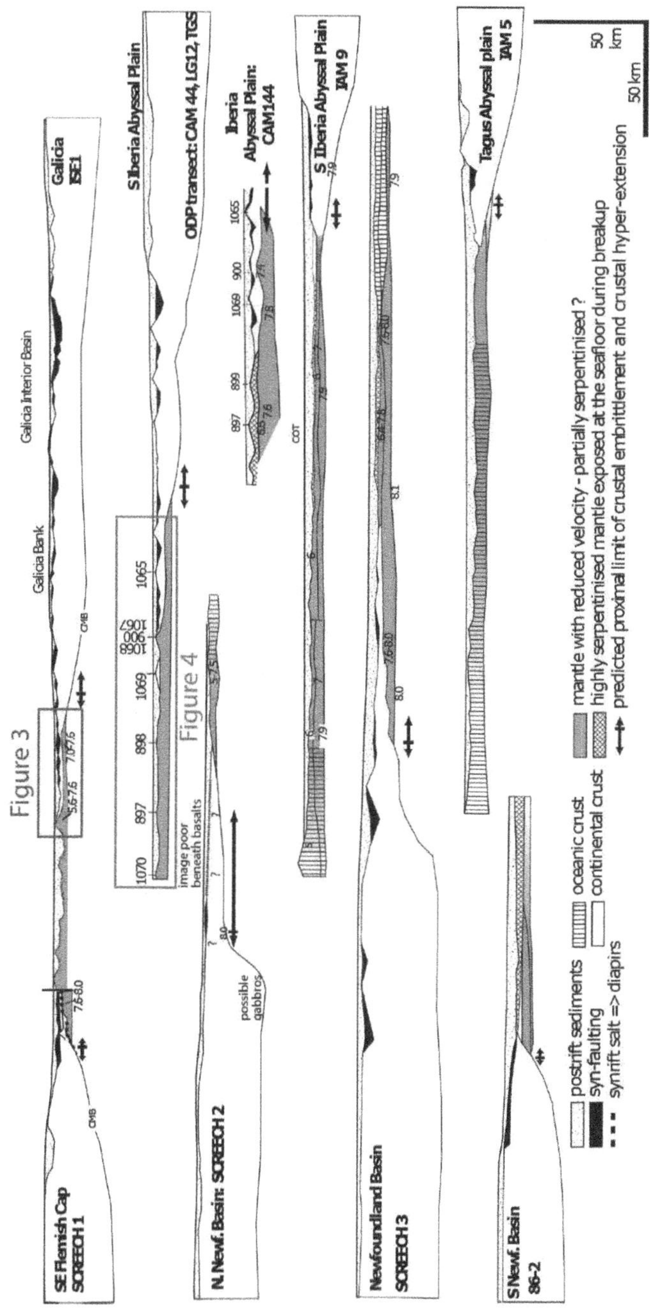

Figure 1.2. *Cross-sections through the Iberian-Newfoundland conjugated rifted margins*

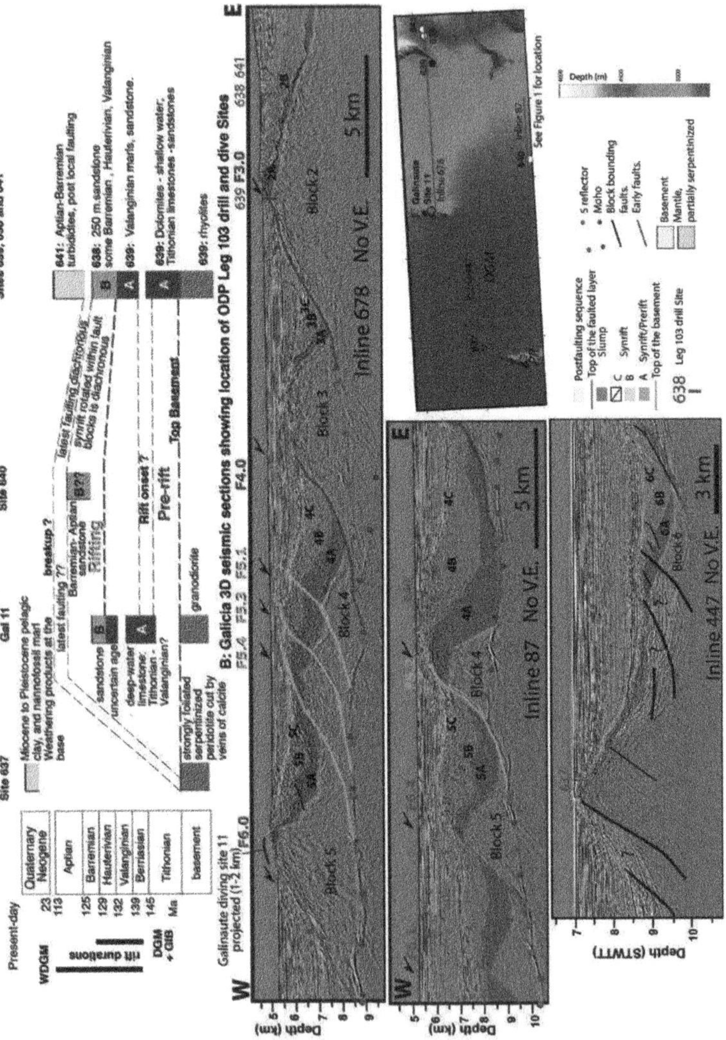

Figure 1.3. Structure of the Deep Galicia Margin. Top: summary of drilling results from ODP Leg 103 (source: adapted from Reston 2005)

CONTINUATION OF CAPTION FOR FIGURE 1.2.– *Adapted from Reston (2009 and 2010) (original sources: SE Flemish Cap: Funck et al. (2003); Galicia: Zelt et al. (2003); South Iberia Abyssal Plain: Chian et al. (1999); North Newfoundland Basin: Van Avendonk et al. (2006); South Iberia Abyssal Plain, IAM9: Dean et al. (2000); Newfoundland Basin (SCREECH3): Lau et al. (2006); South Newfoundland Basin: Reid (1994); Tagus Abyssal Plain: Neves et al. (2009). All noted velocities are based on Ocean Bottom Seismometers and Multi-Channel Seismic data. Note the close match between the predicted landward limit of embrittlement during rifting and of the reduced velocity mantle, interpreted as undercrusting serpentinites. The volumes of water required to generate serpentinization can only come from above, by passage through an entirely brittle crustal section. Thick synrift, especially salt, provides a barrier to these fluids and prevents mantle serpentinization.*

CONTINUATION OF CAPTION FOR FIGURE 1.3.– *Aptian-Barremian age of post-faulting units on block 2 (Site 641) is potentially the same age as syn-faulting units on block 4 (Site 640) suggesting that faulting is diachronous. Thickness of the stratigraphic units are not shown as ODP drills only partially sampled each unit; b) Seismic sections across the Galicia 3D volume with location of the ODP Sites. The location of the seismic sections is shown on a bathymetric map (inset) of the DGM generated within the 3D volume (location shown in Figure 1.1c). Color code for the faults is the same as for Figure 1.6. Block numbering from Ranero and Pérez-Gussinyé (2010). Synrift labels and fault numbering from Lymer et al. (2019). The lowest unit A is cut and rotated by the faulting, but does not thicken into the block bounding faults and so is likely prerift or early synrift; Unit B thickens towards the faults within the half-grabens and is considered to be syn-faulting; Unit C is tilted with the top of the faulted blocks, but onlaps unit B and infills the fault-generated topography, thus being interpreted as synrift, but post-local faulting.*

1.2.1. *The Deep Galicia Margin (DGM)*

At the DGM (Figure 1.1, section ISE1 on Figure 1.2 and Figure 1.3), the hyper-extended crust is made of tilted and faulted crustal blocks that decrease abruptly in size west of ODP Site 639 (Figure 1.3), underlined by the S reflector (de Charpal *et al.* 1978; Boillot and Winterer 1988). Mantle rocks forming the partially buried "Peridotite Ridge" (see ridges and ODP Site 637 on Figure 1.1) have been drilled at ODP Site 637, 25 km to the west of Galinaute dredging Site 11 (Boillot *et al.* 1988) that last sampled continental fault block.

The S detachment fault: S is thought to be both a detachment fault and a tectonic CMB (Figure 1.3), separating the overlying tilted blocks and the underlying partially serpentinized mantle (e.g. Reston 1996). The normal faults that bound the crustal blocks appear to root onto S. The distribution of serpentinization below S obtained from tomographic models revealed that the degree of mantle hydration increases with the slip of the overlying fault, resulting in more serpentinization around where the roots of the faults reach S (Bayrakci *et al.* 2016). Observed in 3D, the surface of the S exhibits corrugations, whose orientation change from E-W to ESE-WNW oceanwards, remaining approximately perpendicular to the strikes of the faults (Schuba *et al.* 2018; Lymer *et al.* 2019). The corrugations confirm the nature of S as a detachment surface, and their change of orientation suggests that the direction of extension changed during rifting, remaining parallel to the corrugations. The corrugations on the block-bounding fault planes align and match with the corrugations observed down-dip on S, suggesting the block-bounding faults and S represented a single slip surface when the faults were active and the corrugations formed. S is therefore a composite surface made of juxtaposed root segments of block-bounding faults (Lymer *et al.* 2019). The partitioned nature of S is further supported by its pronounced distortion visible on the depth sections, where S meets the crust-mantle boundary, and also where some block-bounding faults distort and cut across the S (F4.0 and F5.2 on Figure 1.3).

DGM 3D fault network development: at the DGM, Creswell (2018) and Lymer *et al.* (2019) have described a series of fault sets (Figure 1.6), interpreted as having been made of faults that were active simultaneously. Each set is defined by (1) the neighboring faults with similar strikes that are roughly orthogonal to the underlying corrugations, (2) the geometrical linkage and/or overlap of faults along strike and (3) the complementarity of the heaves within each set: as the heave on one fault decreases laterally, the heaves on an overlapping fault increase to accommodate regional extension (Figure 1.6c, d, e, f). Each set is bounded eastward by a fault that offsets S and continues oceanwards as S. Both the limited along strike length of each fault and the complementarity of heaves between linked faults indicate that the faults within each set were active simultaneously.

Following these criteria, within the most proximal fault set 3/4 (Figure 1.6) the main faults (F3.0, F3.1, F3.2 and F4.0) probably developed with limited lateral extent to accommodate the same regional extension, and continued to slip over the same time period as their heaves are complementary, with a steady if slightly northward decreasing heave sum, a result of the general northward propagation of rifting (Whitmarsh and Miles 1995). As extension increased, the faults became hard-linked as they physically merged laterally, until the faults within the set deactivated and a new set of faults developed oceanward (Figure 1.6b). Similar observations also apply to the oceanward fault sets 5 and 6 (Figure 1.6): F5.1 cuts across the S to the east but is continuous with S to the west, marking the start of fault set 5. Faults within set 5 (F5.1, 5.2, 5.3, 5.4) merge directly and have complementary heaves. Fault F6.4 marks the eastern boundary of the most distal fault set 6, trending NNE-SSW parallel to the underlying corrugations, within which the heaves of F6.0, F6.1, F6.4, are complementary again, and the sum of the heaves remains approximately constant across the volume.

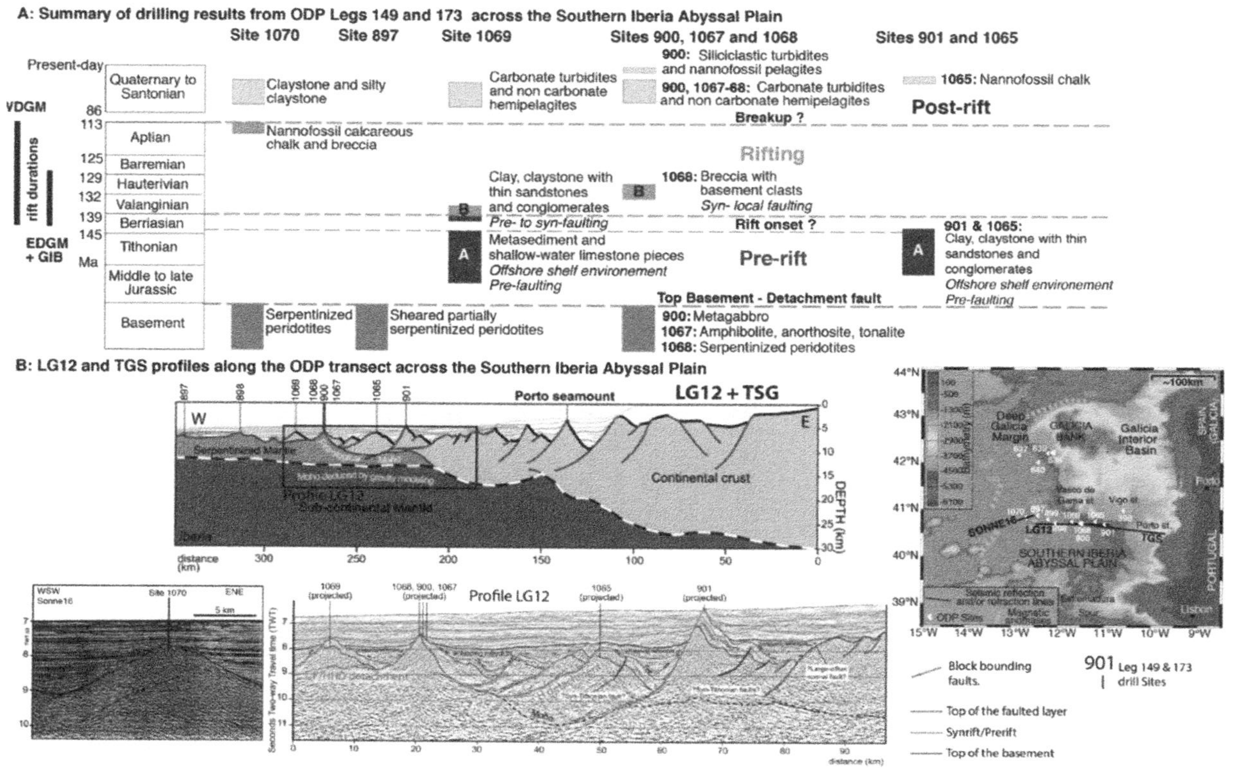

Figure 1.4. *Structure of the Southern Iberia Abyssal Plain (source: adapted from Whitmarsh et al. 1998 and Mohn* et al. *2015)*

CONTINUATION OF CAPTION FOR FIGURE 1.4.– *a) Summary of drilling results from ODP Legs 149 and 173 using a similar nomenclature to that defined at the Deep Galicia Margin (Figure 1.3). Thickness of the stratigraphic units are not shown as ODP drills only partially sampled each unit. b) Interpreted seismic reflection profile (LG12+TGS, Krawczyk et al. 1996 and Sutra and Manatscha, 2012; Sonne 16, Whitmarsh, et al. 1998) across the Southern Iberia Abyssal Plain, with location of ODP sites.*

The faults marking the eastern boundary of each set systematically cut across the S reflector to the east and are continuous with S to the west (Figures 1.3 and 1.6), consistent with an oceanwards migration of the faulting, as each new set cuts across the set immediately landwards. Conversely, the lack of any distortion of S where intersected by other faults within each set (e.g. F5.1, F5.3, F5.4) is interpreted as showing that these faults were active over the same time interval, so that S was simultaneously active beneath these faults.

1.2.2. *The Southern Iberia Abyssal Plain (SIAP)*

Similar to the DGM, the SIAP (Figures 1.1, 1.2 and 1.4) displays a series of tilted blocks bounded by oceanward dipping normal faults that detach along the top of serpentinized mantle. The basement geology across the SIAP is well constrained by numerous seismic experiments and the presence of several ODP boreholes drilled during Legs 149 and 173 (Sawyer *et al.* 1994; Whitmarsh *et al.* 1998). Sites 901 and 1065 sampled the top of two tilted fault-blocks and recovered shallow-water clastics (Figure 1.4), believed to indicate the presence of continental basement at depth (Whitmarsh *et al.* 1998). Sites 1067, 900 and 1068 sampled the same block ~30 km west of Site 1065. Site 1067 recovered anorthosites, amphibolites and tonalites, and Site 900 sampled mafic granulites, all interpreted as lower crustal rocks, while Site 1068 drilled serpentinized peridotites, which were interpreted as serpentinized mantle (Whitmarsh *et al.* 1998). Underlying rotated blocks between Sites 1065 and 1067, the H detachment is warped up and separates Site 900 from Site 1068 (Figure 1.4), so similarly to S, H likely represents a serpentine detachment delineating the CMB (Chian *et al.* 1999). Cutting across and rotating the H detachment, the fault bounding the western side of Site 1068 has been interpreted as marking the development of a newly formed detachment, which has been variously termed F detachment (Krawczyk *et al.* 1996) and HHD detachment (Manatschal *et al.* 2001, pp. 405–428). F/HHD defines the top of seismic basement from this high down to a fault block drilled at Site 1069. This rider block appears to be an allochthonous continental slice (based on the recovered sediment) resting on the same peridotitic basement as the one sampled at Site 1068. Site 899 recovered serpentinite breccias intercalated with sediments, indicating the presence of a nearby serpentinized high

(Whitmarsh *et al.* 1998), and Site 897 recovered highly sheared, partially serpentinized mantle peridotites (Sawyer *et al.* 1994). Finally, Site 1070, located 15 km east of the M0 magnetic anomaly (Figure 1.1), recovered serpentinized mantle rocks and gabbro.

1.2.3. *Conjugated West Iberian and Newfoundland Margins*

The DGM faces the southeast flank of the Flemish Cap at the Newfoundland Margin (Figure 1.1a and sections SCREECH 1/ISE1 on Figure 1.2), whereas the SIAP faces the North Newfoundland Basin (Figure 1.1a and sections SCREECH 2/LG12-TGS on Figure 1.2). Palaeographic reconstructions have proposed that both the Western Iberian and the Newfoundland margins represented a pair of conjugated margins before continental breakup (e.g. Sutra *et al.* 2013). The final root of S observed at the DGM is incidentally believed to be currently located on the Flemish Cap (Reston *et al.* 2007).

A key finding of the drilling at the SIAP was the demonstration of the presence of a wide-zone of unroofed mantle (from Site 900 to 1070 and oceanward, Figure 1.4) that extends between the hyper-thinned continental crust and the first clear magnetic polarity reversals (Figure 1.1, M0), outlining the areas where seafloor spreading occurred. This observation west of Iberia has underpinned the concept of mantle unroofing on a large-scale at other rifted margins (Whitmarsh *et al.* 1998; Manatschal *et al.* 2001). It has also led to the concept of margin pair asymmetry (e.g. Pérez-Gussinyé and Reston 2001), as comparison of sections across both margin segments (Figure 1.2) clearly shows the asymmetry of the unroofed mantle (e.g. Reston 2009). The mantle extends across ~250 km within the SIAP (sections LG12-TSG, IAM-9, Figure 1.2), versus ~50 km west of the COT offshore the Northern Newfoundland Basin (section SCREECH2, Figure 1.2). Asymmetry between the margin pair also exists when considering the thickness of the continental crust: comparison between the two margins still shows a "normal" crustal thickness of ~30 km at the Flemish Cap, whereas at the Galicia Margin, the GIB and the GB are characterized by relatively thin crust, of about ~10 km (e.g. Druet *et al.* 2018). This asymmetry between the WIM and its Newfoundland conjugate margin has been interpreted as having developed once the crust had become completely brittle (e.g. Pérez-Gussinyé and Reston 2001; Reston 2009).

1.3. Synrift tectono-stratigraphy and age and evolution of extension across the West Iberian Margin

The general evolution of the WIM has been derived from an extensive network of 2D seismic reflection profiles, calibrated by scientific drilling acquired during

ODP Legs 103 at the DGM (Boillot and Winterer 1988), and Legs 149 and 173 at the SIAP (Sawyer *et al.* 1994; Whitmarsh *et al.* 1998) (Figures 1.3 and 1.4). The recovered cores and geophysical logging data provided constraints on the lithology, biostratigraphy, palaeomagnetics, geochemistry, bulk density, thermal conductivity, and compressional seismic velocities on the rocks forming the basement and the synrift sediments. ODP results have also provided sparse information on the ages of the synrift sequences, mainly restricted to the top of some tilted blocks and basement highs.

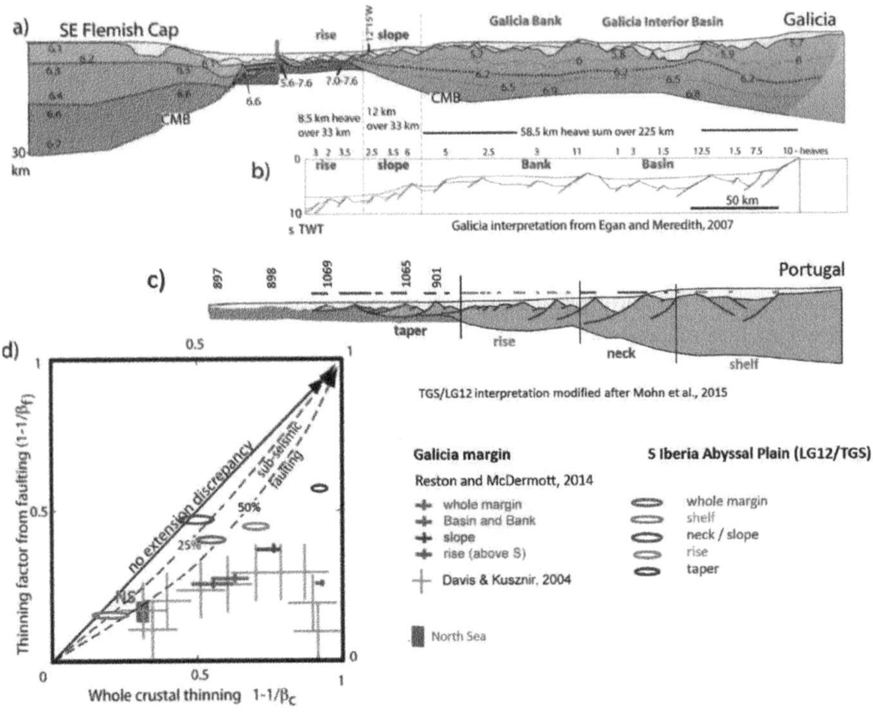

Figure 1.5. *The extension discrepancy across the Iberian margin. a) The Flemish Cap/Galicia transect. Here the amount of measurable extension b) across the Galicia Margin is too slight to explain the observed thinning. c) The same is true, but to a lesser extent, across the SIAP (composite profile LG12/TGS, interpretation modified from Mohn* et al. *2015): there is still an extension discrepancy at the deep margin despite the very large amount of extension associated with different generations of detachments. d) Cross-plot of crustal thinning versus thinning from faulting for these two margins showing the relative magnitude of the extension discrepancy*

The rifting west of Iberia affected Variscan crystalline continental basement that has been sampled from two tilted blocks at ODP Site 639 and at GAL11 diving site (Boillot and Winterer 1988; Boillot *et al.* 1988; Figure 1.3), and interpreted more widely from seismic velocities (Bayrakci *et al.* 2016; Davy *et al.* 2018). Based on a 3D fault analysis, Lymer *et al.* (2019) subdivided the synrift stratigraphy at the DGM in terms of age relative to the local faulting (e.g. the synrift of block 3 is divided into the packages 3A, 3B, 3C, Figure 1.3). Thus, the lowest unit A is likely prerift or early synrift, unit B is considered to be synfaulting, and unit C to be synrift but post-local faulting (see Figure 1.3 for details). Following the Variscan Orogeny, multiple phases of extension affected the WIM over 100 Myr, from the Triassic until the mid-Cretaceous, when eventual northward propagating breakup led to diachronic opening of the North Atlantic (e.g. Srivastava *et al.* 1990; Tucholke *et al.* 2007). Late Triassic to Early Jurassic early rifting affected a wide zone (Tucholke *et al.* 2007), producing large intracontinental basins (Figure 1.1 – LP, PB; Wilson *et al.* 1989, pp. 341–361; Murillas *et al.* 1990). Subsequent Jurassic extension thinned the crust to ≥10 km and led to the development of the GIB (e.g. Pérez-Gussinyé *et al.* 2003). ODP Site 639 recovered shallow water Tithonian carbonates at the western edge of the GB (Boillot and Winterer 1988) (Figures 1.1 and 1.3), that could correspond on seismic sections to relatively thin, poorly reflective packages (unit A on Figure 1.3). At the Early Cretaceous, the DGM developed due to extreme crustal thinning and mantle exhumation processes (Boillot *et al.* 1987), with hyper-thinning of the crust from ~10 km at the western edge of the Galicia Bank to zero in areas of exhumed serpentinized mantle (Figure 1.3). Syn-tectonic sediments (unit B) recovered at the top of a fault block at ODP Sites 638 and 639 have been dated as Valanginian–Barremian, whereas possible syn-tectonic units sampled at more oceanward Site 640 are Barremian–Aptian in age (Applegate and Bergen 1988). The westernmost synrift sample has been obtained by submersible from the top of a crustal block outcropping at the seafloor (GAL-11, Figure 1.3), consisting of sandstones whose age was not readily dateable (Reston 2005). The faulted blocks above S are consequently capped by a deep-water clastic sequence (Boillot and Winterer 1988), rather than true prerift sediments, such as the shallow water carbonates recovered at ODP Site 639. In addition, the ages of the recovered synrift sediments indicate that the top of the synrift units becomes younger towards the distal DGM (Boillot and Winterer 1988; Figure 1.3, compare ages at Sites 641 and 640). The age of the synrift sediments across the Galicia Margin is therefore diachronous (Reston 2005), which is consistent with the oceanwards migration of the rifting, as suggested from 3D observations on the kinematic development of the S detachment (Lymer *et al.* 2019). ODP Leg 103 clearly constrained the age of the extensional faulting at only one fault block (Sites 638, 639 and 641), where the prerift, early synrift and late synrift sequences were recovered. **The age distribution of rifting across the Galicia Margin is therefore uncertain due to the punctual availability of time constraints on the age of synrift sequences.**

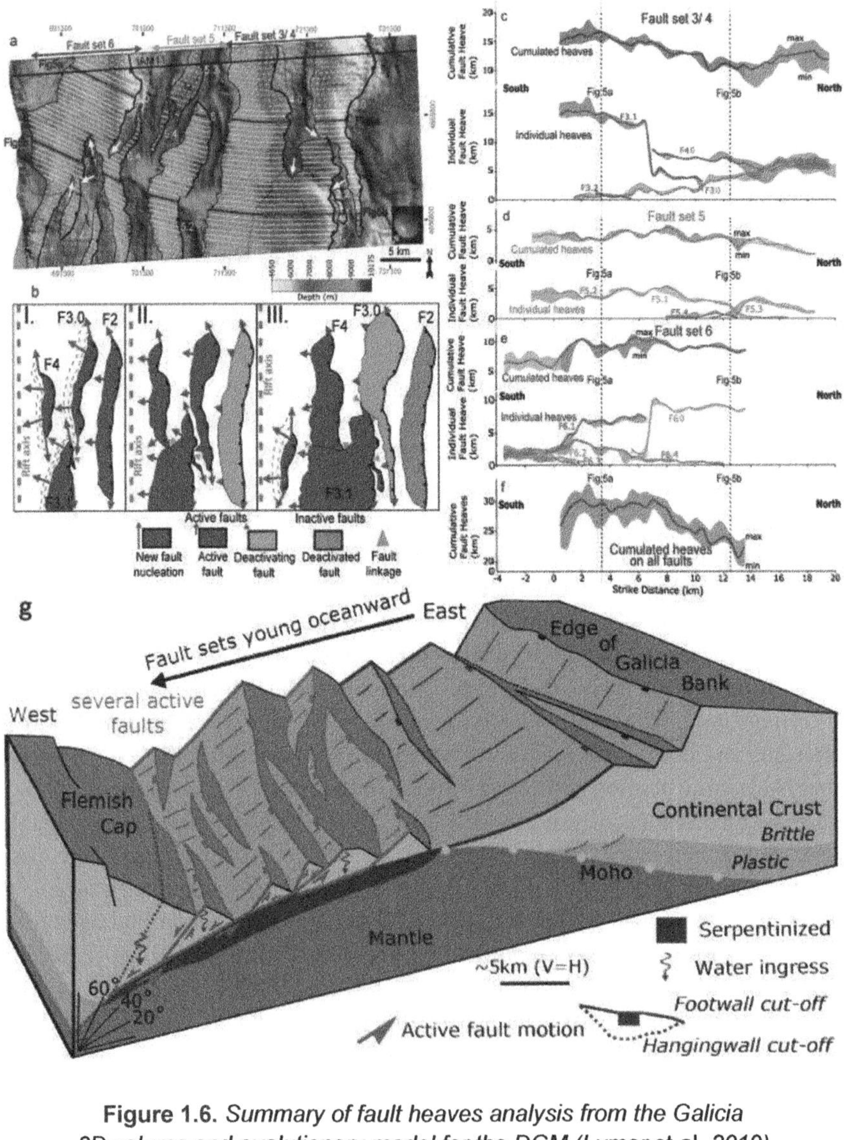

Figure 1.6. *Summary of fault heaves analysis from the Galicia 3D volume and evolutionary model for the DGM (Lymer* et al. *2019)*

CONTINUATION OF CAPTION FOR FIGURE 1.6.– *a) Map of the top basement level (depth shown by scale bar in meters) generated within the Galicia 3D volume (location in Figure 1.1) on which the flowlines identified from corrugations on S are used to measure heaves on the block-bounding faults. b) 3D model of fault development based on the geometry of faults within set 3/4 showing that more than one fault is needed to accommodate extension at any one time as faults have limited along strike length. The flowline heaves (c, d and e) are grouped into three sets in which the heaves are complementary: as one fault dies out the displacement is taken up on another in the same set. Overall, the cumulated heaves f) show a northward decrease in extension, consistent with the propagation of rifting from the south. g) Block diagram summarizing the model of Lymer et al. (2019) for the evolution of the DGM based on the 3D fault analysis shown in a) to f). After a period of extension and thinning, the blue faults cut across the CMB and allow the ingress of water to serpentinize the mantle. Subsequent faults propagate up through the hanging wall of preceding faults as in a rolling hinge model, but one in which the root can slip at low-angle due to serpentinization and in which multiple spatially overlapping faults are active at any one time (color coded by fault set).*

In the SIAP, Legs 149 and 173 recovered synrift stratigraphy through deeper settings and thicker postrift sequence than at the DGM (Figure 1.4). Although the main focus of these legs was to characterize the basement type by specifically targeting structural highs, the SIAP drilling transect provided limited constraints on the age of some individual faults. Tilted Tithonian clays and claystones mixed with clastics and sandstones predating the faulting were recovered at Sites 1065 and 901, interpreted as having been deposited in enclosed, relatively shallow and anoxic basins, close to vegetated land (Whitmarsh *et al.* 1998). Benthic foraminifera recovered at Site 901 indicate a neritic zone (Collins *et al.* 1996), that is continental shelf environment, and were interpreted as marking the onset of the rifting (Whitmarsh and Wallace 2001). A late Berriasian–early Valanginian age for the beginning of significant extension is supported by the pre- to syn-rift nature of the sediments recovered at Site 1069 (Figure 1.4), that are constituted of shallow-water limestones, interpreted as predating the main extensional phase, overlaid by conglomerates and sandstones, interpreted as marking the onset of extension at this location (Whitmarsh *et al.* 1998). Later synrift units have not been sampled but the timing of crustal extension is constrained by 40Ar/39Ar cooling data for samples recovered at Site 1067 and 900, indicating that these deep crustal rocks were exhumed from >18 km (0.6Gpa) through ~4 km depth by 137 Ma (Wilson *et al.* 2001, pp. 429–452). Thus, in the ~10 Myr between the onset of rifting in the Tithonian and the unroofing of these rocks, the crust appears to have been thinned by a factor of ~5. The final phase of crustal extension, which produced the current observed geometry of the margin probably occurred shortly after this (Tucholke *et al.* 2007). In the distal zone of exhumed mantle, oceanward of Site 900

(Figures 1.1 and 1.4), a series of basement highs were drilled: the sediment atop the highs represents a minimum age for the unroofing of that particular portion of peridotitic basement. Drillings have sampled mixtures of serpentine breccias and mass flow deposits of Aptian–Barremian age that may be related to the breakup process (Tucholke *et al.* 2007).

To summarize, in the SIAP, the Tithonian–early Berriasian sequence is tilted and part of the pre-faulting units (Figure 1.4). The late Berriasian–early Albian sequence either onlaps previous sequences/basement, or locally thickens towards the block bounding faults. The evidence thus points towards major rifting and crustal extension occurring over perhaps 10 Myr between the Tithonian and the Valanginian, followed by a phase of mantle unroofing over a further 10 Myr, before the onset of seafloor spreading during the Aptian–Barremian. As within the DGM, extension appears to have migrated oceanwards, but culminates in the unroofing of a far wider zone of mantle rocks. Although the overall time between the onset of rifting and the spreading of seafloor is similar at the two margin segments, approximately half of that time at the SIAP only involved the unroofing and subsequent extension of mantle rocks. Hence, rift duration leading to complete crustal separation may have only been about 10 Myr. Both the DGM and SIAP exhibit tilted fault blocks, detachment faulting, tectonic CMB, serpentinization beneath hyper-extended crust, and a wide zone of mantle exposed at the seafloor during rifting. At both segments of the margin, ODP results suggest that extension during the rifting may have migrated oceanwards. **However, although the WIM has been extensively studied, the details of the timing of rifting remain largely unconstrained due to the fact that most scientific drillings have targeted basement highs**. This strategy, while revolutionizing our understanding of the structures of rifted margins and introducing the concept of mantle unroofing, has not recovered the complete synrift sequences needed to determine the timing of motion on individual faults and the across-strike temporal distribution of fault activity during the rifting. Distinctive modes of fault development and emplacement have resulted into three main models proposed to explain the migration of the extension suggested by the available data. In the following section, we introduce these models designed from observations at the West Iberian Margin, and likely explain how magma-poor margins might have formed.

1.4. Current models of development of the West Iberian Margin

Exploration of the WIM has led to the definition of three major models of continental breakup (Figure 1.7) that appear to offer a solution to the extension discrepancy (Figure 1.5) and which have subsequently applied to the rifting and breakup of other magma-poor margins. These models are: crustal depth-dependent stretching (DDS), cross-cutting polyphase faulting, and migrating faulting models, which include the sequential faulting. They imply fundamentally different mechanisms for thinning the continental crust, the latter two being strongly diachronous in terms of fault activity across the WIM, although ODP Legs 103, 149 and 173 only sampled the complete synrift in a single place (Site 639). For instance, the relative age of the syn-tectonic packages in different sub-basins (e.g. between blocks 3 and 5, Figure 1.7) depends on the mode of extension: simultaneous, focusing or migrating.

Model M1, crustal depth-dependent thinning. Driscoll and Karner (1998) and Davis and Kusznir (2004, pp. 92–136) suggested that both the amount and the mechanisms of extension vary with depth across the crust, due to the different rheologies of the upper and the lower crust. In this way, the extension occurs by faulting in the brittle upper crust, whereas the lower crust deforms through ductile processes, not just stretching but also displacing or flowing, thus thinning the crust much more than indicated by the faulting (Figure 1.5). In some cases, the interpretation of eroded fault block crests and the dominance of shallow water sediments has led to the idea that much of the DDS occurred from the displacement of the lower crust late during the rifting process (Clift and Lin 2001, pp. 489–510; Davis and Kusznir 2004, pp. 92–136). In this interpretation, all the brittle deformation is imaged and occurred during a single phase, so the synrift sediments have the same age across the margin (Figure 1.7).

Model M2, cross-cutting polyphase faulting. An alternative perspective is that the crust thins by the same amount at all levels, but that not all the upper crustal extension is correctly identified. In this polyphase faulting model, extension might be underestimated where young faults cut and displace old faults (Reston 2005; Reston and McDermott 2014) as extension occurs through distinctive phases of deformations that focus with time into the rift axis. This model results in complex structural and stratigraphic geometries which means most of the earlier faults cannot be identified from seismic reflection profiles, thus explaining the extension discrepancy (Reston 2009; Figure 1.5). Seismic modeling of the resulting complex geometries confirms that earlier faulting is indeed likely to be misinterpreted, for example as eroded fault block crests (McDermott and Reston 2015). As faulting may have started at the same time across much of the margin, but focused during rifting towards the rift axis, the earliest synrift should be present across the margin, but the youngest synrift should be present only near the axis of breakup, being of the same age as local post-faulting sediment on the flanks (McDermott and Reston 2015). The initial top basement and pre-rift are dismembered and only preserved locally, as much of the top basement surface comprises early fault surfaces (Figure 1.7).

Model M3, migrating faulting. The migrating faulting models, for example rolling hinge (Buck 1988) and sequential faulting (Ranero and Pérez-Gussinyé 2010; Brune *et al.* 2014), are polyphase but with the locus of the extension migrating oceanwards, and the faulting migrating into the hanging wall of the previous fault, rather than focusing towards the future oceanic spreading axis. Extension occurs on a succession of individual faults active one at a time, with each fault that forms as a steep structure, flexurally rotates, and is abandoned when the angle becomes too low to allow for slip along the fault plane. Only then does a new fault form in the hanging wall of the abandoned fault, and slices ever farther into one side of the rift, with the same process continuously repeating until the crust has been thinned to zero (Ranero and Pérez-Gussinyé 2010; Brune *et al.* 2014). As each fault avoids cutting earlier faults, all the extension should be recognized. Extension discrepancy should only result if combined with another process (e.g. subseismic faulting, Reston and McDermott 2014), or if the migrating locus of faulting overlaps with the preceding ones, producing an element of cross-cutting and unrecognized geometries. A consequence of this model is the systematic oceanwards decreasing of the age of all of the syn-faulting sediments accumulated within each newly-formed half-graben (Figure 1.7).

The sequential faulting, as an inherently asymmetric process, might also explain the development of asymmetry between conjugate margin pairs. As proposed by Ranero and Pérez-Gussinyé (2010), sequential faulting might develop during early rifting when the continental crust is still >20 km thick. However, this early development conflicts with the apparent link between asymmetry development and crustal embrittlement, which requires asymmetry to develop only once the crust is thinner than 10 km (Reston and Pérez-Gussinyé 2007; Reston 2009). However, dynamic models propose a later onset of the development of asymmetry (Brune *et al.* 2014), when the crust becomes thinner than 20 km. The latter scenario is consistent with 3D seismic-derived interpretations showing that extension migrated on a 3D multi-fault variant of the rolling hinge model, only once slip at low-angle was allowed (Lymer *et al.* 2019). Brune *et al.* (2014) proposed hot ductile lower crust as a rooting zone, while Lymer *et al.* (2019) suggested that the recognized serpentinized mantle beneath S (Bayrakci *et al.* 2016) might represent the weak root zone.

The fundamental disparities between the three models – in terms of timing of faulting, number of faulting phases and rheologies – demonstrate that despite decades of exploration of the WIM, our knowledge of rifting and breakup remains fundamentally incomplete, as long as detailed timing of the geological events at rifted margins remains undefined.

1.5. Remaining questions at the West Iberian Margin and other magma-poor margins

The diversity of existing current models developed at the WIM shows that we still fundamentally do not know how the continental crust thins to zero, exposing the Earth's mantle to the surface during rifting, leading to eventual continental breakup. The models presented in the previous section (Figure 1.7) are all potentially viable but imply fundamental differences in the mechanisms of thinning of the crust. The differences between these models can be summarized in three major, yet unanswered, questions, emerging from our current knowledge of the structures of the Galicia Margin, and more globally related to the development and evolution of magma-poor rifted margins.

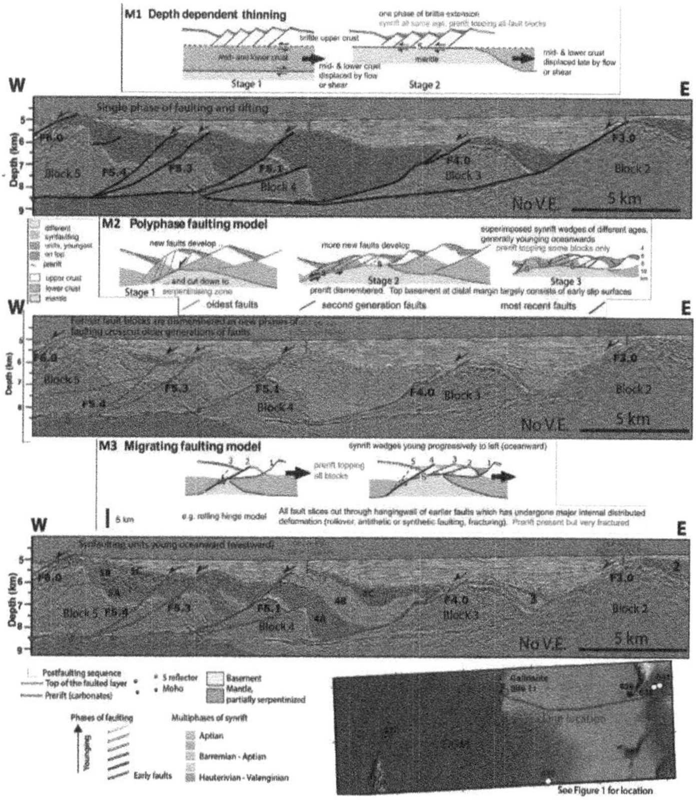

Figure 1.7. *Models proposed to explain fault development and the evolution of the West Iberian Margin, highlighting different predictions in terms of the age distribution of synrift sequences, the different role of detachments and different distributions of prerift/top basement*

CONTINUATION OF CAPTION FOR FIGURE 1.7.– *Although based on observations made at the West Iberian Margin, all of these models have been proposed as explanations for commonly-observed characteristics of magma-poor rifted margins worldwide, including the extension discrepancy. For each model, a simplified interpretation of part of a Galicia 3D is presented, flowline illustrating the different possible resulting stratigraphic geometries of deposition of the syn-tectonic sequences. The same dogleg line through the Galicia 3D volume is shown in each case. Model M1: all faults active over the same time period, before the crust was thinned by late displacement of the lower crust beneath a detachment. M1 predicts widespread prerift sediments (light blue), the same timing of fault onset and abandonment across the margin, and see rapid deepening in the postrift succession. Model M2: Polyphase cross-cutting faults, with extension focusing towards the line of eventual breakup. We would expect limited occurrence of prerift sediments (most top basement being the footwall of earlier faults, showing considerable exhumation), similar ages for the onset of faulting across the margin but very different timing of the cessation of faulting, and complex distribution of synfaulting sequences with a larger age range than predicted by sequential faulting. Model M3: sequential faulting predicts widespread prerift sediments, a pronounced age progression in both the onset and ending of active faulting across the margin (Unit B), and thus also in the oldest post-faulting sequences (Unit C). Block numbering from Ranero and Perez-Gussinyé (2010); fault and synrift unit numbering from Lymer et al. (2019).*

Question 1: how is the crust thinned from 20 km to zero km?

Beneath rift basins such as the North Sea, the crust thins from ~30 km to ~20 km (e.g. Wood and Barton 1983). Together with the presence of normal faults and the subsidence history, the thinning indicates that the basin formed by lithospheric extension (McKenzie 1978) with normal faults accommodating extension of the upper crust, ductile flow extension in the lower crust and lower lithosphere, and possible shear zones (Reston 1993) accommodated the extension of the uppermost mantle. However, early studies suggested that the amount of fault-controlled extension was far less than that required to thin the crust (Ziegler 1983): there is an extension discrepancy (Figure 1.5). Improvement in imaging has reduced the discrepancy under such basins to amounts expected in the presence of subseismic deformation (10%–50% of the fault-controlled extension, Marrett and Allmendinger 1992), that is the deformation occurring between faults and accommodating the distortion of faulted blocks, but as extension increases, especially at rifted margins, this discrepancy increases. This is the case at the WIM, where the amount of extension above S that can be deduced from the geometry of the faulted blocks (Figure 1.5) is considerably less than that implied by the subsidence rate (e.g. Sibuet 1992) and by measured crustal thinning (e.g. Reston 2005). Typically, the faults observed at the WIM indicate about 50%–60% extension (stretching factors of 1.5–1.6, that is thinning factors of 0.4 or lower, Figure 1.5), which is too low for the hyper-thinned crust observed at the WIM. The amount of thinning either requires more extension, or the crust to have been thinned by some other method. This is the extension discrepancy which provides fundamental constraints on the process of crustal thinning towards continental breakup.

Model M1 implies that all the brittle faulting is imaged (Figure 1.7); thus, top basement should, except where subject to erosion, be capped by prerift and early synrift rocks.

Model M2 infers that top basement is locally the exhumed slip surface of an earlier fault (Figure 1.7), cut across and displaced by the more recent faults.

In *Model M3*, the timing of the fault movement should systematically get younger oceanwards (Figure 1.7). Only minor extension discrepancy is excepted towards the deep margin (Ranero and Pérez-Gussinyé 2010), where the crust cut by later faults has locally been prethinned by movement on preceding faults.

Question 2: what is the role of detachments in rifting to breakup and how do they develop?

Large-offset, apparently low-angle normal faults, commonly referred to as detachment faults, accommodate crustal thinning in some areas of high factors of extension, such as at the DGM (S, Figure 1.3) and at the SIAP (H, Figure 1.4). The end-member models M1, M2 and M3 (Figure 1.7) imply different mechanisms of development of detachments.

In *Model M1*, the detachment zone develops late in the rifting history and corresponds to the interface at the top of the mantle and the base of the upper crust, along which the lower crust has been laterally displaced by flow during rifting.

In *Model M2*, the detachment consists of an early fault that cut through the mantle, allowing for serpentinization, which makes it possible for the detachment fault to remain active during all of the latest stages of rifting due to weakening of the underlying rocks. The top of the mantle thus acts as a detachment as extension focuses through subsequent faulting.

In *Model M3*, the detachment is a composite structure that comprises the root segments of successive steep faults which rotated to low angle before being abandoned and a new fault formed in the hanging wall of the former fault, propagating the detachment oceanwards and resulting in an apparently continuous sub-horizontal surface.

Detachment faults have been imaged at a variety of margins and hyperextended basins (e.g. de Charpal *et al.* 1978; Reston *et al.* 2001), but are best known from the GM (e.g. Reston *et al.* 2007). There, 3D analysis (Lymer *et al.* 2019) suggests that S formed from a combination form of models M2 and M3, with an early fault that cut through the CMB and led to mantle serpentinization (M2), but evolved into a 3D rolling hinge onto which multiple faults rooted at once, once the crust had thinned to <10 km (variant of M3). Thus, sequential faulting may develop as one of the later phases of extension in a polyphase mode.

Question 3: what is the evolution of the lithospheric rheology, strain distribution, melt production and serpentinization during rifting?

In contrast to magma-rich margins, magma-poor margins such as the WIM display few and small magmatic bodies atop the well-defined faulted blocks. Several hypotheses have been advanced to explain the lack of magma at such margins: very slow rifting (Pérez-Gussinyé *et al.* 2006), pre-depletion of the asthenosphere following earlier melting events (Muentener and Manatschal 2006), lithospheric (as opposed to crustal) DDS in which crustal breakup occurred before lithospheric separation (Huismans and Beaumont 2011) and cool sub-lithospheric mantle (Reston and Phipps-Morgan 2004). A key to understanding the lack of melt is to constrain the timing and rate, both absolute and relative, of crustal rifting and of mantle thinning (Brune *et al.* 2014, 2017; Ros *et al.* 2017).

To summarize the above section, the remaining key unknowns at the WIM are as follows:

– *How are the lithosphere and the crust thinned in space and time from 20 km to zero?* This can be tested by addressing the extension discrepancy (Figure 1.5) and determining the detailed rift history of the WIM by sampling the age of the synrift in each half-graben to calibrate the existing excellent seismic database at this margin.

– *How and when do detachments develop and control the structural evolution of distal margins?* Determining the detailed rift history will place the role of detachments into a well-constrained framework to test the current models of development of the WIM, and to determine the timing of any switch to migrating, detachment-controlled extension.

– *How does evolving lithospheric rheology control strain localization, necking, mantle upwelling, mantle exhumation, melt production and serpentinization?* The key here is determining the rate of rifting, a natural product of determining the detailed rift history.

These key questions are global in scope, apply to the breakup of all continents and center around the fourth dimension: time. Despite the recent progress of the scientific community in understanding the processes of development of rifted margins, mostly favored by international cooperation (e.g. the "International Ocean Drilling Program", see also Peron-Pinvidic *et al.* (2019)), the *timing* of the key steps at rifted margins remains mostly unconstrained. Observations from seismic data have revealed the geometry of rifted margins, but not the absolute timing and rate of the key steps of rifting, thus the ages of the syn-tectonic units remain interpretive (Figure 1.7). Onshore, observations from analogues of margins cannot fully constrain the processes of rifting because they are discontinuous and overprinted by later, often strong, deformations. Other approaches, such as numerical models, are dependent on time input parameters, for example the velocity of plate divergence, that are often too poorly constrained. **The time aspect of the geological events at rifted margins can only be addressed by scientific drilling**. Given the amount of previous work, the presence of a 3D dataset and the key role the GM has played in framing these questions, the logical place to carry out such an experiment would be the Galicia Margin.

1.6. Conclusions and outlook

The WIM has historically been, and remains, at the forefront of fundamental research on the breakup and development of oceanic basins. Three ODP legs (103, 149 and 173), many geophysical campaigns, and a recent 3D seismic volume place the WIM among the best-documented continental margins. These data allowed the definition of the earliest concepts of oceanic basin development, and still today provide the foundation to the definition of the most recent models of continental breakup. They have provided a glimpse of how extension of the continental lithosphere occurs at rifted margins, demonstrating that rifting is a diachronous process across margins, and suggesting that extension focused/migrated towards the future oceanic spreading center during the rifting history. The latest concepts can be summarized through three main models (section 1.4): depth dependent stretching, cross-cutting polyphase faulting and sequential faulting.

These models imply fundamental differences in terms of the timing and mechanisms of faulting and thinning of the continental crust, of the number of fault phases involved in rifting, of the location of the locus of extension through time, of the rate, temperature and amount of magma involved during lithospheric separation. However, because of the technical, scientific and human implications in recovering direct data from deep present-day rifted margins, these models remain fundamentally unconstrained. Indeed, while the vast amount of data available West of Iberia has provided precise images of the structures of the margin, they currently miss the temporal control to understand the kinematics and the evolution of rifting processes. To provide further time constraints during rifting of the WIM, we need to sample the pre-, syn- and post-rift sediments, as well as the basement, at successive basins across the margins to determine the detailed rift history by:

1) characterizing the nature of the basement and the oldest synrift/prerift unit;

2) dating the sequences deposited during slip on the block-bounding faults above extensional detachment faults, thus recording time constraints on the slip of individual faults across the distal margin;

3) constraining the age and the nature of the youngest synrift units deposited after local faulting to define the mechanisms of post-faulting uplift and faulted-blocks rotation;

4) defining the later subsidence of the margin from discrete palaeoenvironmental data (e.g. the local CCD and depth of the bathyal sediments) obtained through the post-rift sequence.

FURTHER READING.– The above descriptions are abbreviated and often simplified. If interested in reading and learning further, the reader is referred to the following list of publications and references.

– General: (Biari *et al.* 2021; Cresswell 2018; Epin and Manatschal 2018; Gómez-Romeu 2019; King *et al.* 2020; Reston and McDermott 2014; Somoza *et al.* 2019; Sutra *et al.* 2013; Whiting *et al.* 2021).

1.7. References

Applegate, J.L. and Bergen, J.A. (1988). Cretaceous calcareous nannofossil biostratigraphy of sediments recovered from the Galicia margin, ODP Leg 103. *Proceedings of the Ocean Drilling Program, Scientific Results*, 103, 293–348.

Autin, J., Leroy, S., Beslier, M.O., d'Acremont, E., Razin, P., Ribodetti, A., Bellahsen, N., Robin, C., Al Toubi, K. (2010). Continental break-up history of a deep magma-poor margin based on seismic reflection data (northeastern Gulf of Aden margin, offshore Oman). *Geophysical Journal International*, 180(2), 501–519.

Ball, P., Eagles, G., Ebinger, C., McClay, K., Totterdell, J. (2013). The spatial and temporal evolution of strain during the separation of Australia and Antarctica. *Geochemistry, Geophysics, Geosystems*, 14(8), 2771–2799.

Bayrakci, G., Minshull, T.A., Sawyer, D.S., Reston, T.J., Klaeschen, D., Papenberg, C., Ranero, C., Bull, J.M., Davy, R.G., Shillington, D.J., Pérez-Gussinyé, M., Morgan, J.K. (2016). Fault-controlled hydration of the upper mantle during continental rifting. *Nature Geoscience*, 9, 384–388.

Boillot, G. and Malod, J. (1988). The north and north-west Spanish continental margin. *Revista de la Sociedad Geológica de España*, 1(3–4), 295–316.

Boillot, G. and Winterer, E. (1988). Drilling on the Galicia margin: Retrospect and prospect. *Proceedings of the Ocean Drilling Program, Scientific Results*, 103, 809–828.

Boillot, G., Grimaud, S., Mauffret, A., Mougenot, D., Kornprobst, J., Mergoil-Daniel, J., Torrent, G. (1980). Ocean-continent boundary of the Iberian margin: A serpentinite diapir west of the Galicia Bank. *Earth and Planetary Science Letters*, 48, 23–24.

Boillot, G., Recq, M., Winterer, E.L., Applegate, J., Baltuck, M., Bergen, J.A., Comas, M.C., Davies, T.A., Dunham, K., Evans, C.A., Girardeau, J., Goldberg, G., Haggerty, J., Jansa, L.F., Johnson, J.A., Kasahara, J., Loreau, J.P., Luna-Sierra, E., Moullade, M., Ogg, J., Sarti, M., Thurow, J., Williamson, M. (1987). Tectonic denudation of the upper mantle along passive margin: A model based on drilling results (Ocean Drilling Program leg 103, western Galicia margin, Spain). *Tectonophysics*, 132(4), 335–342.

Boillot, G., Comas, M.C., Girardeau, J., Kornprobst, J., Loreau, J.-P., Malod, J., Mougenot, D., Moullade, M. (1988). Preliminary results of the Galinaute Cruise: Dives of the submersible nautile on the western Galicia margin. *Proceedings of the Ocean Drilling Program, Scientific Results*, 103, 37–51.

Boillot, G., Feraud, G., Recq, M., Girardeau, J. (1989). "Undercrusting" by serpentinite beneath rifted margins: The example of the west Galicia margin (Spain). *Nature*, 431, 523–525.

Brune, S., Heine, C., Pérez-Gussinyé, M., Sobolev, S.V. (2014). Rift migration explains continental margin asymmetry and crustal hyper-extension. *Nature Communications*, 5, 4014.

Brune, S., Heine, C., Clift, P., Pérez-Gussinyé, M. (2017). Rifted margin architecture and crustal rheology: Reviewing Iberia-Newfoundland, Central South Atlantic, and South China Sea. *Marine and Petroleum Geology*, 79, 257–281.

Buck, W.R. (1988). Flexural rotation of normal faults. *Tectonics*, 7, 959–973.

de Charpal, O., Guennoc, P., Montadert, L., Roberts, D.G. (1978). Rifting, crustal attenuation and subsidence in the Bay of Biscay. *Nature*, 275, 706–711.

Chian, C., Louden, K.E., Minshull, T.A., Whitmarsh, R.B. (1999). Deep structure of the ocean-continent transition in the southern Iberia Abyssal Plain from seismic refraction profiles: Ocean Drilling Program (Legs 149 and 173) transect. *Journal of Geophysical Research*, 104, 7443–7462.

Clift, P.D. and Lin, J. (2001). Patterns of extension and magmatism along the continent–ocean boundary, South China Margin. In *Non-Volcanic Rifting of Continental Margins: A Comparison of Evidence from Land and Sea*, Wilson, R.C.L., Whitmarsh, R.B., Taylor, B., Froitzheim, N. (eds). Geological Society, London.

Collins, E.S., Kuhnt, W., Scott, D.B. (1996). Tithonian benthic foraminifers from Hole 901A. In *Proceedings of the Ocean Drilling Program*, Whitmarsh, R.B., Sawyer, D.S., Klaus, A. and Masson, D.G. (eds). Scientific Results, College Station, TX.

Cresswell, D.J.F. (2018). Continental break-up at the Galicia Margin. PhD thesis, University of Birmingham [Online]. Available at: http://etheses.bham.ac.uk/id/eprint/8712.

Davis, M. and Kusznir, N.J. (2004). Depth-dependent lithospheric stretching at rifted continental margins. In *Proceedings of NSF Rifted Margins Theoretical Institute*, Karner, G.D. (ed.). Columbia University Press, New York.

Davy, R.G., Minshull, T.A., Bayrakci, G., Bull, J.M., Klaeschen, D., Papenberg, C., Reston, T.J., Sawyer, D.S., Zelt, C.A. (2016). Continental hyperextension, mantle exhumation, and thin oceanic crust at the continent–ocean transition, West Iberia: New insights from wide-angle seismic. *Journal of Geophysical Research: Solid Earth*, 121, 3177–3199.

Davy, R.G., Morgan, J.V., Minshull, T.A., Bayrakci, G., Bull, J.M., Klaeschen, D., Reston, T.J., Sawyer, D.S., Lymer, G., Cresswell, D. (2018). Resolving the fine-scale velocity structure of continental hyperextension at the Depp Galicia Margin using full-waveform inversion. *Geophysical Journal International*, 212(1), 244–263.

Dean, S.M., Minshull, T.A., Whitmarsh, R.B., Louden, K.E. (2000). Deep structure of the ocean–continent transition in the southern Iberia Abyssal Plain from seismic refraction profiles: The IAM–9 transect at 40°20′N. *Journal of Geophysical Research*, 105(B3), 5859–5885.

Driscoll, N.W. and Karner, G.D. (1998). Lower crustal extension across the Northern Carnarvon basin, Australia: Evidence for an eastward dipping detachment. *Journal of Geophysical Research*, 103, 4975–4991, doi.org/10.1029/97JB03295.

Druet, M., Muñoz-Martín, A., Granja-Bruña, J.L., Carbó-Gorosabel, A., Acosta, J., Llanes, P., Ercilla, G. (2018). Crustal structure and continent-ocean boundary along the Galicia continental margin (NW Iberia): Insights from combined gravity and seismic interpretation. *Tectonics*, 37, 1576–1604, doi.org/10.1029/2017TC004903.

Eagles, G., Pérez–Diaz, L., Scarselli, N. (2015). Getting over continent ocean boundaries. *Earth–Science Reviews*, 1–56, doi.org/10.1016/j.earscirev.2015.10.009.

Funck, T., Hopper, J.R., Larsen, H.C., Louden, K.E., Tucholke, B.E., Holbrook, W.S. (2003). Crustal structure of the ocean–continent transition at Flemish Cap: Seismic refraction results. *Journal of Geophysical Research*, 108, 2531, doi:10.1029/2003JB002434.

Gillard, M., Autin, J., Manatschal, G., Sauter, D., Munschy, M., Schaming, M. (2015). Tectonomagmatic evolution of the final stages of rifting along the deep conjugate Australian-Antarctic magma-poor rifted margins: Constraints from seismic observations. *Tectonics*, 34(4), 753–783.

Hoffmann, H.-J. and Reston, T.J. (1992). Nature of the S reflector beneath the Galicia Banks rifted margin: Preliminary results from prestack depth migration. *Geology*, 20(12), 1091–1094, doi.org/10.1130/0091-7613(1992)020<1091: NOTSRB>2.3.CO;2.

Hopper, J.R., Funck, T., Tucholke, B.E., Larsen, H.C., Holbrook, W.S., Louden, K.E., Shillington, D., Lau, H. (2004). Continental breakup and the onset of ultraslow seafloor spreading off Flemish Cap on the Newfoundland rifted margin. *Geology*, 32, 93–96.

Huismans, R. and Beaumont, C. (2011). Depth-dependent extension, two-stage breakup and cratonic underplating at rifted margins. *Nature*, 473(7345), 74–78.

Krawczyk, C.M., Reston, T.J., Beslier, M.-O., Boillot, G. (1996). Evidence for detachment tectonics on the iberia abyssal plain margin. *Proc. ODP, Sci. Results*, 149, 603–615.

Lau, K.W.H., Louden, K.E., Deemer, S., Hall, J., Hopper, J.R. Tucholke, B.E., Holbrook, W.S., Larsen, H.C. (2006). Crustal structure across the grand banks – Newfoundland basin continental margin. II. Results from a seismic reflection profile. *Geophysical Journal International*, 167(1), 157–170, doi.org/10.1111/j.1365-246X.2006.02989.x.

Le Pichon, X. and Barbier, F. (1987). Passive margin formation by low-angle faulting within the upper crust: The northern Bay of Biscay margin. *Tectonics*, 6, 133–150.

Lymer, G., Cresswell, D., Reston, T.J., Bull, J.M., Sawyer, D.S., Morgan, J.K., Stevenson, C., Causer, A., Minshull, T.A., Shillington, D.S. (2019). The development of detachment faulting and asymmetry during 3D continental breakup. *Earth Planetary Science Letters*, 515, 90–99.

Malod, J.A. and Mauffret, A. (1990). Iberian plate motions during the Mesozoic. *Tectonophysics*, 184(3–4), 261–278, doi.org/10.1016/0040-1951(90)90443-C.

Manatschal, G., Froitzheim, N., Rubenach, M., Turrin, B.D. (2001). The role of detachment faulting in the formation of an ocean-continent transition: Insights from the Iberia Abyssal Plain. In *Non-Volcanic Rifting of Continental Margins: A Comparison of Evidence from Land and Sea*, Wilson R.C.L., Whitmarsh, R.B., Taylor, B., Froitzheim, N. (eds). Geological Society, London.

Manatschal, G., Lavier, L., Chenin, P. (2015). The role of inheritance in structuring hyperextended rift systems: Some considerations based on observations and numerical modeling. *Gondwana Research*, 27(1), 140–164.

Marrett, R. and Allmendinger, R.W. (1992). Amount of extension on “small’ faults: An example from the Viking graben. *Geology*, 20(1), 47–50.

Mauffret, A. and Montadert, L. (1987). Rift tectonics on the passive continental margin off Galicia (Spain), *Marine and Petroleum Geology*, 4(1), 49–70, doi.org/10.1016/0264-8172(87)90021-3.

McDermott, K. and Reston, T. (2015). To see, or not to see? Rifted margin extension. *Geology*, 43(11), 967–970, doi.org/10.1130/G36982.1.

McKenzie, D.P. (1978). Some remarks on the development of sedimentary basins. *Earth and Planetary Science Letters*, 40, 20–32.

Miles, P., Bouysse, P., De Souza, K. (2012). *Structural Map of the Atlantic Ocean*. CCGM-CGMW, Paris.

Mohn, G., Karner, G.D., Manatschal, G., Johnson, C.A. (2015). Structural and stratigraphic evolution of the Iberia–Newfoundland hyper-extended rifted margin: A quantitative modelling approach, *Geological Society, London, Special Publications*, 413, 53–89, doi.org/10.1144/SP413.9.

Muentener, O. and Manatschal, G. (2006). High degrees of melt extraction recorded by spinel harzburgite of the Newfoundland margin: The role of inheritance and consequences for the evolution of the North Atlantic. *Earth and Planetary Science Letters*, 252, 437–452.

Murillas, J., Mougenot, D.G., Boillot, G., Comas, M.C., Banda, E., Mauffret, A. (1990). Structure and evolution of the Galicia interior basin (Atlantic western Iberian continental margin). *Tectonophysics*, 184, 297–319.

Neves, M.C., Terrinha, P., Afilhado, A., Moulin, M., Matias, L., Rosas, F. (2009). Response of a multi-domain continental margin to compression: Study from seismic reflection–refraction and numerical modelling in the Tagus Abyssal Plain. *Tectonophysics*, 468(1–4), 113–130, doi.org/10.1016/j.tecto.2008.05.008.

Osmundsen, P.T., Peron-Pinvidic, G., Ebbing, J., Erratt, D., Fjellanger, E., Bergslien, D., Syvertsen, S.E. (2016). Extension, hyperextension and mantle exhumation offshore Norway: A discussion based on 6 crustal transects. *Norwegian Journal of Geology*, 96(4), 343–372.

Pérez-Gussinyé, M. and Reston, T.J. (2001). Rheological evolution during extension at passive non-volcanic margins: Onset of serpentinization and development of detachments to continental break-up. *Journal of Geophysical Research*, 106, 3691–3975.

Pérez-Gussinyé, M., Ranero, C.R., Reston, T.J., Sawyer, D. (2003). Mechanisms of extension at nonvolcanic margins: Evidence from the Galicia interior basin, west of Iberia. *Journal of Geophysical Research*, 108(B5), 2245, doi:10.1029/2001JB000901.

Pérez-Gussinyé, M., Phipps Morgan, J., Reston, T.J., Ranero, C.R. (2006). The rift to drift transition at non-volcanic margins: Insights from numerical modelling. *Earth and Planetary Science Letters*, 244, 458–473.

Peron-Pinvidic, G., Manatschal, G., Osmundsen, P.T. (2013). Structural comparison of archetypal Atlantic rifted margins: A review of observations and concepts. *Marine and Petroleum Geology*, 43, 21–47, doi.org/10.1016/j.marpetgeo.2013.02.002.

Peron-Pinvidic, G., Manatschal, G., The "IMAGinING RIFTING" Workshop Participants (2019). Rifted margins: State of the art and future challenges. *Frontiers in Earth Science*, 7, 218–225.

Pickup, S., Whitmarsh, R.B., Fowler, C.M.R., Reston, T.J. (1996). Insight into the nature of the ocean-continent transition off West Iberia from a deep multichannel seismic reflection profile. *Geology*, 24, 1079–1082, doi.org/10.1130/0091-7613(1996)024<1079:IITNOT>2.3.CO;2.

Prada, M., Watremez, L., O'Reilly, B., Minshull, T.A., Chen, C., Reston, T.J., Shannon, P., Klaeschen, D., Wagner, G., Gaw, V. (2017). Crustal strain-dependent serpentinisation in the Porcupine Basin, offshore Ireland. *Earth and Planetary Science Letters*, 474, 148–159.

Ranero, C.R. and Pérez-Gussinyé, M. (2010). Sequential faulting explains the asymmetry and extension discrepancy of conjugate margins. *Nature*, 468, 294–297.

Reid, I.D. (1994). Crustal structure of a nonvolcanic rifted margin east of Newfoundland. *Journal of Geophysical Research*, 99, 15161–15180.

Reston, T.J. (1993). Evidence for extensional shear zones in the mantle, offshore Britain, and their implications for the extension of the continental lithosphere. *Tectonics*, 12(2), 492–506.

Reston, T.J. (1996). The S reflector west of Galicia: The seismic signature of a detachment fault. *Geophysical Journal International*, 127, 230–244.

Reston, T.J. (2005). Polyphase faulting during the development of the west Galicia rifted margin. *Earth and Planetary Science Letters*, 237(3–4), 561–576.

Reston, T.J. (2009). The structure, evolution and symmetry of the magma-poor rifted margins of the north and central Atlantic: A synthesis. *Tectonophysics*, 468(1–4), 6–27.

Reston, T.J. (2010). The opening of the central segment of the south Atlantic: Symmetry and the extension discrepancy. *Petroleum Geoscience*, 16, 199–206.

Reston, T.J. and McDermott, K. (2014). An assessment of the cause of the "Extension Discrepancy" with reference to the west Galicia margin. *Basin Research*, 26, 135–153.

Reston, T.J. and Morgan, J.P. (2004). Continental geotherm and the evolution of rifted margins. *Geology*, 32(2), 133–136, doi.org/10.1130/G19999.1.

Reston, T.J. and Pérez-Gussinyé, M. (2007). Lithospheric extension from rifting to continental breakup at magma-poor margins: Rheology, serpentinisation and symmetry. *International Journal of Earth Sciences*, 96(6), 1033–1046, doi.org/10.1007/s00531-006 - 0161-z.

Reston, T.J. and Phipps-Morgan, J. (2004). Continental geotherm and the evolution of rifted margins. *Geology*, 32(2), 133–136, doi.org/10.1130/G19999.1.

Reston, T.J., Pennell, J., Stubenrauch, A., Walker, I., Pérez-Gussinyé, M. (2001). Detachment faulting, mantle serpentinization, and serpentinite- mud volcanism beneath the Porcupine Basin, southwest of Ireland. *Geology*, 29(7), 587–590.

Reston, T.J., Leythaeuser, T., Booth-Rea, G., Sawyer, D., Klaeschen, D., Long, C. (2007). Movement along a low-angle normal fault: The S reflector west of Spain. *Geochemistry, Geophysics, Geosystems*, 8, Q06002.

Ros, E., Pérez-Gussinyé, M., Araujo, M., Thoaldo Romeiro, M., Andrés-Martinez, M., Morgan, J.P. (2017). Lower crustal strength controls on melting and serpentinization at magma-poor margins: Potential implication for the South Atlantic. *Geochemistry, Geophysics, Geosystems*, 18, 4538–4557, doi.org/10.1002/2017GC007212.

Sawyer, D.S., Whitmarsh, R.B., Klaus, A. *et al.* (1994). Leg 149. *Proceedings of the Ocean Drilling Program, Initial Reports* [Online]. Available at: http://www-odp.tamu.edu/publications/149_IR/149TOC.HTM.

Schuba, N.C., Gray, G.G., Morgan, J.K., Sawyer, D.S., Shillington, D.J., Reston, T.J., Bull, J.M., Jordan, B.E. (2018). A low-angle detachment fault revealed: Three-dimensional images of the S-reflector fault zone along the Galicia passive margin. *Earth and Planetary Science Letters* [Online]. Available at: https://doi.org/10.1016/j.epsl.2018.04.012.

Shillington, D.J., Holbrook, W.S., Van Avendonk, H.J.A., Tucholke, B.E., Hopper, J.R., Louden, K.E., Larsen, H.C., Nunes, G.T. (2006). Evidence for asymmetric nonvolcanic rifting and slow incipient oceanic accretion from seismic reflection data on the Newfoundland margin. *Journal of Geophysical Research*, 111, B09402.

Sibuet, J.-C. (1992). New constraints on the formation of the non-volcanic continental Galicia-Flemish Cap conjugate margins. *Journal of the Geological Society*, 149, 829–840.

Sibuet, J.–C., Srivastava, S., Manatschal, G. (2007). Exhumed mantle-forming transitional crust in the Newfoundland–Iberia rift and associated magnetic anomalies, *Journal of Geophysical Research*, 112, B06105.

Srivastava, S.P., Roest, W.R., Kovacs, L.C., Oakey, G., Lévesque, S., Verhoef, J., Macnab, R. (1990). Motion of Iberia since the Late Jurassic: Results from detailed aeromagnetic measurements in the Newfoundland Basin. *Tectonophysics*, 184(3–4), 229–260, doi.org/10.1016/0040-1951(90)90442-B.

Stica, J.M., Zalán, P.V., Ferrari, A.L. (2014), The evolution of rifting on the volcanic margin of the Pelotas Basin and the contextualization of the Paraná–Etendeka LIP in the separation of Gondwana in the South Atlantic. *Marine and Petroleum Geology*, 50, 1–21.

Sutra, E. and Manatschal, G. (2012). How does the continental crust thin in a hyper-extended rifted margin? Insights from the Iberia margin. *Geology*, 40, 139–142.

Sutra, E., Manatschal, G., Mohn, G., Unternehr, P. (2013). Quantification and restoration of extensional deformation along the western Iberia and Newfoundland rifted margins. *Geochemistry, Geophysics, Geosystems*, 14, 2575–2597, doi.org/10.1002/ggge.20135.

Thinon, I., Fidalgo-González, L., Réhault, J.-P., Olivet, J.-L. (2001). Déformations pyrénéennes dans le golfe de Gascogne, Comptes Rendus de l'Académie des Sciences – Series IIA. *Earth and Planetary Science*, 332(9), 561–568, doi.org/10.1016/S1251-8050(01)01576-2.

Tucholke, B.E., Sawyer, D.S., Sibuet, J.C. (2007). Breakup of the Newfoundland-Iberia rift. In *Imaging, Mapping and Modelling Continental Lithosphere Extension and Breakup*, Karner, G.D., Manatschal, G., Pinheiro, L.M. (eds). Geological Society, London.

Tugend, J., Manatschal, G., Kusznir, N.J. (2015). Spatial and temporal evolution of hyperextended rift systems: Implication for the nature, kinematics, and timing of the Iberian-European plate boundary. *Geology*, 43(1), 15–18, doi.org/10.1130/G36072.1.

Turner, J.P. and Wilson, P.G. (2009). Structure and composition of the ocean–continent transition at an obliquely divergent transform margin, Gulf of Guinea, West Africa. *Petroleum Geoscience*, 15, 305–311.

Van Avendonk, H.J., Holbrook, W.S., Nunes, G.T., Shillington, D.J., Tucholke, B.E., Louden, K.E., Larsen, H.C., Hopper, J.R. (2006). Seismic velocity structure of the rifted margin of eastern Grand Banks of Newfoundland, Canada. *Journal of Geophysical Research*, 111(B11404), doi.org/10.1029/2005JB004156.

Vegas, R., Vázquez, J.T., Olaiz, A.J., Medialdea, T. (2016). Tectonic model for the latest Triassic-Early Jurassic extensional event in and around the Iberian Peninsula. *Geogaceta*, 60, 23–26.

Welford, K., Hall, J., Sibuet, J.-C., Srivastava, S.P. (2010). Structure across the northeastern margin of Flemish Cap, offshore Newfoundland from Erable multichannel seismic reflection profiles: Evidence for a transtensional rifting environment. *Geophysical Journal International*, 183(2), 572–586, doi.org/10.1111/j.1365-246X.2010.04779.x.

Whitmarsh, R.B. and Miles, P.R. (1995). Models of the development of the West Iberia rifted continental margin at 40°30'N deduced from surface and deep-tow magnetic anomalies. *Journal of Geophysical Research*, 100, 3789–3806.

Whitmarsh, R.B. and Wallace, P.J. (2001). The rift-to-drift development of the west Iberia nonvolcanic continental margin: A summary and review of the contribution of Ocean Drilling Program Leg 173. *Proceedings of the Ocean Drilling Program: Scientific Results*, 173, 1–36.

Whitmarsh, R.B., Beslier, M.-O., Wallace, P.J. *et al.* (1998). Leg 173. *Proceedings of the Ocean Drilling Program, Initial Reports* [Online]. Available at: http://www-odp.tamu.edu/publications/173_IR/173TOC.HTM.

Whitmarsh, R.B., Dean, S.M., Minshull, T.A., Tompkins, M. (2000). Tectonic implications of exposure of lower continental crust beneath the Iberia Abyssal Plain, Northeast Atlantic Ocean: Geophysical evidence. *Tectonics*, 19(5), 919–942.

Whitmarsh, R.B., Manatschal, G., Minshull, T. (2001). Evolution of magma-poor continental margins from rifting to seafloor spreading. *Nature*, 413, 150–154, doi.org/10.1038/35093085.

Wilson, R.C.L., Hiscott, R.N., Willis, M.G., Gradstein, P.M. (1989). The Lusitanian basin of west-central Portugal: Mesozoic and tertiary tectonic stratigraphic and subsidence history. In *Extensional Tectonics and Stratigraphy of the North Atlantic Margins*, Tankard, A.J. and Balkwill, H.R. (eds). American Association of Petroleum Geologists, Tulsa.

Wilson, R.C.L., Manatschal, G., Wise, S. (2001). Rifting along non-volcanic passive margins: Stratigraphic and seismic evidence from the Mesozoic successions of the Alps and western Iberia. In *Non-volcanic Rifting of Continental Margins: A Comparison of Evidence from Land and Sea*, Wilson, R.C.L., Whitmarsh, R.B., Taylor, B., Froitzheim, N. (eds). Geological Society, London.

Wood, R. and Barton, P. (1983). Crustal thinning and susbsidnce in the North Sea. *Nature*, 302, 134–136.

Zalán, P.V., Severino, M.G., Rigoti, C.A., Magnavita, L.P., Oliveira, J.A.B., Viana, A.R. (2011). An entirely new 3-D view of the crustal and mantle structure of a south Atlantic passive margin – Santos, Campos and Espírito Santo Basins, Brazil. *Proceedings of the AAPG Annual Conference and Exhibition, Volume Making the Next Giant Leap in Geosciences*, Houston, TX, 10–13.

Zelt, C.A., Sain, K., Naumenko, J.V., Sawyer, D.S. (2003). Assessment of crustal velocity models using seismic refraction and reflection tomography. *Geophysical Journal International*, 153(3), 609–626, doi.org/10.1046/j.1365-246X.2003.01919.x.

Ziegler, P.A. (1983). Comment on: Crustal thinning and subsidence in the North Sea. *Nature*, 304, 561.

2

Tectonic Evolution of the Mid-Norwegian Rifted Margin

Per Terje OSMUNDSEN[1,2] and Gwenn PERON-PINVIDIC[3,1]

[1] *Department of Geoscience and Petroleum, Norwegian University of Science and Technology, Trondheim, Norway*

[2] *Department of Geosciences, University of Oslo, Norway*

[3] *Geological Survey of Norway (NGU), Trondheim, Norway*

The Norwegian rifted margin consists of an abandoned rift (the North Sea), a hyperextended margin (offshore Mid-Norway) and a sheared margin (the Western Barents Sea). Following a prolonged phase of Devonian orogenic "collapse", rifting peaked in Permian-Early Triassic, Mid-Jurassic into Early Cretaceous and Late Cretaceous to Paleogene times. Whereas the earliest phase of rifting was characterized by the reactivation of orogenic collapse structures, and the truncation of mid-crustal detachments by steeper faults, which heralded the sequence of successive deformation modes that eventually led to breakup. In the North Sea, rifting waned after the Late Jurassic, whereas extension continued in the Mid-Norwegian margin until most of the crystalline crust had been excised. With linked, seawards-dipping mega-faults and extensional detachments bounding morphologically distinct margin domains, parts of the Mid-Norwegian margin structurally resembled magma-poor margins elsewhere. Substantial volumes of magma did, however, accompany breakup in the Early Eocene at c. 54 Ma, and the line of breakup appears as oblique to the Norwegian-Greenland rift system. Based on these characteristics, the Mid-Norwegian margin classifies as a type of hybrid margin where prolonged,

---

For a color version of all the figures in this chapter, see www.iste.co.uk/peronpinvidic/margins2.zip.

Continental Rifted Margins 2,
coordinated by Gwenn PERON-PINVIDIC.

apparently magma-poor extension and hyperextension pre-dated a distinctly magmatic breakup. Progressive focusing of deformation is shown from the earliest Late Paleozoic-Early Triassic rift, which formed inside the North Atlantic area of orogenic collapse, via more localized, lithospheric scale necking in the Late Jurassic and Early Cretaceous to continued Cretaceous hyperextension and Paleogene breakup. As extensional faulting thinned the crust in a series of successive deformation events, an array of syn-rift basins formed within the progressively extended crust. Whereas three generations of syn-kinematic basins are stacked in the proximal domain, and seawards migration of syn-rift depocenters started with the localization of the necking breakaway complexes. Following lithospheric-scale necking, hyperextension affected the margin in the Early Cretaceous. It is uncertain how much crust remained in the outer margin at the time of breakup, and whether mantle exhumation, an important process in magma-poor margins, took place at the Norwegian margin. The breakup phase was characterized by the intrusion of numerous sills and dykes, as well as volcanism producing lava flows, hyaloclastic deltas and distinct packages of seawards-dipping reflectors. ODP drill cores have confirmed that these reflector packages represent basaltic flows and subordinate intercalated sediments.

2.1. Introduction

The Norwegian part of the northeast Atlantic rift system comprises the abandoned North Sea rift, the Mid-Norwegian rifted margin, the Lofoten-Vesterålen margin and the transtensional NW-SE-trending west Barents Sea margin (e.g. Faleide *et al.* 2010, Figure 2.1a). The Mid-Norwegian margin is located between the narrow Lofoten margin and the abandoned North sea rift. Extensive hydrocarbon exploration and production from the 1980s onwards has supplied the industry, as well as academia with seismic reflection data of ever-increasing quality and stratigraphic information from a large number of exploration and production wells. Regional refraction profiles have been constructed through large parts of the margin based on OBS data (e.g. Mjelde *et al.* 2004, 2009; Breivik *et al.* 2010; Kvarven *et al.* 2014). Potential field data in terms of gravimetric and magnetic surveys have been collected from onshore and offshore areas by the Geological Survey of Norway (NGU). The modeling of potential field data, anchored in the refraction profiles, has resulted in regional 2D and 3D crustal models (e.g. Ebbing and Olesen 2010). Thus, the margin has been studied at a range of scales and serves as a principal natural laboratory for the evolution of rifted margins (e.g. Skogseid *et al.* 1992; Brekke 2000; Eldholm *et al.* 2002; Osmundsen *et al.* 2002; Gernigon *et al.* 2004; Tsikalas *et al.* 2005; Faleide *et al.* 2008, 2010; Osmundsen and Ebbing 2008; Peron-Pinvidic *et al.* 2013; Abdelmalak *et al.* 2017; Zastrozhnov *et al.* 2020). A continued supply of state-of-the-art geophysical data ensures the continuation of scientific contributions from this margin, as well as continued discussions on its evolution.

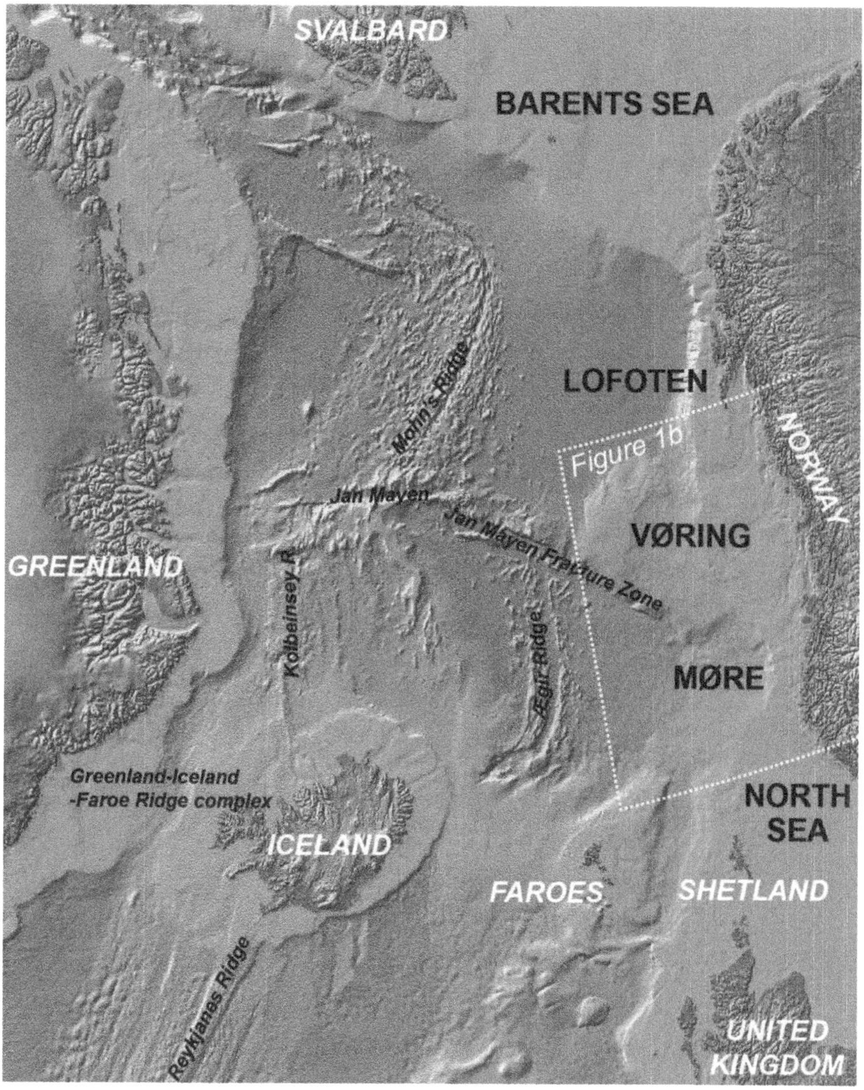

Figure 2.1a. *Map of the North Atlantic with locations of the Vøring and Møre segments of the Mid-Norwegian margin*

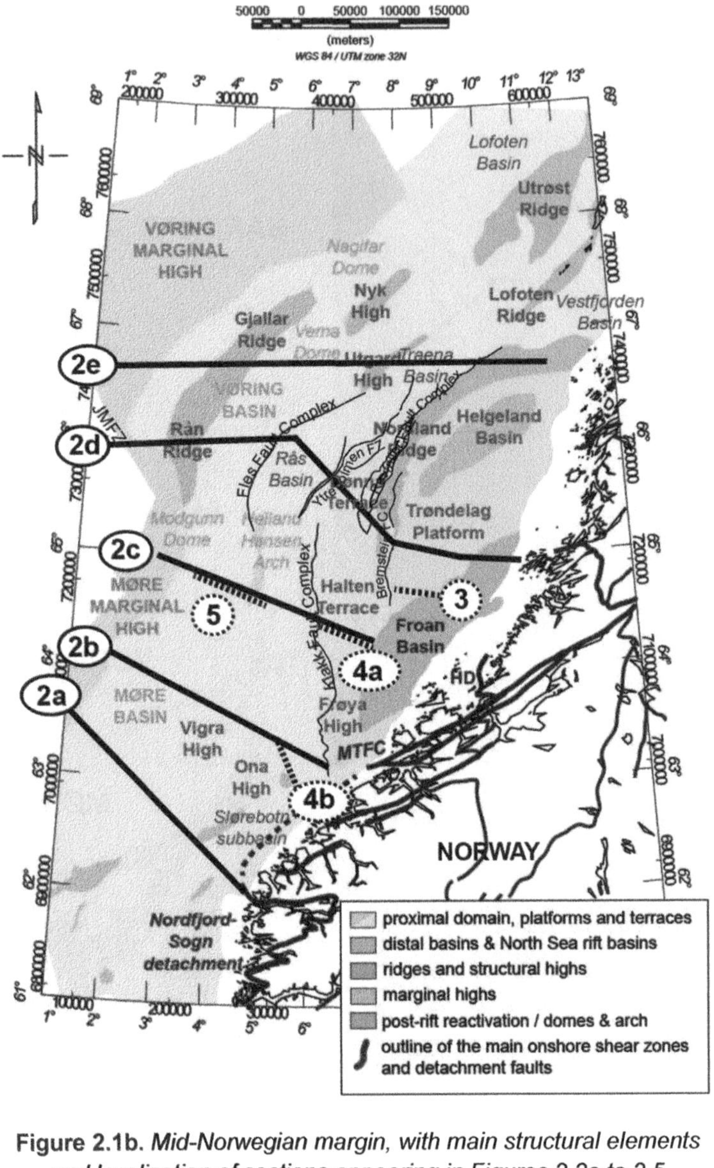

Figure 2.1b. *Mid-Norwegian margin, with main structural elements and localization of sections appearing in Figures 2.2a to 2.5*

Following the acquisition of long-offset industrial seismic reflection data in the mid- and late 1990s, the deep structure of the margin could be addressed to a degree not permitted before. Based on such data, some workers proposed that the structural configuration of the Mid-Norwegian margin showed similarities to magma-poor margins elsewhere, notably the hyperextended Iberian margin (Osmundsen *et al.* 2002; Osmundsen and Ebbing 2008; Lundin and Doré 2011; Peron-Pinvidic *et al.* 2013). The role of large-magnitude extensional detachment faults where the morphology of the margin changed and, in particular, in the areas of most pronounced crustal thinning, indicated that crustal thinning had, in part, occurred in a way similar to that observed at magma-poor margins (op. cit.). However, contrary to the Iberian margin, the Mid-Norwegian margin has not been drilled to crystalline, non-volcanic basement in its distal, most highly extended areas, and the question remains of how much the crust was actually extended prior to magmatic breakup (e.g. Rüpke *et al.* 2013; Nirrengarten *et al.* 2014; Osmundsen *et al.* 2016; Zastrozhnov *et al.* 2020). As different, and sometimes opposing views are largely based on the same datasets, the question largely relies on the philosophy of interpretation and, in particular, the emphasis on seismic reflection versus refraction data and associated potential field models (e.g. Peron-Pinvidic *et al.* 2016).

Recently, renewed exploration activity in the proximal Norwegian margin has led to the acquisition of a new generation of long-offset 2D and high-resolution 3D seismic reflection data, in areas where most older datasets failed to image the deep structure. This has inspired new ideas about the proximal margin, in particular regarding how the orogenic collapse process affected the early phases of rifting through structural reactivation (e.g. Phillips *et al.* 2016; Lenhart *et al.* 2019; Osmundsen *et al.* 2020; Peron-Pinvidic *et al.* 2020) and how the reactivated structures became variably incised and abandoned in subsequent rift stages (Osmundsen and Peron-Pinvidic 2018; Muñoz-Barrera *et al.* 2020; Osmundsen *et al.* 2020).

2.2. Large-scale structure

Based on its large-scale structure and general morphology, the Mid-Norwegian margin was originally subdivided into platform, terrace, deep basin and marginal high areas (Blystad *et al.* 1995; Brekke 2000), as reflected by the formal names used in official maps and communication (see maps and fact pages at www.npd.no). Peron-Pinvidic *et al.* (2013) suggested a subdivision of rift margin architecture into proximal, necking, distal and outer domains, in order to compare the individual components of different margins more specifically to each other. This type of subdivision is closely linked to an understanding of the margin-forming processes and provides a basis to discuss these processes in a global framework. Osmundsen and Peron-Pinvidic (2018) furthermore described the large-magnitude faults that define the boundaries between the main domains at the Mid-Norwegian margin, and that combined into persistent breakaway complexes through fault growth and interaction.

2.2.1. *The proximal margin and the proximal breakaway complex*

For the Mid-Norwegian margin, the proximal margin domain corresponds to the Trøndelag Platform and a very narrow platform area inboard of the Møre margin (Figure 2.1b). In addition, an onshore proximal domain can be defined that encompasses the area between the coast and the line along which normal crustal thickness is reached. In practice, the latter is located a few tens of km west of the present-day watershed. This area contains arrays of extensional shear zones and faults that formed after the peak of the Silurian Caledonian orogeny and that dissected and extended the Caledonian nappe pile in Early and Middle Devonian times as part of a transtensional orogenic "collapse" process, which also involved sinistral strike-slip along the future Møre Trøndelag Fault Complex (e.g. Séranne 1992; Braathen *et al.* 2000, 2002; Osmundsen *et al.* 2003, 2006). The inner boundary for the extended area (innermost boundary fault of Mosar 2003) was termed the proximal breakaway complex and assigned to a margin-wide hierarchy of domain-bounding faults by Osmundsen and Peron-Pinvidic (2018). The widespread transtension facilitated by Devonian extensional and strike-slip structures resulted in the disintegration of the overthickened Caledonian crust and the eventual reduction of crustal thickness to (nearly) normal (e.g. Andersen *et al.* 1999). Devonian transtension also resulted in the deposition of an array of "Old Red" supradetachment basins (see Figures 2.2a to 2.2c).

The offshore part of the proximal domain appears to contradict the common view that proximal margins are only moderately extended and dominated by low-β half-graben basins (Peron-Pinvidic *et al.* 2020). The proximal domain offshore Mid-Norway is significantly extended, and contains deep basins juxtaposed with detachment faults and extensional culminations (Osmundsen *et al.* 2020; Peron-Pinvidic *et al.* 2020). The ages of these basins are uncertain, but they are most likely Paleozoic to Early Triassic in age and are assumed to have developed above deep structures inherited from the preceding phase of orogenic "collapse". This deformation mode has been termed the "reactivation mode", as it was characterized by the reactivation and modification of structures related to orogenic "collapse". The "reactivation" mode of deformation facilitated crustal thinning to considerably less than 30 km thickness (Osmundsen *et al.* 2020; Peron-Pinvidic *et al.* 2020). The reactivated structures and associated basins were, in turn, truncated by steeper and more deeply rooted, landwards- and seaward-dipping faults that bound regional half-graben basins of Late Permian (?) to Early Triassic age (Osmundsen *et al.* 2020). The proximal faults and basins were draped by an Upper Triassic to Lower Jurassic inter-rift succession to record thermal subsidence after Paleozoic-Early Triassic rifting (e.g. Ravnås *et al.* 2014).

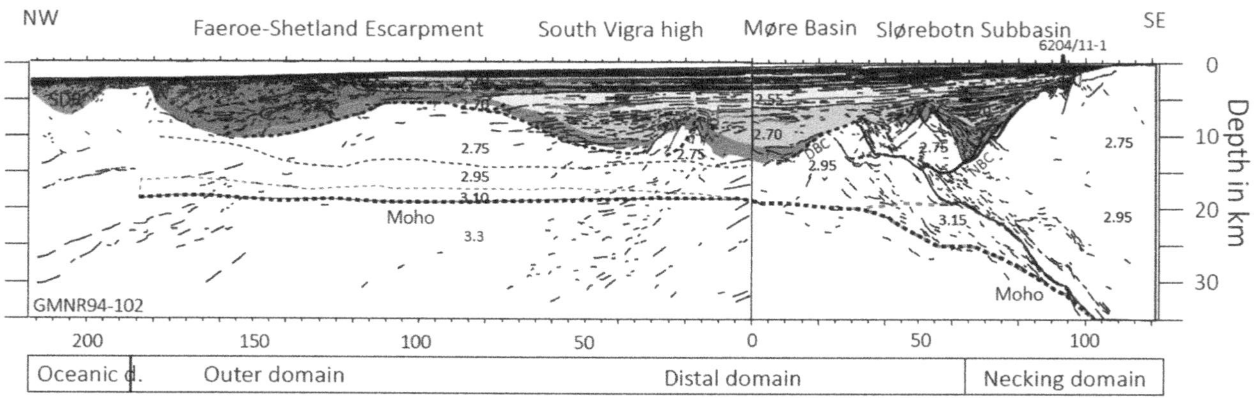

Figure 2.2a. *Section from the NE Møre margin in the area of the Slørebotn Subbasin. Location in* Figure 2.1. *Note the sharply tapering crystalline crust between the top of seismic basement and a set of northwestwards rising deep reflectors that likely represent the Moho or strongly deformed lower crust just above the Moho. The modeled Moho attains a subhorizontal orientation west of the sharpest crustal taper, at an angle with sets of dipping reflectors interpreted from seismic reflection data. Also note high-density (3.15 kg/dm*^{3}*) body under sharply tapered basement, as identified by Kvarven (2014). Also note apparent northwestwards increase in the thickness of seismic basement above the Moho, in an area where reflectors are horizontal. In the northwest, volcanics dominate the outer margin and no interpretation of deep structure can be made (source: adapted from Osmundsen* et al. *2016)*

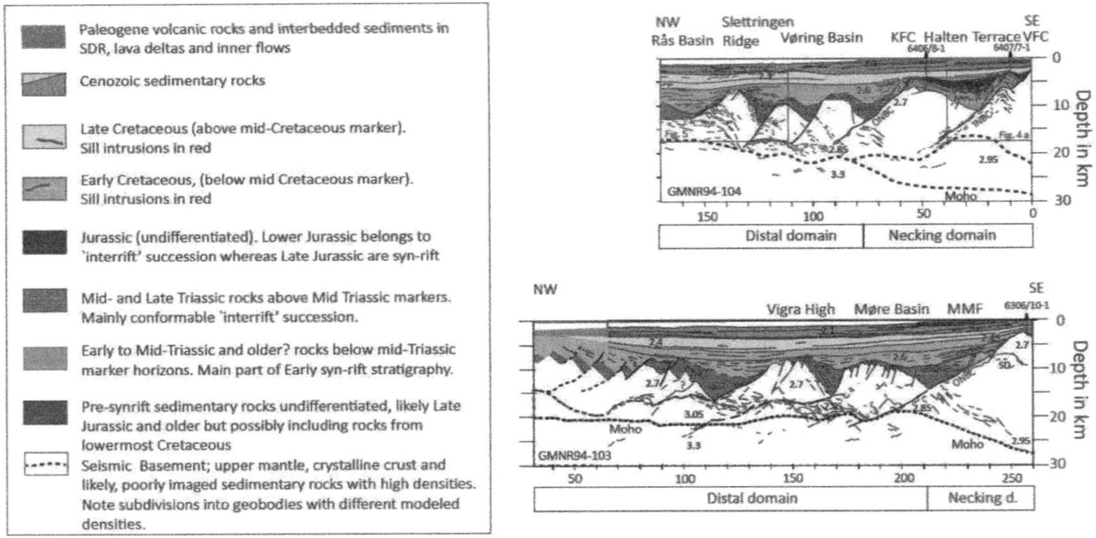

Figure 2.2b. *Bottom right: section from the NE Møre basin. Location in Figure 2.1. The Main Møre Boundary fault (MMF) constitutes the outer necking breakaway complex in this area, separating relatively shallow parts of the Sløbrebotn Subbasin (see also Figure 2.4b) from the deep Møre Basin. Note the array of fault-blocks west of the MMF, characterized by thicknesses of 3–12 km for the seismic basement. This region has been described as a hyperextended subdomain in the distal margin (Peron-Pinvidic* et al. *2013), positioned between the outer necking and distal breakaway complexes. Northwest of the distal breakaway complex, the structure rises northwestwards towards the area of breakup (source: adapted from Osmundsen and Ebbing 2008)*

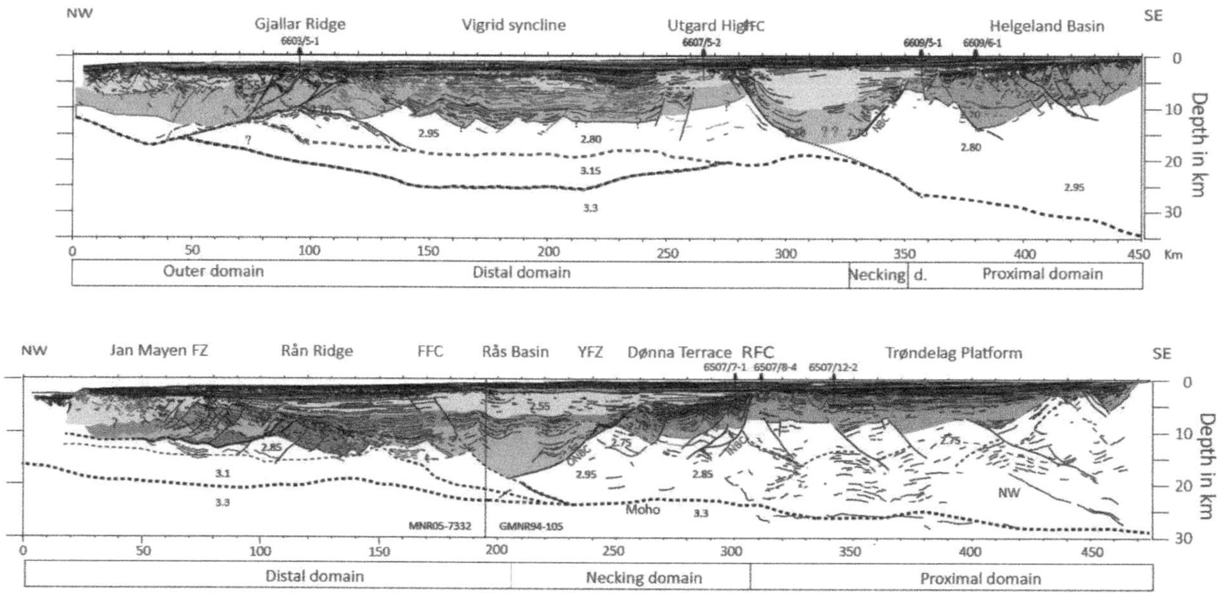

Figure 2.2c. *Top: section through central parts of the Vøring Basin from the Trøndelag Platform across the Rån Ridge. Note southeastwards-dipping faults in Early Triassic (and older?) strata under draping late Triassic and younger strata under the Trøndelag platform.*

CONTINUATION OF CAPTION FOR FIGURE 2.2b.– *Top right: section through the south Halten Terrace and parts of the Vøring Basin. Red rectangles mark areas of seismic clippings in Figures 2.4b and 2.5. Note ramp-flat shape of the inner necking breakaway complex, represented in this area by the Vingleia Fault Complec (VFC). Also note the fault-controlled, broadly synclinal shape of the downfaulted Halten Terrace and lower crustal culmination under the VFC. West of the Slettringen Ridge, the distal breakaway complex (DBC) KFC – Klakk Fault Complex.*

CONTINUATION OF CAPTION FOR FIGURE 2.2c.– *Reflector geometries in the upper parts of the "seismic basement" indicate they represent pre-Triassic sedimentary rocks and the deep structure of the proximal margin shows considerable variation. The inner and outer necking breakaway complexes that facilitated Jurassic-Cretaceous crustal-scale necking of the seismic basement are mainly west- and northwest-dipping and are represented in this section by the Revfallet and Ytreholmen fault complexes (RFC and YFC, respectively). Note the landwards-dipping Fles Fault Complex (FFC).*

Large-magnitude extensional faulting characterizes the trailing edge of the Rån Ridge, as well as the area northwest of the ridge, which is occupied by a very deep Cretaceous basin associated with a deep-seated detachment fault. Bottom: section through the northern Vøring basin from the Helgeland Basin to the Gjallar Ridge (Figure 2.1). The section is oblique to the structural grain. The northeastwards continuation of the FFC facilitated landwards downfaulting in the Træna basin in the Early Cretaceous. The Vigrid syncline is densely intruded, effectively masking the deep structure. In the south Gjallar Ridge, faulting affected Late Cretaceous strata and the Cretaceous succession was drilled to 5 km depth.

Figures 2.2a–c show geoseismic sections from the Møre and Vøring segments of the Mid-Norwegian rifted margin, which have been adapted from Osmundsen *et al.* (2016). The sections illustrate the large-scale structure of these areas well, but they also show some of the problems associated with the interpretation of structure and lithology at depth. Whereas most of the sections represent depth-converted interpretations of long-offset seismic reflection data, stippled lines in the lower parts of the sections outline geobodies extracted from updated versions of 2D and 3D potential field models for the Norwegian margin, developed at the Geological Survey of Norway (Ebbing and Olesen 2010), as well as the interpretation of the Møre margin by Kvarven *et al.* (2014).

It is important to note the position of the Moho and geobodies (commonly referred to as Lower Crustal Bodies or LCB), with calculated densities between 3.0 and 3.3 kg/dm^3. In general, these may represent high-density lower crust, mafic intrusions, serpentinized mantle or a mixture of these. Although some authors may argue for one or the other of these alternatives with admirable confidence, we maintain that the lithological content of these levels remains speculative. Moreover, they are likely to be highly deformed, limiting the validity of methods such as Vp/Vs ratios in the assessment of lithology.

In fact, significant uncertainty also concerns rocks of densities lower than 3.0 kg/dm^3, for instance where deeply buried sediments have lost their porosity at great depth (e.g. Peron-Pinvidic *et al.* 2016). The main faults that separate margin domains are the inner and outer necking breakaway complexes (INBC and ONBC, respectively) and the distal breakaway complex (DBC). In areas where an inner necking breakaway complex cannot be defined, the main boundary to the deep basin is termed the necking breakaway complex (NBC)

2.2.2. *The necking domain and the necking breakaway complexes*

The necking domain corresponds to the areas of the Dønna and Halten terraces and the Slørebotn Subbasin, plus kilometers-wide, tectonically defined basement slopes between the terraces and the deep basins. It is framed by inner and outer breakaway complexes, respectively, that comprise an array of N-S to ENE-WSW-trending, large-magnitude normal faults such as the Bremstein, Klakk and Møre-Trøndelag fault complexes (MTFC; Figure 2.1). The areas that constitute the necking domain are characterized by the strong thinning of the crystalline crust, down-stepping of the top of basement and rise of the Moho in the direction of the taper break, which denotes the boundary between the necking and distal domains (Osmundsen and Redfield 2011).

Inboard of the Vøring Basin, the faults that bound the necking domain on its landwards side vary in geometry from planar in the north to ramp-flat along the south margin of the Halten Terrace (Figure 2.3a). They represent the Mid-Late Jurassic reactivation of deeply rooted faults inherited from Early Triassic rifting and a focusing of deformation on seawards-dipping faults. Jurassic reactivation varied strongly in magnitude, and was moderate in parts of the central Halten terrace where an extensionally forced monocline formed, but was not broken, above the reactivated Bremstein Fault Complex. Farther south, reactivation increased along the BFC, re-using Triassic ramp-flat geometries and becoming responsible for the formation of a rollover antiform-synform pair in the south Halten terrace area (see Figure 2.3a and Ehrlich and Gabrielsen 2004). Farther south in the Møre margin, the main strand of the MTFC provides the landward boundary for the necking domain in the form of the Slørebotn Subbasin, where fault blocks rotated up to 50° on an underlying detachment fault in the Late Jurassic (see Figure 2.3b and Jongepier *et al.* 1996).

The outer boundary for the necking domain is constituted by an array of faults that incised into earlier structures and linked to form the outer necking breakaway complex (Osmundsen and Peron-Pinvidic 2018). These were active in the Early Cretaceous, as evidenced by truncation and offset of the "BCU", as well as by onlap onto exhumed footwalls and footwall uplifts by Early Cretaceous strata (Blystad *et al.* 1995; Jongepier *et al.* 1996; Osmundsen and Peron-Pinvidic 2018). The linkage of faults in the outer necking breakaway complex controlled the width of the necking domain. The north-south trending Klakk Fault complex (KFC) linked ENE-WSW-trending large-magnitude faults outboard of the Trøndelag Platform with the ENE-WSW-oriented main Møre boundary fault (MMF, see Figure 2.3b). This isolated the Halten and Dønna terraces as up to 200 km wide morphological units in the necking domain, whereas to the south, the necking domain consists of the relatively narrow Slørebotn subbasin and the basement slope of the MMF.

In the north, the necking domain mainly consists of basinwards dipping slopes associated with the Ytreholmen and Revfallet fault complexes of Blystad *et al.* (1995). The geometry of the necking domain was also affected by the linkage of large-displacement normal faults. For instance, in the area of widely spaced necking breakaway complexes, the cross-sectional geometry of the necking domain corresponds to that of a regional half-graben or hanging wall of a ramp-flat extensional fault. The necking geometry in the Møre area, where the necking breakaways are much more tightly spaced, varies between that of an uplifted core complex with allochthonous blocks and a faulted, very sharply tapered crustal wedge (see Figures 2.3b and 2.2a).

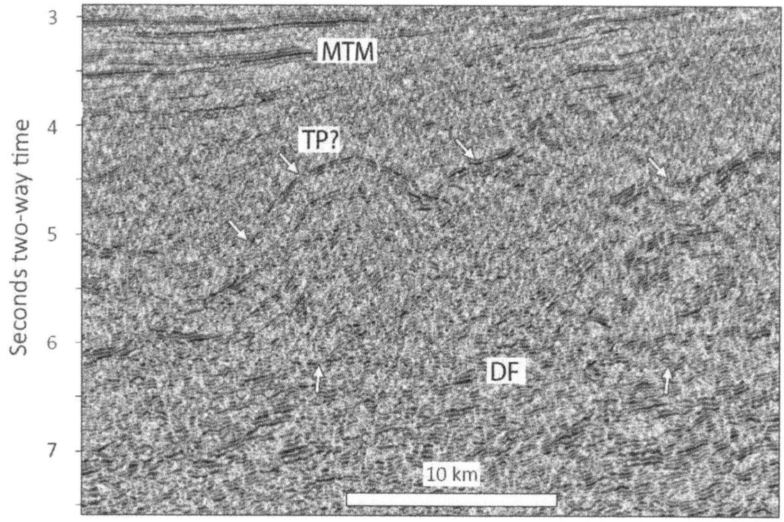

Figure 2.3. *Rotated fault-blocks with pre-Middle Triassic sedimentary wedges on a more gently dipping reflector package representing a detachment zone under the Trøndelag Platform. Such supradetachment fault-blocks and basins occur under substantial parts of the Trøndelag Platform, including the Froan and Helgeland basin areas. MTM – Mid-Triassic markers; TP – Top Paleozoic? DF – Detachment fault*

The outer boundary for the necking domain is the taper break (Osmundsen and Redfield 2011), which marks a persistent and margin-wide morphological change associated with reduction in crustal thickness to less than c. 10 km, most commonly related to hanging wall cutoffs at the foot of the basement slopes associated with the outer necking breakaway complex. The value of 10 km is, in turn, considered critical for the evolution of the distal margin as in the principle, embrittlement may occur at this point and cause truncation of the entire crust by faults (Peréz-Gussinyé and Reston 2001). In some areas, the fault that defines the taper break is assumed to cut directly into the mantle (e.g. Osmundsen and Ebbing 2008). In other areas, the geometry at depth is more complex and involves ramp-flat geometries in the reflective lower crust (see Figure 2.3b).

The combination of wide versus narrow platform and corresponding necking domains give rise to variations in bulk structure corresponding to sharply and more gently tapered margin segments, respectively, inboard of the taper break (Osmundsen and Redfield 2011). The variable sharpness in crustal taper appears to have controlled variations in onshore topography in the post-rift phase (op. cit.).

2.2.3. *The distal domain and the distal breakaway complex*

The basement and deep structure of the distal domain are buried under up to 15 kilometers or more of sedimentary rocks, and the high number of sills intruded into the Cretaceous Møre and Vøring basins reduces the quality of seismic imaging. However, interpretation of modern long-offset seismic reflection data and updated potential field models have provided new insight into these areas. The distal margin domain offshore Norway can be divided into two main parts, based on interpretations of crustal thickness, structural style and morphology (Osmundsen and Ebbing 2008). These were assigned to a hyperextended and an exhumed subdomain, respectively, by Peron-Pinvidic *et al.* (2013). The hyperextended subdomain, lozenge-shaped in the map view (Figure 2.1), has been mapped in the northern Møre and southern Vøring basins, where it is constituted by an array of deep-seated fault blocks in crust thinned to less than 10 km on the average according to depth-converted sections (e.g. Figure 2.2b, c). The fault-blocks were divided into three distinct seismic facies; an upper layer that represents sedimentary strata preserved in deep-seated half-graben, an underlying layer that is transparent or characterized by short, discontinuous reflectors and a lower layer of strongly reflective material, commonly defining sigmoidal shapes between block-bounding faults (Osmundsen and Ebbing 2008). The hyperextended subdomain is best defined in the south Vøring and northeast Møre basin; southwards, the tilted fault blocks become less well-defined and eventually invisible in the seismic data and northwards, the subdomain pinches out against the northwest side of the Trøndelag Platform. This is likely due to linkage and interaction between major faults and convergence of the outer necking and distal breakaway complexes in the north (Osmundsen and Peron-Pinvidic 2018). However, data quality tends to be poorer to the south and north of the mapped subdomain, hampering our interpretation of the transition between the hyperextended subdomain and the deep basins to the north and south.

Outboard of the hyperextended subdomain, a marked change in margin morphology occurs across the faults of the distal breakaway complex, which tend to flatten with depth at levels between 9 and 10 s twt (see Figure 2.4). A corridor at least 30–50 km wide in the south Vøring and north Møre basin is thus underlain by a system of flat-lying or very gently seawards dipping linked detachment faults that truncate landwards-dipping reflectors in the underlying lithosphere (Osmundsen and Ebbing 2008). Based on magma-poor margin analogues such as the Iberian margin and remains of Tethys margins preserved in the Alps (e.g. Manatschal 2004), it has been suggested as a possibility that the footwall of this structure may comprise rocks exhumed from the lithospheric mantle (Osmundsen and Ebbing 2008; Lundin and Doré 2011; Peron-Pinvidic *et al.* 2013). However, as independent evidence for mantle exhumation have not been obtained at the mid-Norwegian margin, the lithological content of the basement that underlies these most highly extended parts is highly uncertain and remains debated (e.g. Rüpke *et al.* 2011; Nirrengarten *et al.* 2014; Gernigon *et al.* 2015; Osmundsen *et al.* 2016; Peron-Pinvidic *et al.* 2016; Abdelmalak *et al.* 2017; see Figure 2.4). Thus, it is uncertain whether this area represents a highly thinned domain of hyperextended crust west and northwest of the distal breakaway complex or an exhumed subdomain with mantle rocks directly underneath the footwall of the regional detachment. In the hanging wall of this detachment, allochthonous blocks with wedge-shaped syn-kinematic basins occur below subhorizontal reflector packages of interpreted Early Cretaceous age (Osmundsen and Peron-Pinvidic 2018). In the central Vøring Basin, the Rån Ridge appears as a larger, highly rotated block above the regional detachment fault and is associated with smaller allocthonous blocks at its trailing edge (see Figure 2.2d and Osmundsen *et al.* 2016; Peron-Pinvidic and Osmundsen 2016; Osmundsen and Peron-Pinvidic 2018).

In the northern Vøring basin, large rotated fault-blocks of the type that characterizes the hyperextended subdomain farther south are not well represented in the seismic reflection data. It is possible that they are simply absent, consistent with a merging of the outer necking and distal breakaway complexes close to the northwest corner of the Trøndelag platform. Large parts of the distal margin do, however, belong to the so-called Vigrid syncline (Blystad *et al.* 1995), which is infested by a dense mesh of sill intrusions that effectively masks the underlying structure (e.g. Planke *et al.* 2005). Uncertainties are thus large regarding the deep structure of the north Vøring basin. In addition, areas such as the Nyk High (Blystad *et al.* 1995) represent areas uplifted in the Cenozoic, thus at a much later stage than the deeply buried highs that occur in the distal margin farther south.

From the Møre to the North Vøring basin, the structure rises towards the line of breakup. The deep structures in these areas are masked by intrusions and are poorly imaged, if imaged at all. Based on combinations of seismic reflection interpretation, refraction studies and potential field modeling, most workers interpret a gradual thickening to c. 10–15 km of the crystalline crust from the deepest parts of the Møre and Vøring basins towards the line of breakup, indicating that breakup occurred in a crust that was thicker than the crust preserved under the deepest parts of the basins.

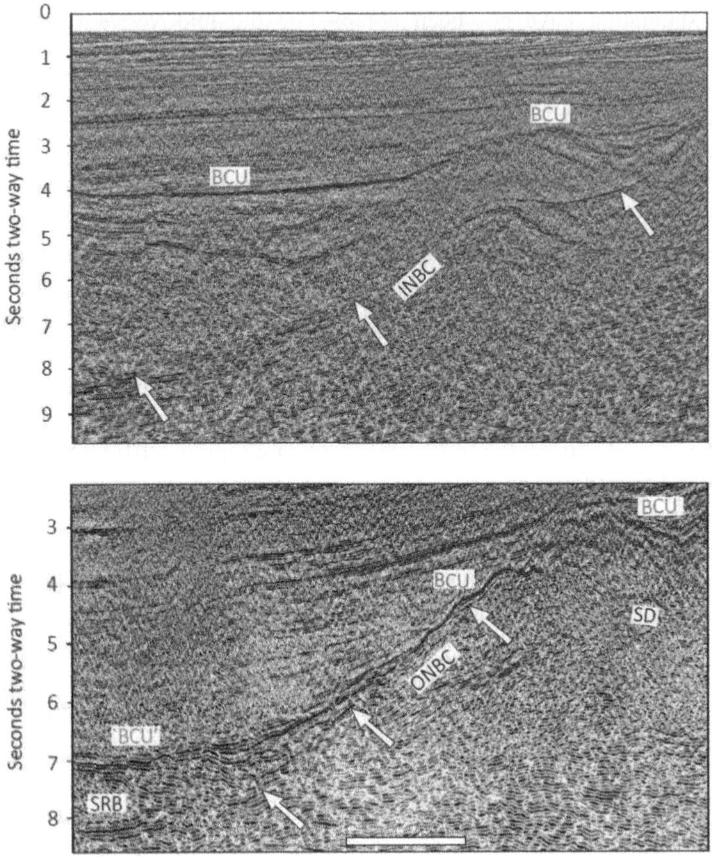

Figure 2.4. *Seismic images of faults that constitute the inner and outer necking breakaway complexes, respectively, at the Mid-Norwegian margin*

CONTINUATION OF CAPTION FOR FIGURE 2.4.– *Top: the Vingleia Fault complex, south Halten Terrace. Compare to depth-converted version in right-hand part of Figure 2.2 c. The ramp-and-flat geometry of this impressive extensional fault strongly affected the geometry of its overlying hanging wall in the southern Halten Terrace area), leading to Late Jurassic depocenter localization in synclinal and downfaulted areas in the hanging wall (e.g. Ehrlich and Gabrielsen 2004; Elliott et al. 2015). The Vingleia fault was also active in the Triassic (Bunkholt et al. in press), indicating necking was heralded by truncation of older mid-crustal detachment faults already in the Triassic (Osmundsen et al. submitted). BCU – Base Cretaceous; INBC – Inner necking breakaway complex, represented here by the Vingleia Fault Complex. Bottom: the Slørebotn detachment with overlying rotated fault blocks and the incising outer necking breakaway complex constituted by the MMF, in the Gossa High area of the Slørebotn subbasin. In this area, a metamorphic core complex is assumed to have developed in Late Jurassic into Early/Mid-Cretaceous time, as the Slørebotn detachment was deactivated and incised by the outer necking breakaway complex, represented here by the main boundary fault for the Møre basin (MMF; Osmundsen and Ebbing 2008). Note the "bumpy" appearance of MMF where it is directly onlapped by Cretaceous strata, likely indicating erosion into the fault plane. A rotated syn-rift basin is preserved below the BCU in the lower left and in the upper right, strongly rotated pre-to syn-rift Triassic into Late Jurassic strata are preserved below the BCU, on top of the Slørebotn Detachment (SD). Less complex fault geometries are associated with the necking breakaway complexes in other parts of the margin, where constituent faults have more planar geometries.*

2.2.4. *The outer domain*

The outer margin comprises a series of outer highs such as the Gjallar Ridge (Figure 2.2e) and the marginal high areas. The Gjallar Ridge contains Cretaceous sediments to great depth, as shown by the 6603/5-1 well (see https://factpages.npd.no), which penetrated to 5000 meters, with Aptian strata at the termination depth. As the well was drilled with a Jurassic target (see https://factpages.npd.no), this showed that the Cretaceous succession was deeper than previously thought, supporting high values of extension in the outer margin. Several more kilometers of sediments are believed to reside beneath this depth; however, the position of the BCU remains uncertain. Moreover, the Cretaceous sediments encountered close to the termination depth had densities of c. 2.7 g/cm, consistent with an observation of velocities of 6 km/s in parts of the Cretaceous succession (Peron-Pinvidic *et al.* 2016). Thus, it was pointed out that compaction of deeply buried sediments, as well as serpentinization of mantle may both cause misinterpretation of geophysical datasets, as the rocks in question would acquire densities that overlap with those of crystalline crust (Osmundsen *et al.* 2016; Peron-Pinvidic *et al.* 2016). The result of such misinterpretation may be a thicker crust in the outer margin areas. However, any interpretation of deep structure or basement lithology in this area will be hampered with uncertainty.

2.3. Syn-rift basin configuration and evolution

The Mid-Norwegian margin is associated with a suite of syn-rift basins that generally become younger in the seaward direction (e.g. Doré and Lundin 1997) overlain diachronously by post-kinematic strata. The oldest extensional basins associated with the margin are the mainly Devonian, "Old Red" supradetachment basins associated with orogenic transtensional "collapse". Juxtaposed with kilometers-thick, regional detachment zones that exhumed high-grade metamorphic rocks from the Caledonian root in their footwalls, the basins record the tectonic and erosional disintegration of the Caledonian orogen that followed Silurian mountain building (e.g. Andersen *et al.* 1994). The "Old Red" basins are entirely continental (Steel *et al.* 1985), consisting of coarse breccias, conglomerates and sandstones tilted landwards when viewed in sections parallel to the WNW-ESE to WSW-ENE maximum elongation direction (Séranne and Seguret 1987, Osmundsen *et al.* 1998). Orthogonal shortening related to the transtensional strain field that accompanied deposition resulted in the basins being folded with axes parallel to the extension direction (Krabbendam and Dewey 1998, Osmundsen and Andersen 2001). Basin deposition was thus accompanied by fault growth, rollover formation and orthogonal shortening (Osmundsen *et al.* 2000). Uplift and erosion of the Devonian structural template led to formation of a major unconformity recognized in the northern North Sea (Lenhart *et al.* 2019), as well as offshore Mid-Norway (Osmundsen *et al.* 2021).

Early phases of sedimentation in Paleozoic, post-Middle Devonian into Early Triassic basins preserved in the offshore proximal domain likely took place under continued or renewed detachment faulting and represent a transitional mode of deformation that post-dated Devonian transtension, but pre-dated formation of the set of steeper faults that developed slightly later. Thus, the basins deposited in the "reactivation phase" can also be described as supradetachment basins, but they do not appear to have been significantly shortened normal to the extension direction and rest unconformably on the folded Caledonian-Devonian substrate (op. cit.). Following widespread reactivation and decoupled crustal thinning, the truncation of the detachments by steeper, more deeply rooted normal faults resulted in the formation of large half-graben basins in Late Paleozoic(?) and/or Early Triassic time. The basins in the transition and early rifting phase contain the transition to marine conditions offshore Mid-Norway, and Late Permian to early Triassic basins contain carbonates, shales, sandstones and evaporites (e.g. Müller *et al.* 2005; Bunkholt *et al.* in press). Some magmatic activity accompanied the Late Permian-Early Triassic rifting, as represented onshore by Permian and Early Triassic dykes (Torsvik *et al.* 1997; Fossen and Dunlap 1999). On the conjugate East Greenland margin, Mid-Permian and probably Middle Triassic fault activity has been documented with K-Ar illite dating (Rotevatn *et al.* 2018), and an Early Triassic dyke swarm intruded in Liverpool Land at c. 244 Ma (Seidler *et al.* 2004). In the proximal margin offshore Mid-Norway, two Middle Triassic salt intervals and their lateral equivalents herald Late Triassic and Early Jurassic "interrift" sedimentation. The "interrift" succession records fluctuating shorelines and widespread coastal conditions including fluvial, swamp and shallow marine environments (e.g. Brekke *et al.* 2001; Ravnås *et al.* 2014). The "interrift" stage was characterized by relatively quiet tectonic conditions and likely by isostatic and thermal subsidence after the Late Paleozoic into Early Triassic rifting (Ravnås *et al.* 2014).

Landwards of the inner necking breakaway, Mid- and Late Jurassic deposits are also mostly unaffected by faulting and appear as "post-kinematic" with respect to underlying Permo-Triassic faults (e.g. Brekke 2000). From Mid- and I particular Late Jurassic time, however, renewed rifting and fault-block rotation outboard of the inner necking breakaway resulted in a change to deep marine conditions, with repeated input of coarse debris and formation of thick, syn-tectonic wedges along major controlling faults (Jongepier *et al.* 1996; Brekke 2000; Elliott *et al.* 2015; Bunkholt *et al.* submitted). Late Jurassic syn-rift deposits are likely also present on deeply buried fault-blocks and allochthon in the distal margin domain.

In the Early Cretaceous, both the proximal and necking domains were draped by Early Cretaceous strata. Inherited fault scarps were still forming topographic highs and some faulting activity also occurred in the Cretaceous (Bunkholt *et al.* submitted). Outboard of the outer necking breakaway, however, syn- and post-kinematic Early Cretaceous strata onlapped rotated fault blocks in the hyperextended subdomain and some of the fault-controlled highs were probably subject to erosion (Fjellanger *et al.* 2005). Outboard of the distal breakaway, strata of inferred Early Cretaceous age became involved in the faulting and rotation of allochthonous blocks on regional, flat-lying detachment faults and wedge-shaped reflector packages of unknown age are associated with the allochthonous blocks. The abovementioned 30-50 km wide expanse representing a regional detachment fault is overlain by an equally wide supradetachment basin (Figure 2.4) For the most deeply buried parts of the distal domain, syn-kinematic deposition was likely over by Albian-Aptian times (Osmundsen *et al.* 2016; Peron-Pinvidic and Osmundsen 2016). Zastrozhnov *et al.* (2018) did, however, report evidence for Mid-Cretaceous syntectonic sedimentation in the central Vøring Basin and northwest of the Rån Ridge and on the Gjallar Ridge in the outer Vøring basin, deformation was resumed in the Late Cretaceous and Paleogene, heralding magmatic breakup. Northwest of the Rån Ridge, a densely faulted deep Cretaceous basin developed above a listric detachment fault. Correlation with drilled intervals in the 6603/5-1 S well on the south Gjallar Ridge shows that Albian-Aptian strata are even more deeply buried in the Cretaceous basin west of Rån, indicating that faulting activity and syn-rift sedimentation migrated seawards through much of the Cretaceous. Drilled intervals of the Late Cretaceous in the Møre and Vøring basins were dominated by deep marine background sedimentation interrupted by subordinate sandy basin-floor fan deposits (Fjellanger *et al.* 2005; Sømme and Jackson 2013). In the Vøring Basin, thick sandy successions were deposited as basin-floor fans into the deep marine environment in Campanian and Maastrichtian times, reflecting the erosion of source areas in East Greenland (Fjellanger *et al.* 2005).

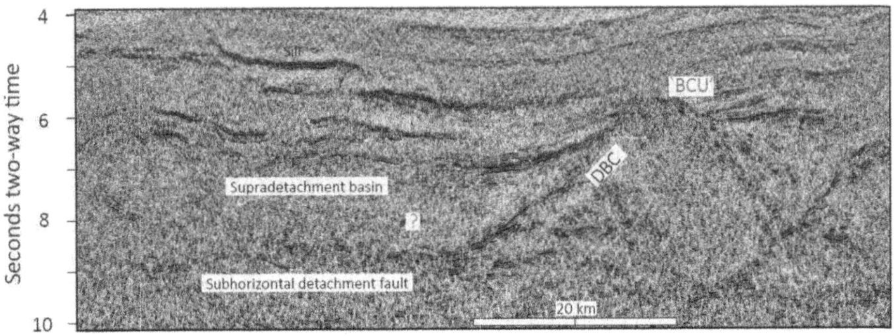

Figure 2.5. *The distal breakaway complex (DBC) as imaged northwest of the Slettringen Ridge in the south Vøring Basin. Compare to the depth-converted version in left-hand part of* Figure 2.2c. *The main fault in the DBC flattens with depth to constitute the top of strongly reflective basement, representing rocks in the lower crust or upper mantle. This "flat zone" with rotated strata resting on a subhorizontal to gently seawards dipping reflector or reflector package is c. 30–50 km wide and can be mapped along strike for 300 km in the northern Møre and south Vøring basins. It is interpreted as a subhorizontal, Early Cretaceous detachment fault (Osmundsen and Peron-Pinvidic 2018) or system of releayed detachment faults overlain by east-to southeastwards rotated strata in a distal supradetachment basin*

The result of the generally seawards-migrating tectonic activity during necking and hyperextension was a highly diachronous depositional pattern in the Late Jurassic and Cretaceous. As faults in more proximal locations became extinct, they were draped by sediments that were post-kinematic in that location, but syn-kinematic farther outboard, where fault-block rotation resulted in wedge-shaped reflector packages. Thus, parts of the stratigraphy of the Mid-Norwegian margin show an overall distribution of syn- and post-kinematic successions that resemble those described by Ribes *et al.* (2019) from the Tethyan margin preserved in the Alps. This pattern is, however, not absolute. Basin configurations in the proximal and outer domains of the Norwegian margin appear quite different from those reported from Tethyan and Iberian analogues. This is likely due to the strong imprint of inherited orogenic collapse structures offshore Norway (e.g. Osmundsen *et al.* 2020; Peron-Pinvidic *et al.* 2020.) and the Early Cenozoic processes that led to magmatic breakup.

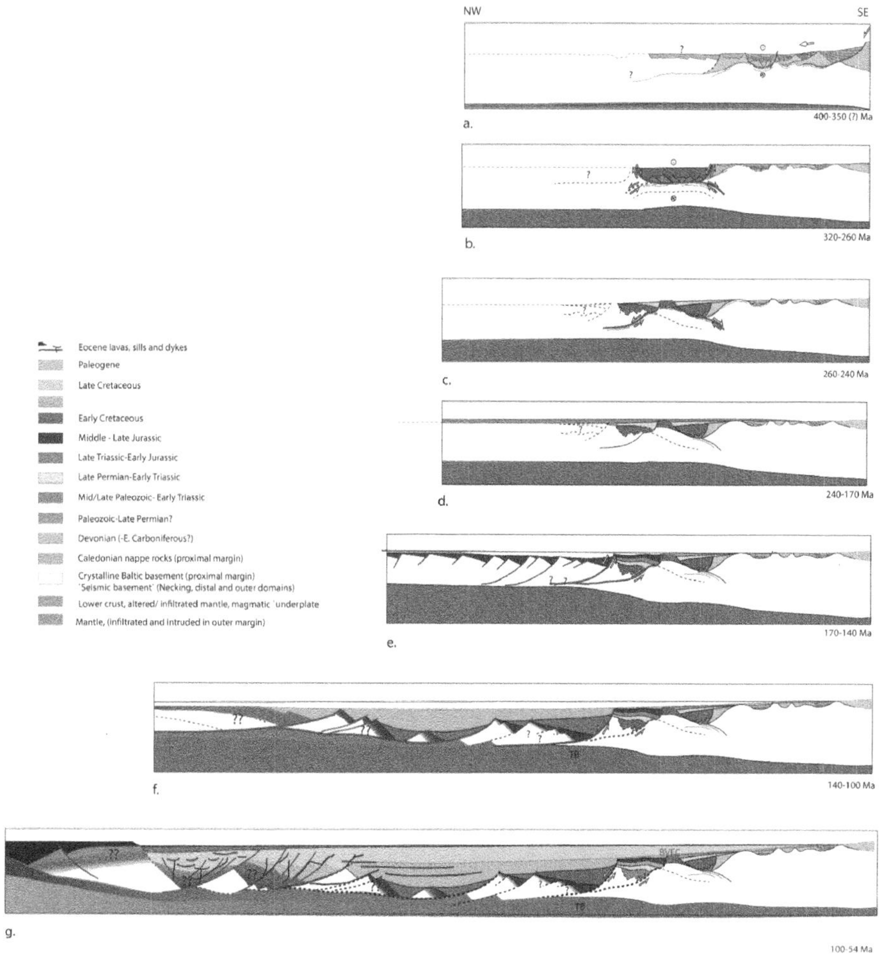

Figure 2.6. *Synthetic, schematic cartoon (not to scale) illustrating the main stages in the tectonostratigraphic evolution of the Mid-Norwegian rifted margin from the Devonian to the Eocene*

CONTINUATION OF CAPTION FOR FIGURE 2.6.– *The active faults are in red, reactivation indicated by stipples. The extinct faults are in black. Please note that the margin architecture varies significantly along strike and that few, if any single 2D section can be expected to image all of the features illustrated in the figure. Age intervals are approximate and provide variably constrained brackets for the successive stages or deformation modes. Note that the proximal margin includes areas that are now part of the onshore Norwegian mainland. Some of these areas were covered in sediments in parts of the Mesozoic and Cenozoic as indicated in the figure, but with the exception of some small Jurassic basin remnants; their extent and thickness distributions are mostly unknown.*

a) The transtensional collapse of the Caledonian mountain belt led to the formation of large-scale extensional detachments and strike-slip zones, and to the formation of a series of Devonian supradetachment basins. The basins are dominated by alluvial, fluvial and lacustrine sediments and range in age from the Early and Middle Devonian to, possibly, Late Devonian and Early Carboniferous. The post-orogenic phase of transtension commenced in overthickened orogenic crust and is thought to have lasted until the crustal thickness had been reduced to (near) normal.

b) Following Devonian transtension, the margin entered a phase of early rifting, characterized by the reactivation and modification of inherited detachment faults, and by significant amounts of crustal thinning. Judging from some previously published plate reconstructions (e.g. Mosar et al. 2003) the strike-slip component of deformation may have still been significant. A suite of Late Paleozoic into Early Triassic basins developed, with rotated fault blocks over mid-crustal detachments and ramp-flat megafaults.

c) The template of core complexes, detachment faults and associated basins became cut in the Early Triassic by steeper, more deeply rooted extensional faults. These bound wedge-shaped basins with large-scale half-graben geometries, such as parts of the Triassic Froan Basin. Some of these face towards the continent (east- and southeastwards), whereas some face west- or southwestwards, towards the future ocean.

d) From the Late Triassic through most of the Early Jurassic, regional subsidence under relatively quiet tectonic conditions led to deposition of generally conformable, draping "interrift" succession most likely related to thermal cooling after Triassic rifting.

e) In the Middle and Late Jurassic, major extensional faulting became localized along west- and northwest-dipping faults that eventually linked up to form the boundaries between the platform, subbasin and terrace areas. These faults contributed to the individualization of the necking domain and controlled the formation of Late Jurassic basins along much of the future margin. The regionally developed "Base Cretaceous Unconformity" formed near the end of this tectonic phase.

f) Localization of large magnitudes of extension along selected west- and northwest-dipping faults in the Earliest Cretaceous led to up to tens of kilometers of normal displacement of the "BCU", to the finalization of the necking domain and to the formation of the taper break (TB), which constitutes the boundary between the necking domain and the distal margin. In the distal margin, continued extension led to the near-excision of continental crust and to the dismemberment of basement and pre-Cretaceous basin rocks into rotated allochthonous fault blocks above low-angle detachment faults. Deeply buried supradetachment basins in the most highly extended parts of the margin are undated, and thus, the exact ages of these basins and associated detachment faults remain hampered with uncertainty. They are tentatively indicated as Early Cretaceous here, but other workers prefer to interpret them as entirely pre-Cretaceous in origin (see text).

g) Extension continued or was resumed in the outer parts of the margin in Late Cretaceous and Paleocene time, resulting in further fragmentation of Early and Mid-Cretaceous basin deposits. Breakup in the Eocene at c. 54 Ma was associated with large volumes of magma in the form of dykes, sills and volcanics. In the outer margin, volcanics accumulated together with subordinate sediments in kilometers-thick wedges recognized in seismic data as seawards-dipping reflectors ("SDR").

2.4. Breakup

Contrary to magma-poor counterparts such as that of Iberia, the Mid-Norwegian margin underwent a strongly magmatic breakup in the Eocene around 54 Ma. The line of breakup appears to cut obliquely through the Norwegian-Greenland rift system, albeit some parts closely follow the inherited structural grain (Figure 2.1). Thus, albeit parts of the pre-breakup architecture of the Mid-Norwegian margin resemble that of a magma-poor margin, the magmatic breakup processes induced some important differences. ODP drilling penetrated a substantial section of the volcanic outer margin (e.g. Eldholm *et al.* 1989). Extensive volcanism accompanied breakup, as recorded by the numerous basaltic flows organized in wedge-shaped seawards-dipping successions, by landwards prograding lava deltas, by several extensive, inner volcanic flow units and by a multitude of sills intruded into the Cretaceous sedimentary basin (Blystad *et al.* 1995; Planke *et al.* 1999, 2000; Berndt *et al.* 2001; Abdelmalak *et al.* 2017). The outer margin comprises both inner and outer wedges of seawards-dipping reflectors (SDR), suggested by some workers to be separated by landwards-dipping normal faults (Geoffroy 2005). Superposed on the previous margin architecture, largely dominated by seawards-dipping Jurassic-Cretaceous faults, this represents a flip of the main fault polarity on the margin.

2.5. Post-rift evolution

The Mid-Norwegian margin is covered by an Eocene and younger post-rift basin, which mainly records thermal subsidence following Mesozoic-Earliest Cenozoic rifting and processes related to episodic uplift of the Norwegian mainland (e.g. Brekke 2000; Faleide *et al.* 2002; Henriksen *et al.* 2005, 2011; Sømme *et al.* 2009; Løseth *et al.* 2017). Several prograding units are assumed to be related to uplift events, including, in particular the so-called Thanetian delta (Paleogene) and the Late Miocene-Early Pliocene or, alternatively, mainly Pliocene Molo Formation (Eidvin *et al.* 2014; Løseth *et al.* 2017). Repeated glaciations over the last 2.7 million years resulted in the progradation of a thick Pliocene-Pleistocene sediment wedge (the Naust Formation) subdivided into several subunits (e.g. Ottesen *et al.* 2009).

The cause(s) of post-rift uplift are much debated and not well-understood. Thermal effects of breakup, ridge push, thermal mantle “fingers”, lithospheric buckling and flexural effects imposed by erosion and deposition have been proposed as fundamental mechanisms behind Cenozoic uplift events (see Faleide *et al.* (2002) and the references therein). Some authors argue that the uplift of Scandinavia is unrelated to margin processes as such, and that its onshore topography represents gradual isostatic adjustments since the 400 Ma old Caledonian orogeny (Nielsen *et al.* 2009; Egholm *et al.* 2009; Steer *et al.* 2012; Pedersen *et al.* 2016). However, one fundamental observation is that the present onshore topographic distribution closely reflects large-scale offshore structure in terms of the inherited Mesozoic crustal thinning gradient (Osmundsen and Redfield 2011). Thus, whatever other mechanisms were involved in the uplift, the rift-related taper of the crystalline crust played an important role long into the post-rift phase and is reflected in Scandinavian topography even at present day (op. cit.). The reactivation of inherited faults in the onshore domain likely accompanied Cenozoic uplift adjacent to sharply tapered margin segments, adding to the asymmetry of onshore Scandinavia and to the alignment of rivers perpendicular and parallel to fault strike (Redfield *et al.* 2005; Osmundsen *et al.* 2010; Redfield and Osmundsen 2013). Likely, the topographic effects of fault reactivation facilitated sediment delivery along parts of the offshore margin (Sømme and Jackson 2013).

The formation of large dome structures led to inversion of Late Cretaceous and Paleogene depocentres and to the formation of an important Mid-Miocene unconformity, and may have evolved above areas with serpentinized mantle in the most highly extended parts of the margin (Lundin and Doré 2011).

2.6. Future challenges

The excellent imaging of large parts of the Norwegian margin is countered by some areas with very poor imaging, most notably close to the area of breakup and in parts of the deepest Cretaceous basin areas, where combinations of sill intrusions and large sediment thicknesses lead to the deterioration of the seismic signal. The deep structure, basin configuration and lithological properties of basement, as well as basin rocks are particularly poorly known at depth in an area between the deepest parts of the distal basins to the area of the SDR. Improved imaging of this area in the future will enable better constraints on its interpretation. In particular, the interpretation of so-called lower crustal bodies ("LCB") with velocities >7 but < 8 km/s and calculated densities $\geq$ 3 but < 3.3 g/cm is debated. Based on geophysical data, several authors have argued that at the Norwegian margin, such bodies should be interpreted as crust, based on their "crustal" velocities and densities (up to values of 3.2 kg/dm^3) and on calculated Vp/Vs ratios (e.g. Zastrozhnov *et al.* 2018). Others have discussed the possibility that rocks from the lithospheric mantle became exhumed or nearly exhumed under allochthonous blocks in the most highly extended areas of the margin, opening for the interpretation of LCB's as serpentinized mantle (Ren *et al.* 1998; Osmundsen and Ebbing 2008; Reynisson *et al.* 2010, Rüpke *et al.* 2011; Peron-Pinvidic *et al.* 2013; Osmundsen *et al.* 2016; Lundin and Doré 2011; Peron-Pinvidic *et al.* 2016). Thus, both the existence and extent of possible serpentinization remains debated, albeit the large-scale structural interpretation appears to have converged over the last few years, with the identification of low-angle extensional detachment faults and strongly rotated fault blocks in the distal Norwegian margin by most authors. Better seismic imaging in the most distal and outer parts of the margin will probably contribute to the clarification of some of these questions. However, solid proof for the presence or absence of serpentinization or for other interpretations of distal structure not yet proposed can only be obtained by deep drilling to the basement in the distal or outer margin. Poor imaging in the area of the Vigrid syncline due to a dense mesh of sills in the Cretaceous section efficiently masks the deep structure and deep basin configuration in the transition between the northern Vøring basin and better imaged areas to the south. The causes of the apparent differences in architecture and, to some degree, ages of deformation between these areas remain enigmatic. Recent leaps in our understanding of the proximal margin offshore Norway (e.g. Lenhart *et al.* 2019) were based on the acquisition and interpretation of state-of-the-art 2D and 3D data. With continued industry interest, we are convinced that the abovementioned challenges in the Norwegian margin will also eventually be resolved.

2.7. Acknowledgments

The authors acknowledge support from the hydrocarbon industry through a series of previous industry-sponsored projects, the Norwegian University of Science and Technology, the Geological Survey of Norway and the University of Oslo. The authors thank colleagues in industry and academics for many inspiring discussions on the geology of rifted margins and Atle Rotevatn and Kim Senger for their constructive reviews.

FURTHER READING.– The above descriptions are abbreviated and often simplified. If interested in reading and learning further, the reader is referred to the following list of publications and references.

– General: (Brekke 2000; Faleide *et al.* 2008; Lenhart *et al.* 2019; Kvarven *et al.* 2014; Osmundsen and Peron-Pinvidic 2018; Osmundsen *et al.* 2020; Peron-Pinvidic *et al.* 2013; Peron-Pinvidic and Osmundsen 2018; Planke *et al.* 2017; Zastrozhnov *et al.* 2020).

2.8. References

Abdelmalak, M.M., Faleide, J.I., Planke, S., Gernigon, L., Zastrozhnov, D., Shephard, G.E., Myklebust, R. (2017). The T-Reflection and the deep crustal structure of the Voring Margin offshore Mid-Norway. *Tectonics* 36, 2497–2523 [Online]. Available at: https://doi.org/10.1002/2017TC004617.

Andersen, T.B. and Jamtveit, B. (1990). Uplift of deep crust during orogenic extensional collapse: A model based on field studies in the Sogn-Sunnfjord Region of western Norway. *Tectonics*, 9, 1097–1111.

Andersen, T.B., Osmundsen, P.T., Jolivet, L. (1994). Dep crustal fabrics and a model for the extensional collpase of the southwest Norwegian Caledonides. *Journal of Structural Geology*, 16, 1191–1203.

Andersen, T.B., Torsvik, T.H., Eide, E.A., Osmundsen, P.T., Faleide, J.I. (1999). Permian and Mesozoic extensional faulting within the Caledonides of central south Norway. *Journal of the Geological Society*, London, 156, 1073–1080.

Berndt, C., Planke, S., Alvestad, E., Tsikalas, F., Rasmussen, T. (2001). Seismic volcanostratigraphy of the Norwegian Margin: Constraints on tectonomagmatic break-up processes. *Journal of the Geological Society*, London, 158, 413–426.

Blystad, P., Brekke, H., Færseth, R.B., Larsen, B.T., Skogseid, J., Tørudbakken, B. (1995). Structural elements of the Norwegian continental shelf, Part II. The Norwegian Sea Region. *Norwegian Petroleum Directorate Bulletin*, 8, 0–45.

Braathen, A., Nordgulen, Ø., Osmundsen, P.T., Andersen, T.B., Solli, A.. Roberts, D. (2000). Devonian, orogen-parallel, opposed extension in the Central Norwegian Caledonides. *Geology*, 28, 615–618.

Braathen, A., Osmundsen, P.T., Nordgulen, Ø., Roberts, D., Meyer, G. (2002). Orogen-parallel extension of the Caledonides in northern Centrel Norway: An overview. *Norwegian Journal of Geology*, 82, 225–241.

Breivik, A.J., Mjelde, R., Raum, T., Faleide, J.I., Murai, Y., Flueh, E.R. (2010). Crustal structure beneath the Trøndelag Platform and adjacent areas of the Mid-Norwegian margin, as derived from wide-angle seismic and potential field data. *Norwegian Journal of Geology*, 90, 141–161.

Brekke, H. (2000). The tectonic evolution of the Norwegian sea continental margin with emphasis on the Vøring and Møre Basins. In *Dynamics of the Norwegian Margin*, Nøttvedt, A., (ed.). Geological Society Special Publication, Geological Society of London, 167, 327–378.

Brekke, H., Sjulstad, H.I., Magnus, C., Williams, R.W. (2001). Sedimentary environments offshore Norway – An overview. In *Sedimentary Environments Offshore Norway – Palaeozoic to Recent*, Martinsen, O.J., Dreyer, T. (eds). Elsevier, Amsterdam.

Bunkholt, H., Oftedal, S.S., Hansen, B.T., Løseth, J.A., Kløvjan, H., Norwegian O.S. (submitted). Sea rifted contiental margin tectono-sedimentary element: Review of the geology of the Trøndelag Platform and Halten and Dønna Terraces. Manuscript submitted to the Geological Society, London.

Dalland, A., Worsley, D., Ofstad, K. (1988). A lithostratigraphic scheme for the Mesozoic and Cenozoic succession offshore Norway north of 62. *Norwegian Petroleum Directorate Bulletin*, 4, 65.

Doré, T., Lundin, E.R., Jensen, L.N., Birkeland, Ø., Eliassen, P.E., Fichler, C. (1999). Principal tectonic events in the evolution of the Northwest European Atlantic margin. *Geological Society, London, Petroleum Geology Conference Series*, 5, 41–61.

Ebbing, J. and Olesen, O. (2010). New compilation of top basement and basement thickness for the Norwegian continental shelf reveals the segmentation of the passive margin system. *Geological Society, London, Petroleum Geology Conference Series*, 7, 885–897.

Egholm, D.L., Nielsen, S.B., Pedersen, V.K., Lesemann, J.E. (2009). Glacial effects limiting mountain height. *Nature*, 460, 884–887.

Ehrlich, R. and Gabrielsen, R.H. (2004). The complexity of a ramp-flat fault and its effect on hanging-wall structuring: An example from the Njord oil field, offshore Mid Norway. *Petroleum Geoscience*, 10, 305–317.

Eidvin, T., Riis, F., Rasmussen, E. (2014). Oligocene to lower Pliocene deposits of Norwegian continental shelf, Norwegian Sea, Svalbard, Denmark and their relation to the uplift of Fennoscandia: A synthesis. *Marine and Petroleum Geology*, 56, 184–221.

Eldholm, O., Thiede, J., Taylor, E. (1989). Evolution of the Voring volcanic margin. In *Proceedings of the Ocean Drilling Program*, Eldholm, O., Thiede, J., Taylor, E. (eds). Scientific Results. Ocean Drilling Program, College Station, TX.

Eldholm, O., Tsikalas, F., Faleide, J.I. (2002). Continental margin off Norway 62–75°N, Palaeogene tectono-magmatic segmentation and sedimentation. In The North Atlantic Igneous Province: Stratigraphy, Tectonic, Volcanic and Magmatic Procresses, Jolley, D.W. and Bell, B.R. (eds). The Geological Society of London.

Elliott, G.M., Jackson, C.A.L., Gawthorpe, R.L., Wilson, P., Sharp, I.R., Michelsen, L. (2015). Late syn-rift evolution of the Vingleia Fault Complex, Halten Terrace, offshore Mid-Norway: A test of rift basin tectono-stratigraphic models. *Basin Research* [Online]. Available at: https://doi.org/101111/bre.12158.

Faleide, J.I., Kyrkjebø, R., Kjennerud, T., Gabrielsen, R.H., Jordt, H., Fanavoll, S., Bjerke, M.D. (2002). Tectonic impact on sedimentary processes during Cenozoic evolution of the North Sea and surrounding areas. *Geological Society of London, Special Publications*, 196, 235–269.

Faleide, J.I., Tsikalas, F., Breivik, A.J., Mjelde, R., Ritzmann, O., Engen, Ø., Wilson, J., Eldholm, O. (2008). Structure and evolution of the continental margin off Norway and the Barents Sea. *Episodes*, 31(1), 82-01.

Faleide, J.I., Bjørlykke, K., Gabrielsen, R. (2010). Geology of the Norwegian continental shelf. In *Petroleum Geoscience*, Bjørlykke, K. (ed.). Springer-Verlag, Berlin Heidelberg.

Fjellanger, E., Surlyk, F., Wamsteeker, L.C., Midtun, T. (2005). Upper cretaceous basin-floor fans in the Vøring Basin, Mid Norway shelf. In *Onshore-Offshore Relationships on the North Atlantic Margin*, Wandaas, B., Nystuen, J.P., Eide, E.A., Gradstein, F.M. (eds). Elsevier, Amsterdam.

Fossen, H. (2010). Extensional tectonics in the North Atlantic Caledonides: A regional view. In *Continental Tectonics and Mountain Building: The Legacy of Peach and Horne*, Law, R.D., Butler, R.W.H., Holdsworth, R.E., Krabbendam, M., Strachan, R.A. (eds). Geological Society, London, Special Publications, 335, 767–793.

Fossen, H. and Dunlap, W.J. (1999). On the age and tectonic significance of Permo-Triassic dikes in the Bergen-Sunnhordland region, southwest Norway. *Norsk Geologisk Tidsskrift*, 79, 169–178.

Fossen, H., Teyssier, C., Whitney, D. (2013). Transtensional folding. *Journal of Structural Geology*, 56, 89–102.

Fossen, H., Fazlikhani, H., Faleide, J.I., Ksienzyk, A., Dunlap, W.J. (2016). Post-Caledonian extension in the west Norway-northern North Sea Region: The role of structural inheritance. In *The Geometry and Growth of Normal Faults*, Childs, C., Holdsworth, R.E., Jackson, C.A.-L., Manzocchi, T., Walsh, J.J., Yielding, G. (eds). The Geological Society, Special Publications, London [Online]. Available at: https://doi.org/10.1144/SP439.6.

Geoffroy, L. (2005). Volcanic passive margins. *Comptes rendus géoscience*, 337, 1395–1408.

Gernigon, L., Ringenbach, J-C., Planke, S., Le Gall, B. (2004). Deep structures and breakup along volcanic rifted margins: Insights from integrated studies along the outer Vøring Basin (Norway). *Marine and Petroleum Geology*, 21, 363–372.

Gernigon, L., Blischke, A., Nasuti, A., Sand, M. (2015). Conjugate rifted margins, sea-floor spreading and microcontinent. Insights from new high-resolution aeromagnetic surveys in the Norway Basin. *Tectonics*, 34, 907–933.

Henriksen, S., Fichler, C., Grønlie, A., Henningsen, T., Laursen, J., Løseth, H., Ottesen, D., Prince, I. (2005). The Norwegian Sea during the Cenozoic. In *Onshore-Offshore Relationships on the North Atlantic Margin*, Wandaas, B, Nystuen, J.P., Eide, E.A., Gradstein, F.M. (eds). Elsevier, Amsterdam.

Henriksen, S., Helland-Hansen, W., Bullimore, S. (2011). Relationships between shelf-edge trajectories and sediment dispersal along depositional dip and strike: A different approach to sequence stratigraphy. *Basin Research*, 23, 3–21.

Jongepier, K., Rui, J.C., Grue, K. (1996). Triassic to early Cretaceous stratigraphic and structural development of the northeastern Møre Basin margin, off Mid-Norway. *Norsk Geologisk Tidsskrift*, 76, 199–214.

Krabbendam, M. and Dewey, J.F. (1998). Exhumation of UHP rocks by transtension in the Western Gneiss region, Scandinavian Caledonides. In *Continental Transpressional and Transtensional Tectonics*, Holdsworth, R.E., Strachan, R.A., Dewey, J.F. (eds). The Geological Society, London, Special Publications, 135, 159–181.

Kvarven, T., Ebbing, J., Mjelde, R., Faleide, J.I., Libak, A., Thybo, H., Flueh, E.R., Murai, Y. (2013). Crustal structure across the Møre margin, Mid-Norway, from wide-angle seismic and gravity data. *Tectonophysics*, 626, 21–40.

Lenhart, A., Jackson, C., Bell, A.-L., Duffy, R.E., Gawthorpe, O.B., Fossen, R.L.H. (2019). Structural architecture and composition of crystalline basement offshore west Norway. *Lithosphere*, 11(2), 273–293.

Løseth, H., Kyrkjebø, R., Hilde, E., Wild, R.J., Bunkholt, H. (2017). 500 m of rapid base-level rise along an inner passive margin – Seismic observations from the Pliocene Molo Formation, mid Norway. *Marine and Petroleum Geology*, 86, 268–287.

Lundin, E.R. and Doré, A.G. (1997). A tectonic model for the Norwegian passive margin with implications for the NE Atlantic: Early Cretaceous to break-up. *Journal of the Geological Society of London*, 154, 545–550.

Lundin, E.R. and Doré, A.G. (2011). Hyperextension, serpentinization, and weakening: A new paradigm for rifted margin compressional deformation. *Geology*, 39, 347–350.

Manatschal, G. (2004). New models for evolution of magma-poor rifted margins based on a review of data and concepts from West Iberia and the Alps. *International Journal of Earth Sciences*, 93, 432–466.

Mjelde, R., Raum, T., Breivik, A., Shimamura, H., Murai, Y., Takanami, T., Faleide, J.I. (2004). Crustal structure of the Vøring Margin, NE Atlantic: A review of geological implicaitons based on recent OBS data. *Geological Society of London, Petroleum Geology Conference Series*, 6, 803–813.

Mjelde, R., Raum, T., Kandilarov, A., Murai, Y., Takanami, T. (2009). Crustal structure and evolution of the outer Møre Margin, NE Atlantic. *Tectonophysics*, 468, 224–243.

Mosar, J. (2003). Scandinavia's North Atlantic passive margin. *Journal of Geophysical Research, Solid Earth*, 108, B8.

Müller, R., Nystuen, J.P., Eide, F., Lie, H. (2005). Late Permian to Triassic basin infill history and palaeogeography of the Mid-Norwegian shelf-East Greenland region. In *Onshore-Offshore Relationships on the North Atlantic Margin*, Wandaas, B, Nystuen, J.P., Eide, E.A., Gradstein, F.M. (eds). Elsevier, Amsterdam.

Muñoz-Barrera, J., Rotevatn, A., Gawthorpe, R.L., Henstra, G.A., Kristensen, T.B. (2020). The role of structural inheritance in the development of high-displacement crustal faults in the necking domain of rifted margins: The Klakk Fault Complex, Frøya High, offshore mid-Norway. *Journal of Structural Geology*, 140, 104163.

Naliboff, J.B., Buiter, S.J.H., Peron-Pinvidic, G., Osmundsen, P.T., Tetreault, J. (2017). Complex fault interaction controls continental rifting. *Nature Communications*, 8(1179) [Online]. Available at: https://doi.org/10.1038/s41467-017-00904-x.

Nielsen, S.B., Gallagher, K., Leighton, C., Balling, N., Svenningsen, L., Holm Jakobsen, B., Thomsen, E., Nielsen, O., Heilmann, C., Clausen, D., Egholm, D.L., Summerfield, M.A., Clausen, O.R., Piotrowski, J.A., Thorsen, M.R., Huuse, M., Abrahamsen, N., King, C., Lykke-Andersen, H. (2009). The evolution of western Scandinavian topography: A review of Neogene uplift versus the ICE (isostasy–climate–erosion) hypothesis. *Journal of Geodynamics*, 47, 72–95.

Nirrengarten, M., Gernigon, L., Manatschal, G. (2014). Lower crustal bodies in the Møre volcanic rifted margin: Geophysical determination and geological implications. *Tectonophysics*, 636, 143–157.

Osmundsen, P.T. and Andersen, T.B. (2001). The middle Devonian basins of western Norway: Sedimentary response to large-scale transtensional tectonics? *Tectonophysics*, 332, 51–68.

Osmundsen, P.T. and Ebbing, J. (2008). Styles of extension offshore mid-Norway and implications for mechanisms of crustal thinning at passive margins. *Tectonics*, 27(6). doi:10.1029/2007TC002242.

Osmundsen, P.T. and Peron-Pinvidic, G. (2018). Crustal-scale fault interaction at rifted margins and the formation of domain-bounding breakaway complexes: Insights from offshore Norway. *Tectonics*, 37(3), 935–964.

Osmundsen, P.T. and Redfield, T. (2011), Crustal taper and topography at passive continental margins. *Terra Nova*, 23(6), 349–361.

Osmundsen, P.T., Andersen, T.B., Markussen, S., Svendby, A.K. (1998). Tectonics and sedimentation in the hanging wall of a major extensional detachment: The Devonian Kvamshesten Basin, western Norway. *Basin Research*, 10, 213–234.

Osmundsen, P.T., Bakke, B., Svendby, A.K., Andersen, T.B. (2000). Architecture of the Middle Devonian Kvamshesten Group, western Norway: Sedimentary response to deformation above a ramp-flat extensional fault. In *New Perspectives on the Old Red Sandstone*, Friend P.F., Williams, B.P.J. (eds). Geological Society, London, Special Publications.

Osmundsen, P.T., Sommaruga, A., Skilbrei, J.R., Olesen, O. (2002), Deep structure of the Mid Norway rifted margin. *Norwegian Journal of Geology*, 82(4), 205–224.

Osmundsen, P.T., Braathen, A., Nordgulen, Ø., Roberts, D., Meyer, G.B., Eide, E. (2003). The Devonian Nesna shear zone and adjacent Gneiss-cored culminations, North-Central Norwegian Caledonides. *Journal of the Geological Society, London*, 160, 137–150.

Osmundsen, P.T., Eide, E.A., Haabesland, N.E., Roberts, D., Andersen, T.B., Kendrick, M., Bingen, B., Braathen, A., Redfield, T.F. (2006). Kinematics of the Høybakken detachment zone and the Møre-Trøndelag Fault Complex, central Norway. *Journal of the Geological Society, London*, 163, 303–318.

Osmundsen, P.T., Redfield, T.F., Hendriks, B.H.W., Bergh, S., Hansen, J-A., Henderson, I.H.C., Dehls, J., Lauknes, T.R., Larsen, Y., Anda, E., Davidsen, B. (2010). Fault-controlled alpine topography in Norway. *Journal of the Geological Society, London*, 167, 83–98.

Osmundsen, P.T., Peron-Pinvidic, G., Ebbing, J., Erratt, D., Fjellanger, E., Bergslien, D., Syvertsen, S.E. (2016). Extension, hyperextension and mantle exhumation offshore Norway: A discussion based on 6 crustal transects. *Norwegian Journal of Geology*, 96, 343–372.

Osmundsen, P.T., Peron-Pinvidic, G., Bunkholt, H. (2021). Rifting of collapsed orogens: Successive incision of continental crust in the proximal margin offshore Norway. *Tectonics* [Online]. Available at: https://doi.org/10.1029/2020TC006283.

Ottesen, D., Rise, L., Andersen, E.S., Bugge, T., Eidvin, T. (2009). Geological evolution of the Norwegian continental shelf between 61oN and 68oN during the last 3 million years. *Norwegian Journal of Geology*, 89, 251–265.

Pedersen, V.K., Huismans, R.S., Moucha, R. (2016). Isostatic and dynamic support of high topography on a North Atlantic passive margin. *Earth and Planetary Science Letters*, 446, 1–9.

Pérez-Gussinyé, M. and Reston, T.J. (2001). Rheological evolution during extension at passive non-volcanic margins: Onset of serpentinization and development of detachments to continental break-up. *Journal of Geophysical Research*, 106, 3691–3975.

Peron-Pinvidic, G. and Manatschal, G. (2009). The final rifting evolution at deep magma-poor passive margins from Iberia-Newfoundland: A new point of view. *International Journal of Earth Sciences (Geologisches Rundschau)*, 98, 1581–1597.

Peron-Pinvidic, G. and Osmundsen, P.T. (2016). Architecture of the distal and outer domains of the Mid-Norwegian rifted margin: Insights from the Rån-Gjallar ridges system. *Marine and Petroleum Geology*, 77, 280–299.

Peron-Pinvidic, G., Manatschal, G., Osmundsen, P.T. (2013). Structural comparison of archetypal Atlantic rifted margins: A review of observations and concepts. *Marine and Petroleum Geology*, 43, 21–47.

Peron-Pinvidic, G., Osmundsen, P.T., Ebbing, J. (2016). Mismatch of geophysical datasets in distal rifted margin studies. *Terra Nova*, 28, 340–347.

Peron-Pinvidic, G., Osmundsen, P.T., Bunkholt, H. (2020). The proximal domain of the Mid-Norwegian rifted margin: The Trøndelag Platform revisited. *Tectonophysics*, 790, 228551.

Phillips, T.B., Jackson, C.A.L., Bell, R.E., Duffy, O.B., Fossen, H. (2016). Reactivation of intrabasement structures during rifting: A case study from offshore southern Norway. *Journal of Structural Geology*, 91, 54–73.

Planke, S., Alvestad, E., Eldholm, O. (1999). Seismic characteristics of basaltic extrusive and intrusive rocks. *The Leading Edge*, 18, 342–348.

Planke, S., Symonds, P., Alvestad, E., Skogseid, J. (2000). Seismic volcanostratigraphy of large-volume basaltic extrusive complexes on rifted margins. *Journal of Geophysical Research*, 105, 19335–19351.

Planke, S., Rasmussen, T., Rey, S.S., Myklebust, R. (2005). Seismic characteristics and distribution of volcanic intrusions and hydrothermal vent complexes in the Vøring and Møre basins. *Geological Society of London, Petroleum Geology Conference Series*, 6, 833–844.

Planke, S., Millett, J.M., Maharjan, D., Jerram, D.A., Abdelmalak, M.M., Groth, A., Hoffmann, J., Berndt, C., Myklebust, R. (2017). Igneous seismic geomorphology of buried lava fields and coastal escarpments on the Voring volcanic rifted margin. *Interpretation*, 5(3), 1–42.

Ravnås, R., Berge, K., Campbell, H., Harvey, C., Norton, M.J. (2014). Halten terrace lower and middle Jurassic inter-rift megasequence analysis: Megasequence structure, sedimentary architecture and controlling parameters. *International Association of Sedimentologists, Special Publication*, 46, 215–252.

Redfield, T.F. and Osmundsen, P.T. (2013). The long-term topographic response of a continent adjacent to a hyperextended margin: A case study from Scandinavia. *Geological Society of America, Bulletin*, 125, 184–200.

Redfield, T.F., Osmundsen, P.T., Hendriks, B.W.H. (2005). The role of fault reactivation and growth in the uplift of western Norway. *Journal of the Geological Society, London*, 162, 1013–1030.

Ren, S., Skogseid, J., Eldholm, O. (1998). Late Cretaceous-Paleocene extension on the Vøring Volcanic Margin. *Marine Geophysical Researches*, 20, 343–369.

Reynisson, F., Ebbing, J., Lundin, E., Osmundsen, P.T. (2010). Properties and distribution of lower crustal bodies on the Mid-Norwegian margin. *Geological Society, London, Petroleum Geology Conference Series*, 7, 843–854.

Ribes, C., Manatschal, G., Ghienne, J.-F., Karner, G.D., Johnson, C.A., Figueredo, P.H., Incerpi, N., Epin, M.-E. (2019). The syn-rift stratigraphic record across a fossil hyper-extended rifted margin: The example of the northwestern Adriatic margin exposed in the Central Alps. *International Journal of Earth Sciences* [Online]. Available at: https://doi.org/10.1007/s00531-019-01750-6.

Rotevatn, A., Kristensen, T., Ksienzyk, A.K., Wemmer, K., Henstra, G.A., Midtkandal, I., Grundvåg, S.-A., Andresen, A. (2018). Structural inheritance and rapid rift-length establishment in a multiphase rift: The East Greenland rift system and its Caledonian orogenic ancestry. *Tectonics*, 37, 1858–1875 [Online]. Available at: https://doi.org./10.1029/2018TC005018.

Rüpke, L.H., Schmid, D., Peréz-Gussinye, M., Hartz, E. (2013). Interrelation between rifting, faulting, sedimentation and mantle serpentinization during continental margin formation – Including examples from the Norwegian Sea. *Geochemistry, Geophysics, Geosystems (G3)*, 14, 4351–4369.

Seidler, L., Steel, R.J., Stemmerik, L., Surlyk, F. (2004). North Atlantic marine rifting in the Early Triassic: New evidence from East Greenland. *Journal of the Geological Society, London*, 161, 583–592.

Séranne, M. (1992). Late Palaeozoic kinematics of the Møre-Trøndelag Fault Zone and adjacent areas, Central Norway. *Norsk Geologisk Tidsskrift*, 72, 141–158.

Séranne, M. and Seguret, M. (1987). The Devonian basins of western Norway: Tectonics and kinematics of an extending crust. In *Continental Extensional Tectonics*, Coward, M.P., Dewey, J.F., Hancock, P.L. (eds). Geological Society, London, Special Publications, 28, 537–548.

Skogseid, J., Pedersen, T., Eldsholm, O., Larsen, B.T. (1992). Tectonism and magmatism during NE Atlantic continental break-up: The Vøring Margin. *Geological Society, London, Special Publications*, 68, 305–320.

Sømme, T.O. and Jackson, C.A.-L. (2013). Source-to-sink analysis of ancient sedimentary systems using a subsurface case study from the Møre-Trøndelag area of southern Norway: Part 2 – Sediment dispersal and forcing mechanisms. *Basin Research*, 25, 512–531.

Sømme, T.O., Martinsen, O.J., Thurmond, J.B. (2009). Reconstructing morphological and depositional characteristics in subsurface sedimentary systems: An example from the Maastrichtian-Danian Ormen Lange system, Møre Basin, Norwegian Sea. *AAPG Bulletin*, 93, 1347–1377.

Steel, R.J., Siedlecka, A., Roberts, D. (1985). The old red sandstone basins of Norway and their deformation: A review. In *The Caledonide Orogen: Scandinavia and Related Areas*, Gee, D.G., Sturt, B.A. (eds). John Wiley and Sons, Chichester.

Steer, P., Huismans, R., Valla, P.G., Gac, S., Herman, F. (2012). Bimodal plio-ouaternary glacial erosion of fjords and low-relief surfaces in Scandinavia. *Nature Geoscience*, 5, 635–639.

Torsvik, T.H., Andersen, T.B., Eide, E.A., Walderhaug, H.J. (1997). The age and tectonic significance of dolerite dykes in western Norway. *Journal of the Geological Society, London*, 154, 961–973.

Tsikalas, F., Faleide, J.I., Eldholm, O., Wilson, J. (2005). Late Mesozoic-Cenozoic structural and stratigraphic correlations between the conjugate mid-Norway and NE Greenland continental margins. *Geological Society, London, Petroleum Geology Conference Series*, 6, 785–801.

Zastrozhnov, D., Gernigon, L., Gogin, I., Abdelmalak, M.M., Planke, S., Faleide, J.I., Eide, S., Myklebust, R. (2018). Cretaceous-Paleocene evolution and crustal structure of the Northern Vøring Margin (Offshore Mid-Norway): Results from integrated geological and geophysical study. *Tectonics*, 37, 497–528.

Zastrozhnov, D., Gernigon, L., Gogin, I., Planke, S., Abdelmalak, M.M., Polteau, S., Faleide, J.I., Manton, B., Myklebust, R. (2020). Regional structure and polyphased Cretaceous-Paleocene rift and basin development of the mid-Norwegian volcanic passive margin. *Marine and Petroleum Geology*, 115, 104269.

3

South and Equatorial Atlantic Margins

Daniel ASLANIAN and Maryline MOULIN
Ifremer, REM/GM/LGS, Plouzané, France

3.1. Introduction

The Atlantic and Indian Oceans are the result of the break-up and dispersion of the mega-continent Pangaea, related to the closure and disappearance of the Tethys paleo-ocean (Figure 3.1). The geodynamics of this break-up can be divided into three main phases: (1) in the Triassic (Sinemurian ~195 Ma), Laurasia separated from Gondwana creating the Central Atlantic Ocean; (2) in the Jurassic (Oxfordian-Kimmeridgian – 165–155 Ma), the supercontinent Gondwana broke into two pieces of similar size, with the formation of oceans in the Somali and Mozambique Basins; and (3) in the Cretaceous, through a global geodynamic reorganization (125–85 Ma, Figure 3.2), all remaining blocks finally broke up to form the North Equatorial and South Atlantic Oceans, as well as the eastern part of the Indian Ocean.

This multiphase evolution implies different geodynamic contexts for each Atlantic passive margins genesis. Moreover, the break-ups generally occurred along old sutures (Figure 3.1). Looking at a simple section along an orogen shows the complexity of such areas and suggests a strong impact of the tectonic heritage on these break-ups and the subsequent passive margins genesis. This complexity must

For a color version of all the figures in this chapter, see www.iste.co.uk/peronpinvidic/margins2.zip.

Continental Rifted Margins 2,
coordinated by Gwenn PERON-PINVIDIC. © ISTE Ltd 2022.

be kept in mind when looking at inherently oversimplified numerical or analogue models

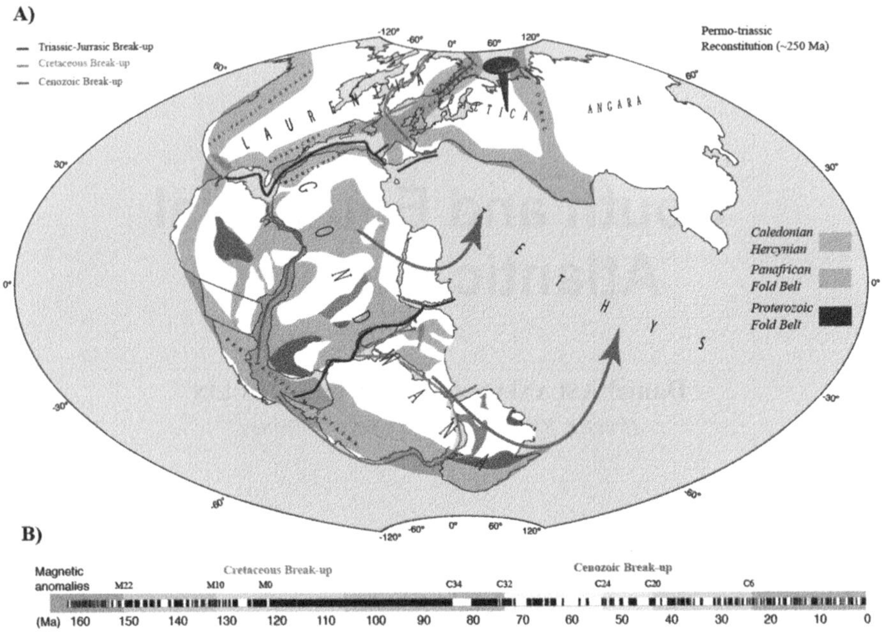

Figure 3.1. *a) Reconstruction at Permo-Triassic time (~250 Ma) (Olivet, Aslanian, pers. comm.) just before the break-up of the mega-continent Pangea. The Upper Triassic-Jurassic episode (in blue, about 195–185 and 165–155 Ma) separated the Pangea into three equivalent blocks along two parallel lines (Laurentia, Baltica and Angara for one side, Africa and South America for another side, Madagascar, India, Antarctica and Australia for the last side). The second episode (in green) occurred during the Early Cretaceous (about 133–120 Ma), and the last episode (in red) during the Tertiary (60–40 Ma). Note that we describe the first movement between two plates, for instance India and Madagascar (about 120 Ma), and not the first oceanic crust (80–90 Ma for this part of Indian Ocean). Note the almost general coincidence between the different break-ups and older orogenic belts. The block is considered as the fixed plate. Hammer Projection. b) Magnetic anomaly inversions. Note the change of the pattern around the Cretaceous and Cenozoic Break-ups (source: modified from Moulin and Aslanian 2010)*

The margins of the North Atlantic Ocean, including the Newfoundland and Iberian Margins, present two distinct episodes of rifting: (1) at Permian–Trias times, distension occurred and aborted in Western Europe and gave rise to the deposition of evaporites (Ziegler (1989), for a general view); (2) in the middle Cretaceous, another distension occurred and broke the system apart. More than 70 Ma separate these two episodes, which strongly complicates the interpretation of the related thinning processes.

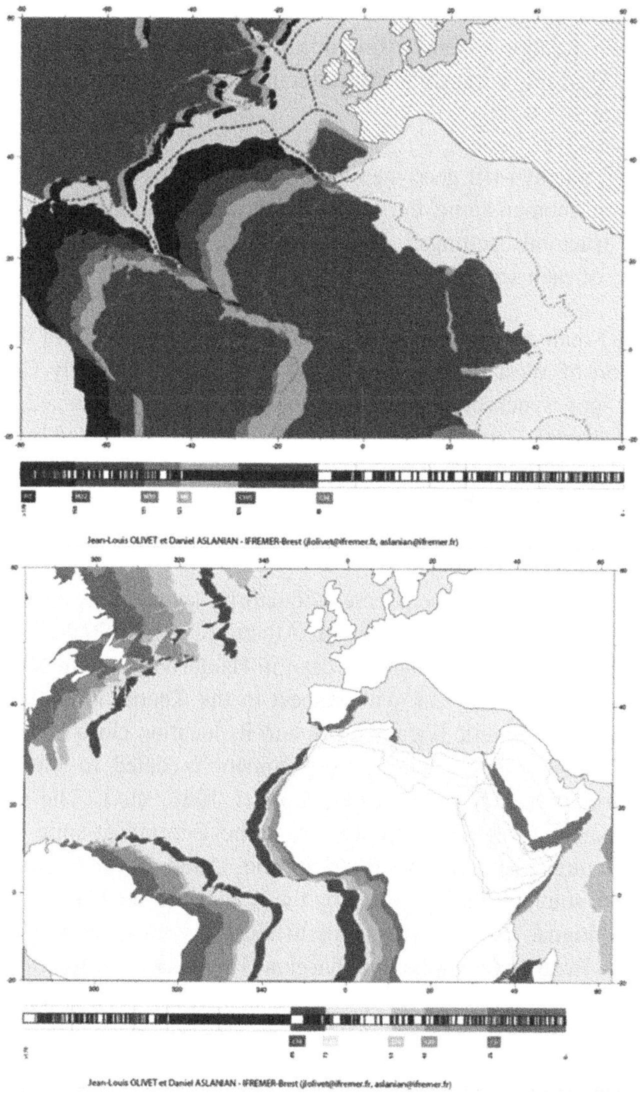

Figure 3.2. *Break-up and dispersion of the Pangea. Top: movement from Sinemurian to about 105 Ma. Bottom: movement from 84 Ma to present day, after the main Cretaceous geodynamic revolution*

In the Central Atlantic, rifting started at end of Sinemurian (Sahabi *et al.* 2004), following the end of the Hercynian orogenic suture, and the margins of the Central Atlantic Ocean seem to have been formed in a single thinning episode, at the Trias–Lias boundary.

In the South Atlantic Ocean, rifting occurred on the location of the Pan-African suture (~600 Ma), more than 450 Ma after its formation, and the two events are clearly dissociated.

The pre-rift and syn-rift geodynamic contexts are thus extremely different from one segment to another along the entire Atlantic Ocean system, which results in contrasting structural settings. We summarize below the main structural characteristics of each segment from north to south.

Whilst the South and Equatorial Atlantic Oceans originated from the rupture and final separation of West Gondwana (Figure 3.1) during the Early Cretaceous, the opening was not synchronous and can be divided into four ~2,000 km-long segments: Equatorial, Central, Austral and Patagonia (as presented by Popoff 1988, and Moulin *et al.* (2010) – Figure 3.3). The geodynamic context of these different 2,000 km-long segments has tremendous influence on the morphology and the genesis of the related passive margins.

Northwards, the strongly segmented Equatorial segment represents a "transfer area" between the opening of the Central Atlantic Ocean, initiated in Sinemurian, and the opening of the South Atlantic Ocean in Hauterivian, about 60 My later, and shifted by about 30° to the east with respect to the Central Atlantic (Figure 3.4). Despite the lack of magnetic lineaments due to its location close to the Equator, the beginning of seafloor spreading in this segment is dated to the upper Albian (Oliveira Marinho 1985; Blarez 1986; Gouyet 1988, etc.). The 2,000 km-long Equatorial Atlantic Ocean is separated from the Central Atlantic Ocean by the Marathon Fracture Zone, to the north of the Demerara and Guinean plateaus (Figure 3.3). Its southern limit, marked by the Romanche and Chain Fracture Zones, north of the Potiguar Basin, corresponds to the Ivory Coast and Ghana transform system and its Brazilian homologous margins (Hayes and Ewing 1970; Le Pichon and Hayes 1971; Mohriak 2004). Inbetween, this ocean can be divided into three 600–800 km-long sub-segments, separated by major fracture zones (Ascension, Chain, Saint Paul and Sierra Leone; Figure 3.3) and presents several pull-apart segments, about 300 km wide, with transform margins at their limits.

The Central Segment is positioned between the Benue Trough to the north, onshore Africa, and the Rio Grande Fracture Zones to the south (Figure 3.3), and is characterized by systems of individual rifts with sag-type basins and deposition of thick layers of evaporites at the end of the rift phase in the Aptian time (Zalán *et al.* 2011; pink polygones on Figure 3.3). The seafloor spreading in this region began after the Barremian, during the quiet magnetic time in the Early Cretaceous time (after Chron M0, see Figure 3.1), so that no magnetic lineament marks the beginning of the oceanic crust accretion.

Between the Central Segment and the Austral Segment, the Santos Basin-São Paulo Plateau System (SSPS) is interpreted as a kinematic buffer with an oblique opening to the global movement (Moulin *et al.* 2010, 2012), corresponding to a relay zone between the plate movements that created the northern Campos and Espirito Santo Basins and the southern volcanic basins of the South Atlantic Ocean. Despite this oblique opening, the SSPS also corresponds to a system of rifts with sag basins and deposition of thick layers of evaporites. This sedimentary-salt margin type corresponds to the most prolific oil-producing province in Brazil and Africa, where the traditional turbidite and carbonate fields of the post-salt and giant to super-giant fields in pre-salt microbialites of Angola, Santos, Campos and Espírito Santo Basins are located.

Southwards, the Rio Grande (or Florianopolis)/Walvis Fracture Zone marks the passage of one system of volcanic rifts in the Austral segment limited by the Falkland-Agulhas Fracture Zones to the south. The seafloor spreading began during the Early Cretaceous (Rabinowitz and LaBrecque 1979), but debate remains about the age of the magnetic anomalies observed in this area. This segment is characterized by the presence of Seaward Dipping Reflectors (SDRs), which indicate the presence of volcanism during the genesis of the margin, both along the Brazilian and Argentine Basins and their conjugates Namibia-South Africa (Hinz 1981; Talwani, Abreu 2000; Franke *et al.* 2007; Stica *et al.* 2014; Clerc *et al.* 2015).

The Patagonia Segment is the junction between the Atlantic ridge and the South Western Indian Ridge (SWIR) (Figure 3.3). The structural setting is complicated by the presence of the fossil Agulhas ridge. The seafloor spreading is presumed to have begun around 134 Ma at M10 (Rabinowitz *et al.* 1976; Goodlad *et al.* 1982; Martin *et al.* 1982; Thompson *et al.* 2019).

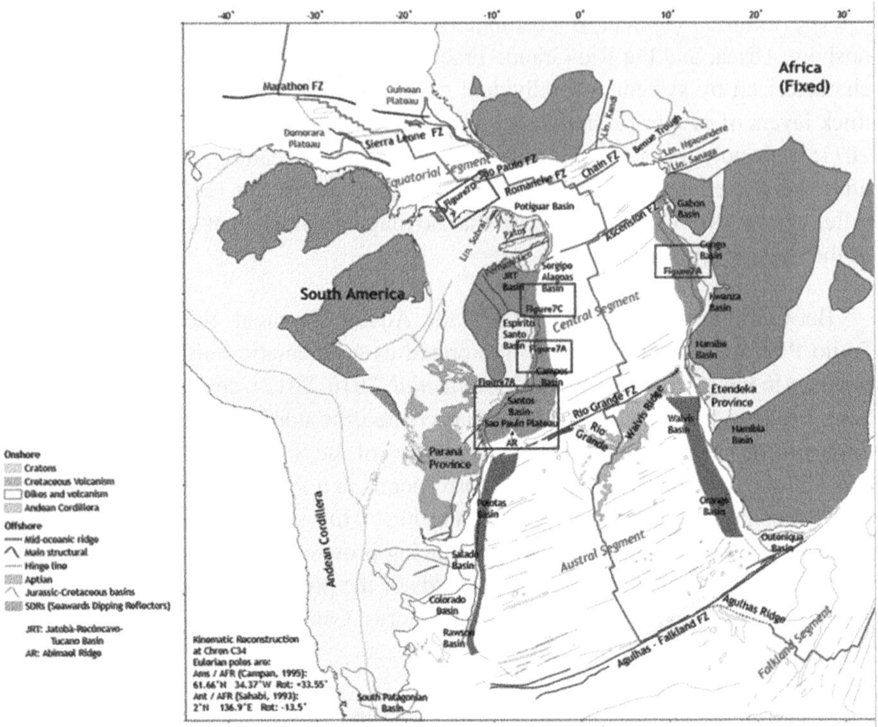

Figure 3.3. *General tectono-structural map of the South Atlantic Ocean at Chron 34 (84 Ma). For this well-constrained reconstruction, we used the Eulerian pole from Campan (1995) for the Equatorial and South Atlantic Oceans and the Eulerian pole from Sahabi (1993) for the South Western Indian Ocean. Fracture zones and seamounts are based on interpretation of satellite-derived gravity 1 mn x 1 mn grid (Sandwell and Smith 2009). The large red line represents the accreting oceanic ridge at Chron 34. Sedimentary Cretaceous basins are represented by green lines. Cratons are represented by brown lines and polygons. Aptian salt extension (pink areas) are based on a compilation of different sources (Moulin* et al. *2010). The location of the Abimael ridge (interpreted by Mohriak 2001, as an aborted oceanic propagator across the Rio Grande Fracture Zone) is given by the letters AR with a black arrow. Seaward Dipping Reflectors (SDRs) are represented in light green areas. The main structural constraints (lineaments and plateau limits) used for the kinematic reconstructions are shown in blue lines. Cretaceous volcanism is depicted in red and triassic volcanism in purple. Hinge lines are depicted with thick red lines. Mercator projection (source: Moulin* et al. *2010)*

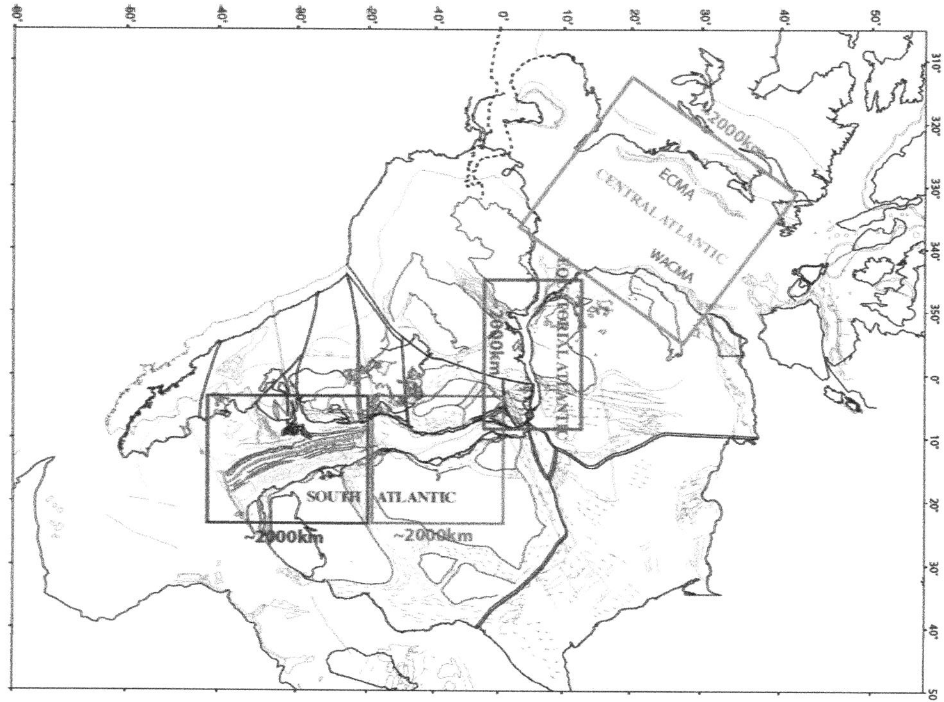

Figure 3.4. *Kinematic reconstruction at M0 (Barremian, Moulin, Aslanian, unpublished). The Central Atlantic Ocean is wide open. The Austral segment of the South Atlantic is already producing oceanic crust (see magnetic anomalies in colored lines). Movements have started in the central segment of the South Atlantic, producing extension and exhumation (Aslanian* et al. *2009). In the equatorial segment, the age of the inset of oceanic crust is interpolated to 106 Ma (source: Moulin* et al. *2010)*

3.2. Geodynamics and understanding passive margins genesis

Both detailed and global kinematic paleo-reconstructions are needed to precisely define conjugate systems in a global context, including the tightest position of the conjugate margins, the location of onshore structures (basins, cratons, etc.), the first direction of horizontal movement (oblique/strike-slip movement, for instance): these parameters may play an important role in thinning processes, as well as and the tightest position of the conjugate margin, in a global context, which constrains the allowed horizontal movement involved in the thinning process. In other words, to understand passive margins genesis and build models that take into account all geophysical and geological, on-land and offshore, parameters, precise kinematic paleo-reconstructions are not only essential but compulsory (Sahabi *et al.* 2004; Aslanian *et al.* 2009; Aslanian and Moulin 2012).

In 1965, on the basis of the work of Carey (1955), Bullard *et al.* (1965) proposed the first South Atlantic Ocean statistical reconstruction. The authors pointed out a geometrical problem: how can we reconstruct the initial pre-opening fit both in the Equatorial and South Atlantic Oceans at the same time? Since then, several authors have tried to solve the problem of the pre-opening fit in the South Atlantic Ocean in the past. Attempts to obtain the pre-opening fit (and evaluate its age) and the early evolution of an ocean are strongly dependent on the hypothesis made for the location of the first typical oceanic crust and often require an assessment of the rigidity of the surrounding plates. This procedure was first undertaken for the South Atlantic Ocean in 1974 for Africa and then for South America. In the central segment of the South Atlantic Ocean, this work is complicated by the lack of well-defined magnetic anomalies in the early history of the ocean. The difference between these models can be attributed to three main uncertainties: the location of intra-continental deformation, the location of the continent–ocean transition zone (and how to correct for its tectonic extension and/or magmatic addition during rifting) and the recognition of the earliest seafloor spreading anomalies.

The existence of several intraplate deformations on both Africa and South America has been well accepted for a long time (Eagles 2006; Torsvik *et al.* 2009; Lawver *et al.* 1999; Seton *et al.* 2012; Verard *et al.* 2012; etc.), but the first global reconstruction presenting a good fit in agreement with both geological and geophysical data was only published in 2010 (Moulin *et al.* 2010). Later, new models based on local and detailed information and common inversion measurements have been proposed to remove uncertainties, but none of these new reconstructions show detailed agreement with the full set of terrestrial and offshore geological and geophysical data all together.

A passive margin is only one piece of the jigsaw. Reconstructing the entire system needs a correct definition of conjugate margins, and therefore a precise kinematic reconstruction.

The figure below presents a zoom on the area of the Reconcavo-South Alagoas-Jequitinhonha margin system in three different palinspastic reconstructions and their implications on the definition of conjugate passive margins system. (A) The Castro (1987) reconstruction (which is very close to the Pindell, Dewey (1982); Mohriak, Rosendahl (2003); Meyers *et al.* (1996) reconstructions) presents the Rio Muni Margin as the conjugate of Sergipe-Alagoas Margins, within an oblique or transform context. These reconstructions present a shift of more than 150 km for the Ascension FZ and cannot be used to define the conjugate margin. (B) The paleo-reconstruction from Moulin *et al.* (2010) constrained by geological and geophysical data of the Equatorial and South Atlantic Oceans. In this reconstruction, the conjugate margin of the Reconcavo-South Alagoas-Jequitinhonha Margin is the South Gabon Margin, in a typical oblique opening system. (C) The "conceptual plate tectonic reconstruction" of Blaich *et al.* (2008), which shifts to the north the African plate with respect to the South American plate (~100 km).

Castro, 1987 in Blaich *et al.*, 2008 | Aslanian *et al.*, 2009 Moulin *et al.*, 2010 | Blaich *et al.*, 2008

Three systems of conjugate margins, inferred by three different palinspastic reconstructions. (A) The conceptual reconstruction of Castro (1987); (B) the reconstruction of Moulin *et al.* (2010), used in Aslanian *et al.* (2009) and (C) the conceptual reconstruction of Blaich *et al.* (2008).

Box 3.1. *Passive margins and the question of its conjugate part*

Whatever their uncertainties and their inability to reflect all data, palinspastic models for the South Atlantic Ocean can be divided into two families according to the degree of tightness they imply in the central segment of the South Atlantic, for the pre-rift position: very tight palinspastic reconstruction family and "opened" palinspastic reconstruction family. These two families of models give different values for the distance between the conjugate hinge lines (green areas in Figure 3.5). In terms of geodynamic evolution, this implies different values for the horizontal movement that occurred between the fit and the first oceanic crust. In terms of crustal structure, this implies different amounts of extension for the initial conjugate passive margins system. In terms of modeling, the possible amount of horizontal movement between the fit and the first oceanic crust will also be very different.

The analysis of passive margin rift processes and evolution should not be carried out on small sketches but only through precise kinematic reconstructions: in other words, conceptual models must be situated on precise paleogeographic maps in order to test their consequences on global models to be validated.

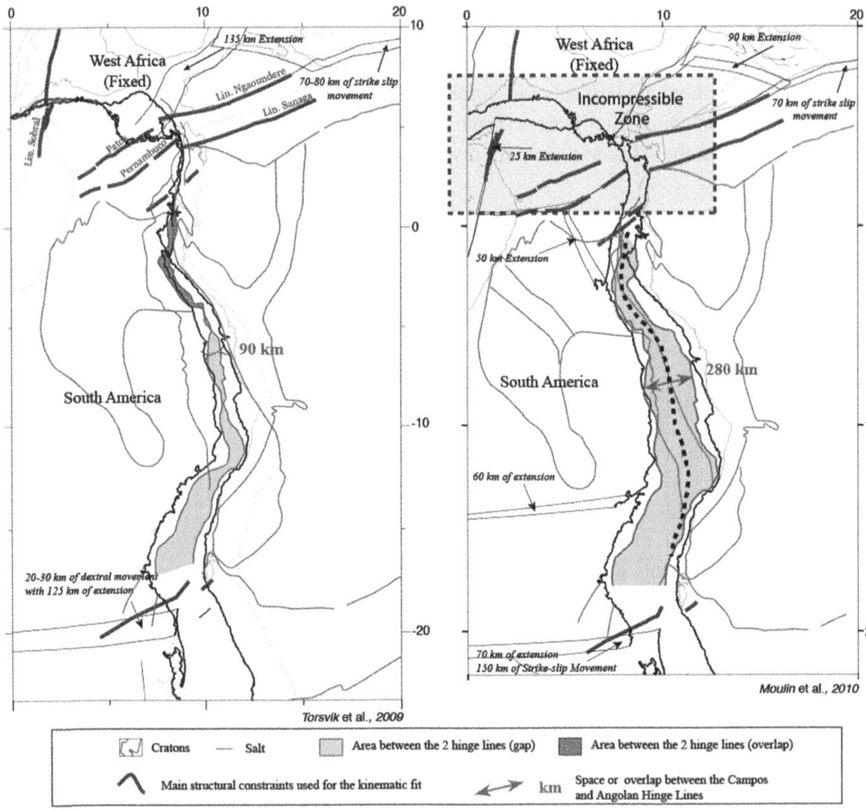

Figure 3.5. *Comparison in the Central Segment of the South Atlantic Ocean of the two families of initial reconstructions (source: Aslanian and Moulin 2010)*

CONTINUATION OF CAPTION FOR FIGURE 3.5.– *West Africa is fixed. In the Equatorial and Central Segments, blue lines represent the Pan-African fault systems*

(Ngaoundéré-Sanaga, Patos-Pernambuco and Kandi-Sobral lineaments). The hinge lines are drawn in red lines. Pink areas represent the Aptian salt extension. The dark green areas represent the overlap, while the light green represents the gap between the 2 hinge lines. (1) The right side presents a very tight palinspastic reconstruction. Note that the hinge lines overlap by more than 90 km in the N-E area. (2) The left side presents a "opened" palinspastic reconstruction. In the Central Segment, this model shows a gap of 280 km between the two hinge lines.

3.3. Crustal structures in the South Atlantic and equatorial margins

Based on the above statements, it is concluded that the analysis of passive margin extensional processes and evolution requires precise kinematic reconstruction models, but on the other hand, kinematic studies need both global and detailed views, and, for the pre-drift period, specific attention on intraplate deformation and passive margins structure and limits.

At first order, the geodynamic segmentation of the South and Equatorial Atlantic Oceans leads to the formation of different types of passive margins, showing a relationship between the regional geodynamic context and the structural architecture of passive margins.

In the Austral Segment of the South Atlantic Ocean, the presence of Seaward-Dipping Reflectors (SDRs) are interpreted as flow lavas and may therefore indicate the presence of volcanism during the genesis of the margin, both along the Brazilian and Argentina Margins and their conjugates Namibia–South Africa Margins. This type of margin is somewhat misleadingly called "volcanic margin" which implies that the other margins have not undergone magmatic events. However, most passive margins show evidence for magmatic events (the SSPP, for example, is full of magmatism), even though the process of their emplacement is probably different.

The typical structure of this category of margins is schematized in Figure 3.6. Whilst there is a good agreement on the description on the seaward bended and dipping reflectors, the nature of the underlying deeper basement is still a matter of debate: is it (proto-)oceanic crust, sediment, upper continental crust or exhumed lower continental crust as seems to be the case for the NW Greenland Margins (Guan *et al.* 2019)? Whatever the interpretation, the continental crust thinning (the necking zone) is rather abrupt and coincident to the emplacement of the SDRs (Guan *et al.* 2019).

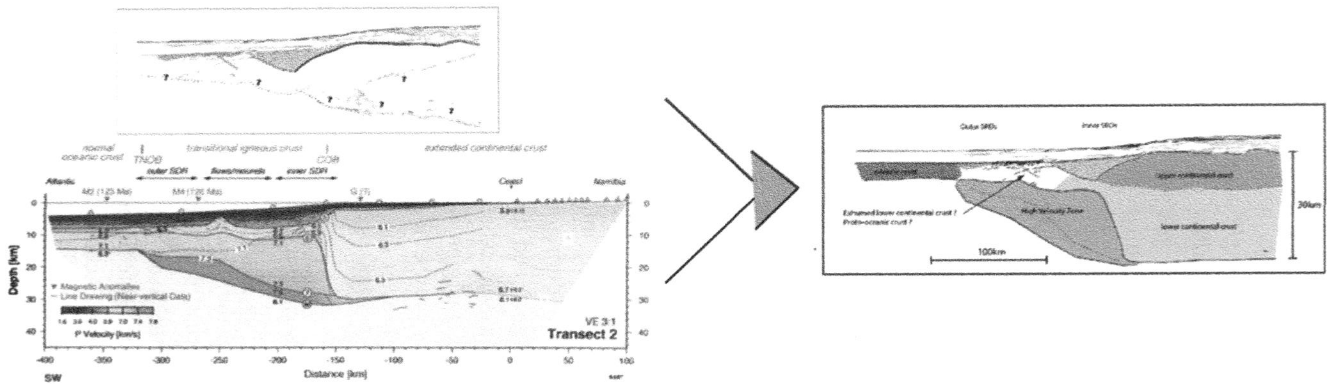

Figure 3.6. *Left top: line drawing based on Industrial MCS data; left bottom: wide-angle model from* Bauer et al. *(2010). Note that these two profiles are not at the same location but on conjugate margins. Right: a tentative of schematic geological interpretation*

Interpretation of the nature and composition of the crust imaged by a seismic experiment is not straightforward. From velocity models based on wide-angle seismic data, velocity versus depth (1D-Vz) profiles can be extracted and compared to other profiles found in the literature. These latter profiles mainly come from specific seismic experiments or compilation of velocity structures from several of these experiments. Interpretations of the nature and composition of the crust imaged during seismic experiments are rarely constrained by direct evidence except when they coincide with drilling experiments. Even in these cases, velocities and lithologies are only directly related for the uppermost crust and only on a very few locations. Laboratory velocity measurements on rock samples confined at high pressure can also provide a direct relationship between velocities and lithologies but seismic experiments are often made in regions that have a specific geodynamic context and history where such a pressure condition most probably does not apply. Finally, there is often a range of possible lithologies for a particular velocity. When comparing and discussing VZ profiles, it is therefore important to know their origins and keep in mind all these limitations, and to replace these profiles in the geodynamic context. Anyway, as shown below on a example in the Western Mediterranean Sea (Moulin *et al.* 2015), these 1D-Vz profiles extracted every 10 or 5 km along the profile exhibit on all margins a strong segmentation with abrupt changes in velocity structure which show a drastic change in the nature of the basement.

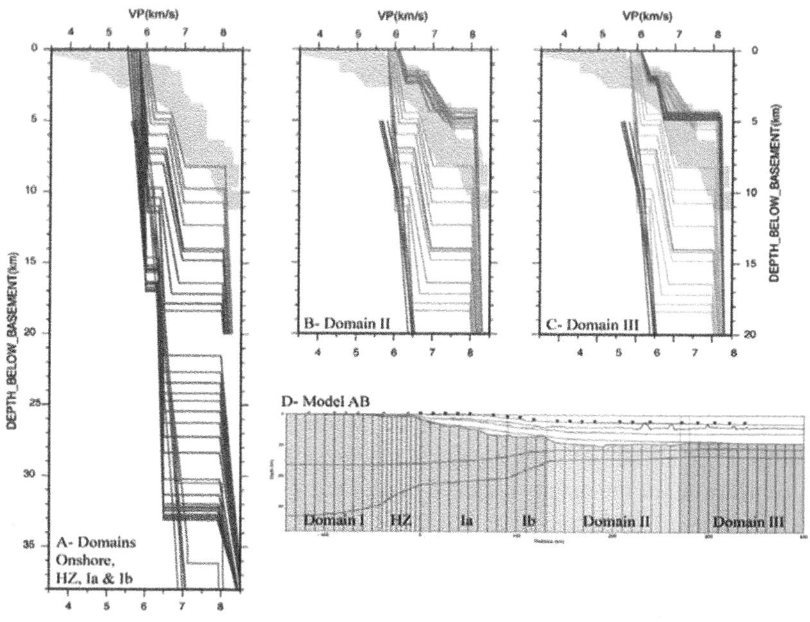

1-D P-wave velocity depth (1D-Vz) profiles below the basement in the three distinct domains defined in this study on a wide-angle seismic profile in the Gulf of Lion (Western Mediterranean Sea). Each domain is compared to the compilations made for continental

crust (Christensen, Mooney 1995) (gray lines) and a "normal oceanic crust" (White et al. *1992) (blue-shaded envelope). Velocity-depth profiles are extracted each 10 km (except in the Hinge Zone, HZ, each 5 km) in (A) the unthinned (in purple) and the continental slope domains (in brown (HZ), blue (Ia) and green (Ib)); (B) the transitional domain (in orange) and (C) the atypical oceanic crust domain (in red). (D) The exact position of the extracted Vz-profiles is illustrated on profile AB with the same colours (source: Moulin* et al. *2015).*

Box 3.2. *Velocity versus depth profiles and the nature and composition of the crust*

The **Central Segment** of the South Atlantic Ocean (Figure 3.3) is characterized by sedimentary basins with pre- and syn-break-up magmatism, and the presence of an approximately 1–2 km-thick salt layer in the so-called continent-ocean transition, overlying a mainly non-marine sequence. Both layers are overlain by marine shale and carbonates (Marton *et al.* 2000; Contrucci *et al.* 2004; Moulin *et al.* 2005; Lentini *et al.* 2010). As shown by Moulin (2003) and Moulin *et al.* (2005) on the Angolan Basin, the presence of a salt cover continuously from the platform to the deep basin, together with the absence of thick marine layers prior to Aptian times and the absence of an erosional surface contemporaneous to salt deposition, indicate that salt is related to the first marine transgression in a shallow lagoon environment. The presence of carbonates overlying the salt layer furthermore shows that this shallow environment lasted after the salt deposition.

Along the central segment, the passive margins on both sides may present different characteristics and morphology due to their different previous tectonic history (i.e. tectonic heritage) and upper mantle segmentation (i.e. thermal segmentation). As already quoted (Aslanian *et al.* 2009), no margin presents all the features needed to support a general model, but each margin supplies pieces of the jigsaw. The main, and the most difficult, trick is to decipher between local and general characteristics. This could only be done by comparing many different margins.

Several combined MCS and wide-angle seismic experiments were carried out in the Central Segment of the South Atlantic Margins, in close collaboration with industries, in an attempt to draw a general model for this area. Figure 3.7 presents a short compilation of the results obtained during the Zaiango cruise (Contrucci *et al.* 2004), Sanba, Salsa (Loureiro *et al.* 2018; Pinheiro *et al.* 2018) and Magic (Aslanian *et al.* 2021, in press) cruises. Figure 3.7a shows the interpretations of what was the first attempt (Aslanian *et al.* 2009) to combine seismic data and analysis (Contrucci *et al.* 2004; Moulin *et al.* 2005) on conjugate margin and kinematic reconstruction (Moulin *et al.* 2010). The main results were: (1) the characterization of a very sharp necking zone; (2) the imaging of a 200 km wide transition zone; (3) a pure vertical and delayed subsidence of the basin. The palinspastic reconstruction shows that it is unlikely that the continental crust would maintain its integrity throughout the thinning process, which invalidates any conservational model as being the sole thinning process.

The Sanba experiment was devoted to the Santos São Paulo Platform (SSPP). The opening direction of this buffer zone between the Central and the Austral Segments of the South Atlantic Ocean is oblique to the general opening motions of the South American and African plates (Moulin *et al.* 2010, 2012 – Figure 3.3). The SSPP presents in the same area both conjugate margins, separated by an aborted ridge. Figure 3.7b shows the strategy used in all of these cruises, involving several cross-cutting profiles in order to cross-correlate the results on each profile and avoid misinterpretation (Klingelhoeffer *et al.* 2014; Evain *et al.* 2015). Beneath the continental shelf, a ~100 km wide necking zone is imaged where the continental crust thins abruptly from ~40 km to less than 15 km. Towards the ocean, most of the SSPS is characterized by velocity values, velocity gradients and a Moho interface considered as typical of thinned mid/lower continental crust. The central domain has, however, a very heterogeneous structure from south to north. This structure is interpreted as atypical oceanic crust, exhumed lower crust, or upper continental crust intruded by mafic material, overlying either altered mantle.

The Salsa experiment was devoted to the lateral segmentation of the NE-Brazilian Margins. Figure 3.7c presents the results on the Jequitinhonha Margin that was imaged by three combined MCS-Wide-Angle profiles. In that margin, the unthinned continental crust was not imaged due to a technical problem with the land-stations. Along these about 200-km long profiles, typical oceanic crust was never imaged and the transitional crust in the Jequitinhonha Basin is modeled as composed of exhumed lower continental crust, underlying an anomalous velocity layer, with high-amplitude undulating reflectors. The seismic propagation velocities and the geometry of the reflectors suggest the existence of a boudin-like feature probably linked to the crustal thinning, with oceanward creep of the lower crust (Loureiro *et al.* 2018).

The Magic project was devoted to the Pará-Maranhão/Barreirinhas Margin, North Brazil, a pull-apart passive margin, with two strike-slip borders, formed during the opening of the Equatorial Atlantic Ocean in Cretaceous times. The combined MCS wide-angle seismic profiles show a very sharp necking zone followed by a 10-km thick sedimentary basin underlined by a 5-thick crust modeled with velocity ranging from 6.2 to 6.9 km/s and interpreted as exhumed lower continental crust (Figure 3.7d) (Aslanian *et al.* 2021). Eastwards, the velocity gradients become characteristic of a proto-oceanic crust, evolving into more standard thin oceanic crust (Figure 3.7d). The proto-oceanic crust presents, in its upper part, seismic velocity higher than "normal" oceanic crust but in continuity with the velocity described in the transitional domain. Following Bott (1971) and Aslanian *et al.* (2009), this suggests a relationship between these two domains, involving lower continental material flowing that fed the first atypical oceanic crust (Aslanian *et al.* 2021, in press), as already observed in the Liguro-Provencal Basin (Afilhado *et al.* 2015; Moulin *et al.* 2015).

Despite their differences, the passive margins of the Central Segment of the South Atlantic and the Equatorial Atlantic Ocean present common characteristics and the recurrence of these general features pleads in favor of general rules regarding the architecture of the related margins:

1) little evidence for large extensional faults, which may point to specific thinning processes;

2) a quite abrupt thinning of the crustal basement (necking zone);

3) a large transitional domain, with a large part of allochthonous material due to lower continental crust exhumation rather than mantle exhumation;

4) a system of conjugate margins which remained in an elevated position, near the sea surface, until, at least, the break-up – the presence of carbonates overlying the salt layer shows that this shallow environment lasted after the salt deposition.

Moreover, according to seismic data and kinematical constraints, it is unlikely that the lower continental crust maintained its integrity throughout the thinning process. Lower/middle continental crust is missing in the palinspastic reconstruction (Aslanian *et al.* 2009; Aslanian and Moulin, 2012).

3.4. Conclusion

This overview of these margins, although very different, shows the essential role of the lower continental crust in the passive margin genesis, from the Equatorial Atlantic Ocean to the Austral Segment of the South Atlantic Ocean.

This makes it possible to propose a three-phase model of evolution (Figure 3.8): the rifting phase occurred on a large area, producing very few tilted blocks (on the top of the platform or at the toe of the continental slope). The mid/lower continental crust exhumation phase is the main thinning phase with the formation of the sag basin and continental to lacustrine sedimentation. The end of this phase is marked in the Central Segment by the first marine sediment (known as the "La Chela" layer), the deposit of the salt layer, in a very shallow environment. The third phase corresponds to the break-up and drift phase. The setup of the oceanic crust is sometimes progressive and involved a part of the lower continental crust.

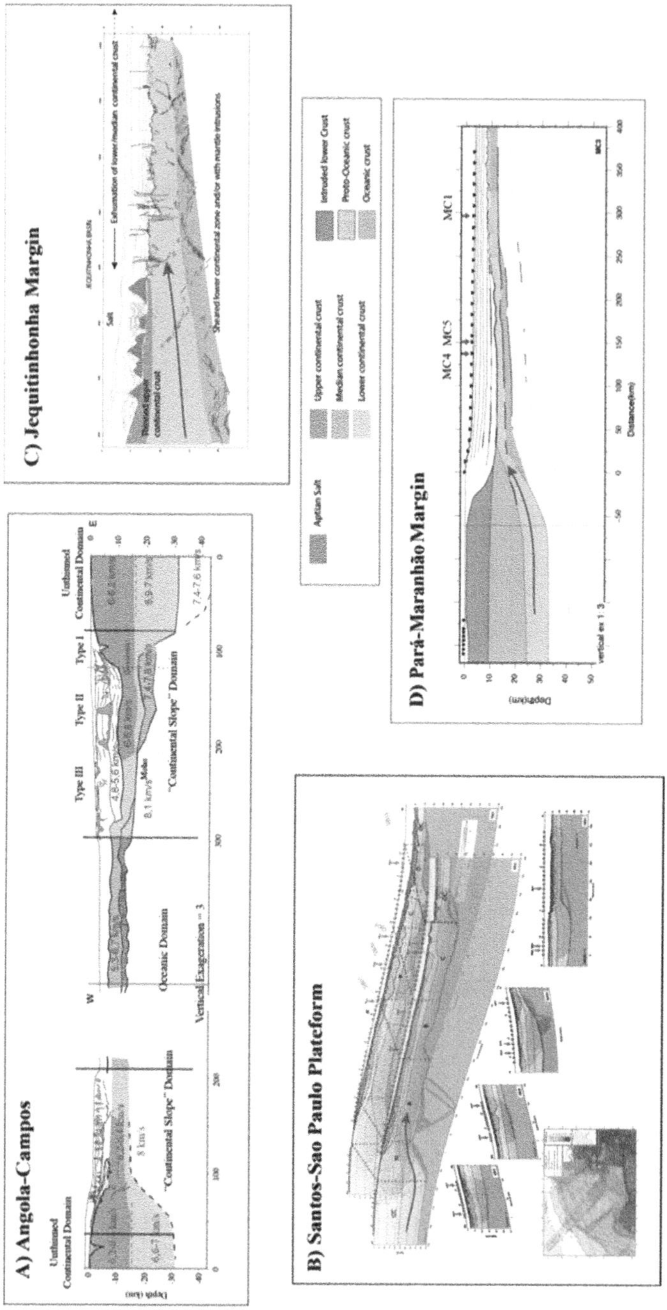

Figure 3.7 *Some results of different combined MCS wide-angle seismic experiments in the Central Segment of the South Atlantic Ocean and the Equatorial Ocean (see locations on Figure 3.3)*

CONTINUATION OF CAPTION FOR FIGURE 3.7.– *(A) the Zaiango experiment in the Angola Basin (Contrucci et al. 2004; Moulin et al. 2005), combined with industrial profile and gravity on the Campos conjugate margin (Aslanian et al. 2009); (B) the Sanba experiment in the Santos São Paulo Platform system (Klingelhoeffer et al. 2014; Evain et al. 2015); (C) the Salsa experiment in the NE Brazilian Margin. Only the results obtained on the Jequitinhonha Margin are presented here (Loureiro et al. 2018); (D) the Magic experiment, on the Pará-Maranhão/Barreirinhas Margin, in the Equatorial Atlantic (source: Aslanian et al. 2021, in press).*

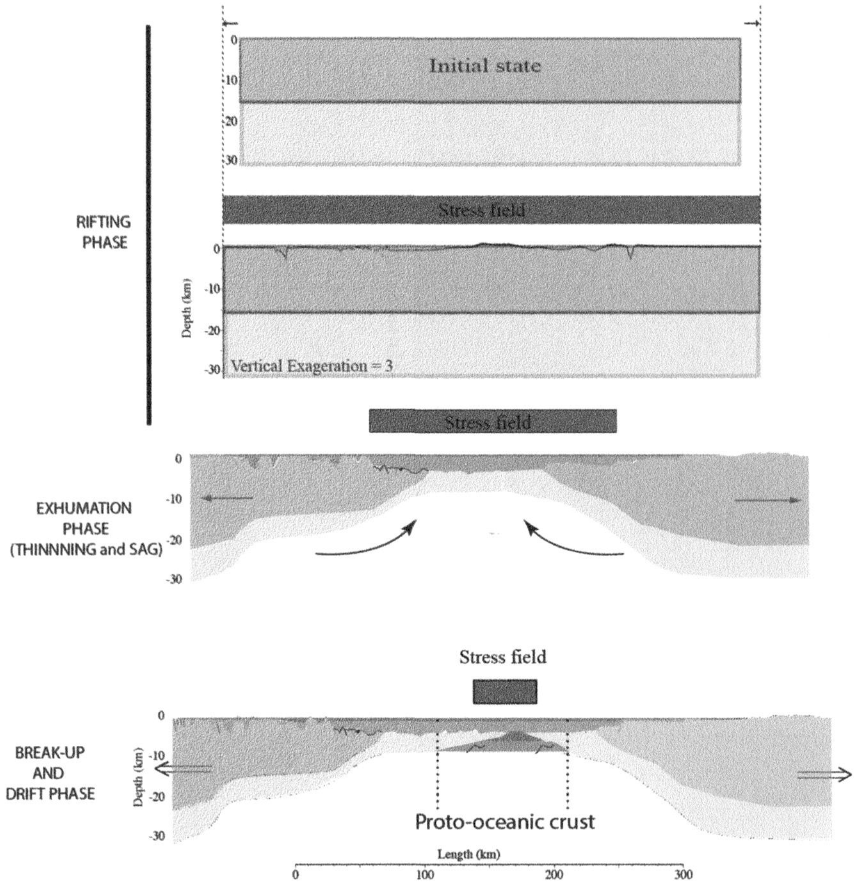

Figure 3.8. *Three-phase model of passive margin evolution showing the primordial role of the median/lower continental crust (source: modified from Aslanian* et al. *(2009))*

3.5. References

Afilhado, A., Moulin, M., Klingelhoefer, F., Aslanian, D., Schnurle, P., Nouzé, H., Rabineau M., Beslier, M.O. (2015). Deep crustal structure across an young passive margin from wide-angle and reflection seismic data (The SARDINIA Experiment) II. Sardinia's margin. *BSGF, ILP Special Volume*, 186, 331–351. doi:10.2113/gssgfbull.186.4-5.331, 2015.

Aslanian, D. and Moulin, M. (2010). Comments on "A new scheme for the opening of the south Atlantic ocean and the dissection of an Aptian salt Basin" from Torsvik, Sonia Rousse, Cinthia Labails and Mark A. Smethurst. *Geophysical Journal International*, 183, 20–28. doi: 10.1111/j.1365-246X.2010.04727.x.

Aslanian, D. and Moulin, M. (2012). Paleogeographic consequences of conservational models in the South Atlantic Ocean. In *Conjugate Divergent Margins. Geological Society*, Mohriak, W.U., Danforth, A., Post, P.J., Brown, D.E., Tari, G.C., Nemcok, M., Sinha, S.T. (eds). Special Publications, London. http://dx.doi.org/10.1144/SP369.5.

Aslanian, D., Moulin, M., Olivet, J.-L., Unternehr, P., Matias, L., Bache, F., Rabineau, M., Nouzé, H., Klingelhoefer, F., Contrucci, I., Labails, C. (2009). Brazilian and Angolan passive margins: The kinematic constraints. *Tectonophysics, Special Issue: Role of the Magmatism*, 468, 98–112.

Aslanian, D., Galais, F., Afilhado, A., Schnurle, P., Moulin, M., Evain, M., Dias, N., Soares, J., Fuck, R., Pessoa Neto, O., Cupertino, J.A., Viana, A., The MAGIC Team (2021). Deep structure of the Pará-Maranhão/Barreirinhas passive margin in the Equatorial Atlantic (NE Brazil). *Journal of South American Earth Sciences*, 110. 10.1016/j.jsames.2021.103322.

Aslanian, D., Schnürle, P., Moulin, M., Evain, M., Pellen, R., Rabineau, M., Afilhado, A., Dias, N. (in press). The Mediterranean Sea: A laboratory to characterize micro-continental drift and oceanic basin formation processes. *Nature*, Springer, Cham.

Bauer, K., Neben, S., Schreckenberger, B., Emmermann, R., Hinz, K., Fechner, N., Gohl, K., Schulze, A., Trumbull, R.B., Weber, K. (2000). Deep structure of the Namibia continental margin as derived from integrated geophysical studies. *Journal of Geophysical Research*, 105, 25829–25853.

Blaich, O.A., Tsikalas, F., Faleide, J.I. (2008). Northeastern Brazilian margin: Regional tectonic evolution based on integrated analysis of seismic reflection and potential field data and modelling. *Tectonophysics*, 458, 51–67.

Blarez, E. (1986). Structure et évolution d'une marge continentale transformante. PhD thesis, Université Pierre et Marie Curie, Paris.

Bott, M.H.P. (1971). Evolution of young continental margins and formation of shelf basin. *Tectonophysics*, 11, 319–337.

Bullard, E.C., Everett, J.E., Smith, A.G. (1965). The fit of the continents around the Atlantic. *Symposium on Continental Drift. Philosophical Transactions of the Royal Society of London. Series A, Mathematical and Physical Sciences*, 258, A, 41–51.

Campan, A. (1995). Analyse cinématique de l'Atlantique Equatorial, implications sur l'évolution de l'Atlantique Sud et sur la frontière de plaque Amérique du Nord/Amérique du Sud. PhD thesis, Université Pierre et Marie Curie, Paris.

Carey, S.W. (1955). Wegener's South America–Africa assembly, fit or misfit? *Geological Magazine*, 92, 196–200.

Castro Jr., A.C.M. (1987). The northeastern Brazil and Gabon basins: A double rifting system associated with multiple crustal detachment surfaces. *Tectonics*, 6(6), 727–738.

Christensen, N.I. and Mooney, W.D. (1995). Seismic velocity structure and composition of the continental crust: A global view. *Journal of Geophysical Research*, 100, 9761. https://doi.org/10.1029/95JB00259.

Clerc, C., Jolivet, L., Ringenbach, J.-C. (2015). Ductile extensional shear zones in the lower crust of a passive margin. *Earth and Planetary Science Letters*, 431, 1–7. http://dx.doi.org/10.1016/j.epsl.2015.08.038.

Contrucci, I., Matias, L., Moulin, M., Géli, L., Klingelhoeffer, F., Nouzé, H., Aslanian, D., Olivet, J.-L., Sibuet, J.-C., Réhault, J.-P. (2004). Deep structure of the West African continental margin, between 5°S and 8°S, from reflection/refraction seismics and gravity data. *Geophysical Journal International*, 158, 529–553.

Eagles, G. (2007). New angles on South Atlantic Opening. *Geophysical Journal International*, 168(1), 353–361, January. doi: 10.1111/j.1365-246X.2006.03206.x.

Evain, M., Afilhado, A., Rigoti, C., Loureiro, A., Klingelhoefer, F., Alves, D., Schnurle, P., Feld, A., Fuck, R., Soares, J., M. Vinicius de Lima, Corela, C., Matias, L., Benabdellouahed, M., Baltzer, A., Rabineau, M., Viana, A., Moulin, M., Aslanian, D. (2015). Deep structure of the Santos Basin-São Paulo Plateau System (SSPS). *Journal of Geophysical Research: Solid Earth*, 120, doi:10.1002/2014JB011561.

Franke, D., Neben, S., Ladage, S., Schreckenberger, B., Hinz, K. (2007). Margin segmentation and volcano-tectonic architecture along the volcanic margin off Argentina/Uruguay, South Atlantic. *Marine Geology*, 244, 46–67. doi:10.1016/j.margeo.2007.06.009.

Goodlad, S.W., Martin, A., Hartnady, C. (1982). Mesozoic magnetic anomalies in the southern Natal Valley. *Nature*, 295, 686–688. https://doi.org/10.1038/295686a0.

Gouyet, S. (1988). Evolution tectono-sédimentaire des marges guyanaises et nord-brésilienne au cours de l'ouverture de l'Atlantique Sud. PhD thesis, Université de Pau et des Pays de l'Adour, Pau.

Guan, H., Geoffroy, L., Gernigon, L., Chauvet, F., Grigné, C., Werner, P. (2019). Magmatic ocean-continent transitions. *Marine and Petroleum Geology*, 104, 438–450. https://doi.org/10.1016/j.marpetgeo.2019.04.003.

Hayes, D.E. and Ewing, M. (1970). North Brazilian ridge and adjacent continental margin. *The American Association of Petroleum Geologists Bulletin*, 54, 2120–2150.

Hinz, K. (1981). A hypothesis on terrestrial catastrophes: Wedges of very thick oceanward dipping layers beneath passive continental margins. *Geological Journal*, E22, 3–28.

Klingelhoefer, F., Evain, M., Afilhado, A., Rigoti C., Loureiro, A., Schnurle, P., Moulin, M., Alves, D., Benhabdelhaoued, M., Baltzer, A., Rabineau, M., Feld, A., Lepretre, A., Viana, A., Aslanian, D. (2014). Imaging proto-oceanic crust off the Brazilian Continental Margin. *Geophysical Journal International*, 200, 471–488.

Lawver, L.A., Gahagan, L., Dalziel, I.W.D. (1999). A tight fit-early Mesozoic Gondwana, a plate reconstruction perspertive. In *Origin and Evolution of Continents: Proceedings of the International Symposium Origin and Evolution of Continents*, Motoyoshi, Y. and Shiraishi, K. (eds). National Institute of Polar Research, Tokyo.

Le Pichon, X. and Hayes, D.E. (1971). Marginal offsets, fracture zones and the early opening of the South Atlantic. *Journal of Geophysical Research*, 7626, 6283–6293.

Lentini, M.R., Fraser, S.I., Scott Sumner, H., Davies, R.J. (2010). Geodynamics of the central South Atlantic conjugate margins: Implications for hydrocarbon potential. *Petroleum Geosciences*, 16, 217–229.

Loureiro A., Schnürle, P., Klingelhöfer, F., Afilhado, A., Evain, M., Gallais, F., Rabineau, M., Baltzer, A., Benabdellouahed, M., Pinheiro, J., Dias, N., Soares, J., Fuck, R., Matias, L., Cupertino, J.A., Viana, A., Moulin, M., Aslanian, D. (2018). Imaging exhumed lower continental crust in the distal Jequitinhonha basin, Brazil. *Journal of South American Earth Sciences*, 84, 351–372. https://doi.org/10.1016/j.jsames.2018.01.009.

Martin, A.K., Goodlad, S.W., Hartnady, C.J.H., du Plessis, A. (1982). Cretaceous palaeopositions of the Falkland Plateau relative to southern Africa using Mesozoic seafloor spreading anomalies. *Geophysical Journal International*, 71, 567–579. https://doi.org/10.1111/j.1365-246X.1982.tb02784.x.

Marton, L.G., Tari, G.C., Lehmann, C.T. (2000). Evolution of the Angolan passive margin, West Africa, with emphasis on post-salt structural styles. In *Atlantic Rifts and Continental Margins*, Mohriak, W.U. and Talwani, M. (eds). American Geophysical Union, Washington, DC.

Meyers, J.B., Rosendahl, B.R., Austin, J.A. (1996). Deep penetrating MCS images of the South Basin: Implications for rift tectonics and post-breakup salt remobilisation. *Basin Research*, 8, 65–84.

Mohriak, W.U. (2001). Salt tectonics, volcanic centers, fracture zones and their relationship with the origin and evolution of the South Atlantic Ocean: Geophysical evidence in the Brazilian and West African margins. *7th International Congress of The Brazilian Geophysical Society*, Salvador, Bahia, Brazil, October 28–31.

Mohriak, W.U. (2003). Bacias Sedimentares da Margem Continental Brasileira. In *Geologia, Tectônica e Recursos Minerais do Brasil*, Bizzi, L.A., Schobbenhaus, C., Vidottie R.M., Gonçalves, J.H. (eds). CPRM, Brasilia.

Mohriak, W.U. (2004). Recursos energéticos associados à ativação tectônica mesozóico-cenozóica da América do Sul. In *Geologia do continente sul-americano: evolução da obra de Fernando Flávio Marques de Almeida*, Mantesso-Neto, V., Bartorelli, A., Carneiro, C.D.R., Brito-Neves, B.B. (eds). Beca Produções Culturais Ltda., São Paulo.

Mohriak, W.U. and Rosendahl, B.R. (2003). Transform zones in the South Atlantic rifted continental margins. In *Intraplate Strike-Slip Deformation Belts*, Storti, F., Holdsworth, R.E., Salvini, F. (eds). Geological Society, London.

Mohriak, W.U., Fernandez, B., Biassussi, A.S. (2004). Salt tectonics domains and structural provinces: Analogies between the South Atlantic and the Gulf of Mexico. In *Salt-Sediment Interactions and Hydrocarbon Prospectivity: Concepts, Applications and Case Studies for the 21st Century*, Post, P.J., Olson, D.L., Lyons, K.T., Palmes, S.L., Harrison, P.F., Rosen, N.C. (eds). SEPM Society for Sedimentary Geology, Houston, TX.

Moulin, M. (2003). Etude géologique et géophysique des marges continentales passives : exemple du Zaïre et de l'Angola. PhD thesis, Université de Bretagne Occidentale, Brest.

Moulin, M. and Aslanian, D. (2010). Corrigendum to: A new starting point for the South and Equatorial Atlantic Ocean (*Earth Science Reviews* 98 (2010), 1–37). *Earth Science Reviews*, 103, 197–198. doi:10.1016/j.earscirev.2010.10.001.

Moulin, M., Aslanian, D., Olivet, J.-L., Contrucci, I., Matias, L,. Géli, L., Klingelhoefer, F., Nouzé, H., Réhault, J.-P., Unternehr, P. (2005). Geological constraints on the evolution of the Angolan margin based on reflection and refraction seismic data (ZaïAngo project). *Geophysical Journal International*, 162, 793–810.

Moulin, M., Aslanian, D., Unternher, P. (2010), A new starting point for the South and Equatorial Atlantic Ocean. *Earth Science Reviews*, 98(1), 1–37.

Moulin, M., Aslanian, D., Rabineau, M., Patriat, M., Matias, L. (2012). Kinematic keys of the Santos – Namibe Basins. In *Conjugate Divergent Margins*, Mohriak, W.U., Danforth, A., Post, P.J., Brown, D.E., Tari, G.C., Nemcok, M., Sinha, S.T. (eds). Geological Society, London. http://dx.doi.org/10.1144/SP369.3.

Moulin, M., Klingelhoefer, F., Afilhado, A., Aslanian, D., Schnurle, P., Nouzé, H., Rabineau, M., Beslier, M.-O., Feld, A. (2015). Deep crustal structure across a young passive margin from wide-angle and reflection seismic data (The SARDINIA Experiment) I. Gulf of Lion's Margin. *BSGF, ILP Special Volume*, 186, 309–330.

Oliveira Marinho, M. (1985). Le plateau marginal de Guinée : transition entre Atlantique Central et Atlantique Equatorial. PhD thesis, Université Pierre et Marie Curie, Paris.

Pindell, J. and Dewey, J.F. (1982). Permo-triassic reconstruction of western pangea and the evolution of the Gulf of Mexico/Caribbean region. *Tectonics*, 1(2), 179–211.

Pinheiro, J.M., Schnurle, P., Evain, M.. Gallais, F., Klingelhoefer, F., Afilhado, A., Loureiro, A., Fuck, R., Soares, J., Cupertino, J.A., Viana, A., Moulin, M., Aslanian, D. (2018). Lithospheric structuration onshore-offshore of the Sergipe-Alagoas passive margin, NE Brazil, based on wide-angle seismic data. *Journal of South American Earth Sciences*, 88, 649–672. https://doi.org/10.1016/j.jsames.2018.09.015.

Popoff, M. (1988). Du Gondwana à l'Atlantique sud : les connections du fossé de la Bénoué avec les bassins du Nord-Est bresilien jusqu'à l'ouverture du golfe de Guinée au Crétacé inférieur. *Journal of African Earth Sciences Special Publication*, 7(2), 409–431.

Rabinowitz, P.D. and La Brecque, J. (1979). The Mesozoic South Atlantic Ocean and evolution of its continental margins. *Journal of Geophysical Research*, 84, 5973–6002.

Rabinowitz, P.D., Cande, S.C., La Brecque, J.L. (1976). The Falkland escarpment and Agulhas Fracture Zone: The boundary between oceanic and continental crust at conjugate continental margins. *Anais da Academia Brasileira de Ciencias*, 48, 241–251.

Sahabi, M. (1993). Modèle général de l'évolution de l'océan Indien. Thesis, Université de Bretagne Occidentale, Brest.

Sahabi, M., Aslanian, D., Olivet, J.-L. (2004). Un nouveau point de départ pour l'histoire de l'Atlantique Central. *C.R. Geoscience*, 336, 1041–1052.

Sandwell, D.T. and Smith, W.H.F. (2009). Global marine gravity from retracked Geosat and ERS-1 altimetry: Ridge segmentation versus spreading rate. *Journal of Geophysical Research*, 114, B01411.

Seton, M, Müller, R.D., Zahirovic, S., Gaina, C., Torsvik, T., Shepard, G., Talsma, A., Gurnis, M., Turner, M., Maus, S., Chandler, M. (2012). Global continental and ocean basin reconstructions since 200 Ma. *Earth-Science Reviews*, 113, 212–270.

Stica, J.M., Zalan, P.V., Ferrari, A.L. (2014). The evolution of rifting on the volcanic margin of the Pelotas Basin and the contextualization of the Paraná-Etendeka LIP in the separation of Gondwana in the south Atlantic. *Journal of Marine and Petroleum Geology*, 50, 1–21.

Talwani, M. and Abreu, V. (2000). Inferences regarding initiation of oceanic crust formation from the U.S. East Coast margin and conjugate South Atlantic margins. In *Atlantic Rifts and Continental Margins*, Mohriak, W.U. and Talwani, M. (eds). American Geophysical Union, Washington, DC.

Thompson, J., Moulin, M., Aslanian, D., Guillocheau, F., de Clarens, P. (2019). New starting point for the Indian Ocean: Second phase of breakup for the Gondwana. *Earth-Science Reviews*, 191, 26–56. https://doi.org/10.1016/j.earscirev.2019.01.018.

Torsvik, T.H., Rousse, S., Labails, C., Smethurst, M.A. (2009). A new scheme for the opening of the South Atlantic Ocean and the dissection of an Aptian salt basin. *Geophysical Journal International*, 177, 1315–1333, doi:10.1111/j.1365-246X.2009.04137.x.

Verard, C., Flores, K., Stampfli, G. (2012). Geodynamic reconstructions of the South America–Antarctica plate system. *Journal of Geodynamics*, 53, 43–60.

White, R.S., McKenzie, D., O'Nions, R.K. (1992). Oceanic crustal thickness from seismic measurements and rare earth element inversions. *Journal of Geophysical Research*, 97(B13), 19683–19715.

Zalán, P.V., Severino, M.C.G., Rigoti, C.A., Magnavita, L.P., de Oliveira, J.A.B., Viana, A.R. (2011). An entirely new 3D-view of the crustal and mantle structure of a South Atlantic passive margin – Santos, Campos and Espírito Santo Basins, Brazil. *AAPG Annual Convention and Exhibition*, Houston, TX, April, [Online]. Available at: http://www.searchanddiscovery.com/abstracts/html/2011/annual/abstracts/Zalan.html.

4

The Continental Margins of the Gulf of Aden

Sylvie LEROY and Chloé NONN
Sorbonne Université, CNRS-INSU, Institut des Sciences de la Terre, ISTeP, Paris, France

4.1. Introduction

The continental margins of the Gulf of Aden are a good example for illustrating and understanding the formation of rifted margins and the transition to an active oceanic ridge. This exceptional geological object allows us to study continental break-up, both in a magmatic context (under the influence of the Afar mantle plume) and in a low magmatic context (associated with the exhumation of sub-continental mantle).

The Gulf of Aden is an oceanic basin that extends for about 1,400 km between the Red Sea and the Indian Ocean (see Figure 4.1). It is the result of an Oligo-Miocene continental rifting process (34–17.6 Myr) that led to the separation of the Arabian and Somali plates. The result of the rifting is the break-up of the continental lithosphere and the establishment of an oceanic ridge in the center of the basin, here the Aden and Sheba ridges. Currently active, they are a site of intense seismicity (Figure 4.1).

For a color version of all the figures in this chapter, see www.iste.co.uk/peronpinvidic/margins2.zip.

Continental Rifted Margins 2,
coordinated by Gwenn PERON-PINVIDIC.

The Gulf of Aden's relatively recent post-rift history (less than 20 Myr) also enables the observation of a continental rift undergoing thermal and compositional rebalancing. Less than 300 km to 650 km apart, the Gulf of Aden's continental margins define a narrow rift, thus promoting correlations between the conjugate segments. The relatively thin sedimentary cover facilitates the observation of the structures by indirect geophysical methods. In addition, the obliquity of the Gulf (the direction of extension is oblique to the general direction of the continental rift, not perpendicular) results in significant segmentation of the margins, which are offset by transfer fault zones that evolve over time to transform fault zones of variable offset (from ~20 to 180 km, see Figure 4.1).

The present-day structure of the Gulf of Aden's continental margins is closely related to the regional geological history.

4.2. Geological and geodynamic context and heritage

Bordered by the Afar mantle plume to the west and the oceanic lithosphere of the Indian Ocean to the east, the Gulf of Aden opened within the Arabian-African lithosphere, which is made up of terrains of different geological natures (Figure 4.1). These include Neo-Proterozoic bedrock formed by the combination of various arcs and accreted "terranes", and different sedimentary series of Permo-Triassic, Jurassic or Cretaceous ages.

The metamorphic bedrock has undergone several episodes of rifting at different time periods: in the Permo-Triassic, Jurassic and Upper Cretaceous. These extensional episodes were expressed in the form of grabens and basins distributed on the eastern Afro-Arabic platform in the Permo-Triassic, near the future Gulf of Aden region in the Upper Jurassic and on the Arabian platform in the Upper Cretaceous (see Figure 4.1).

In the Late Eocene (~35 Myr), the extension was localized in the Red Sea-Aden system when the collision started west of the Zagros and the Afar mantle plume was formed (Bellahsen *et al.* 2003; Koptev *et al.* 2018). At the paroxysm of activity of the Afar plume (~30 Myr), the rifting of the Gulf of Aden continues. Its expression is clearly volcanic in the west and non-volcanic in the east (Leroy *et al.* 2012).

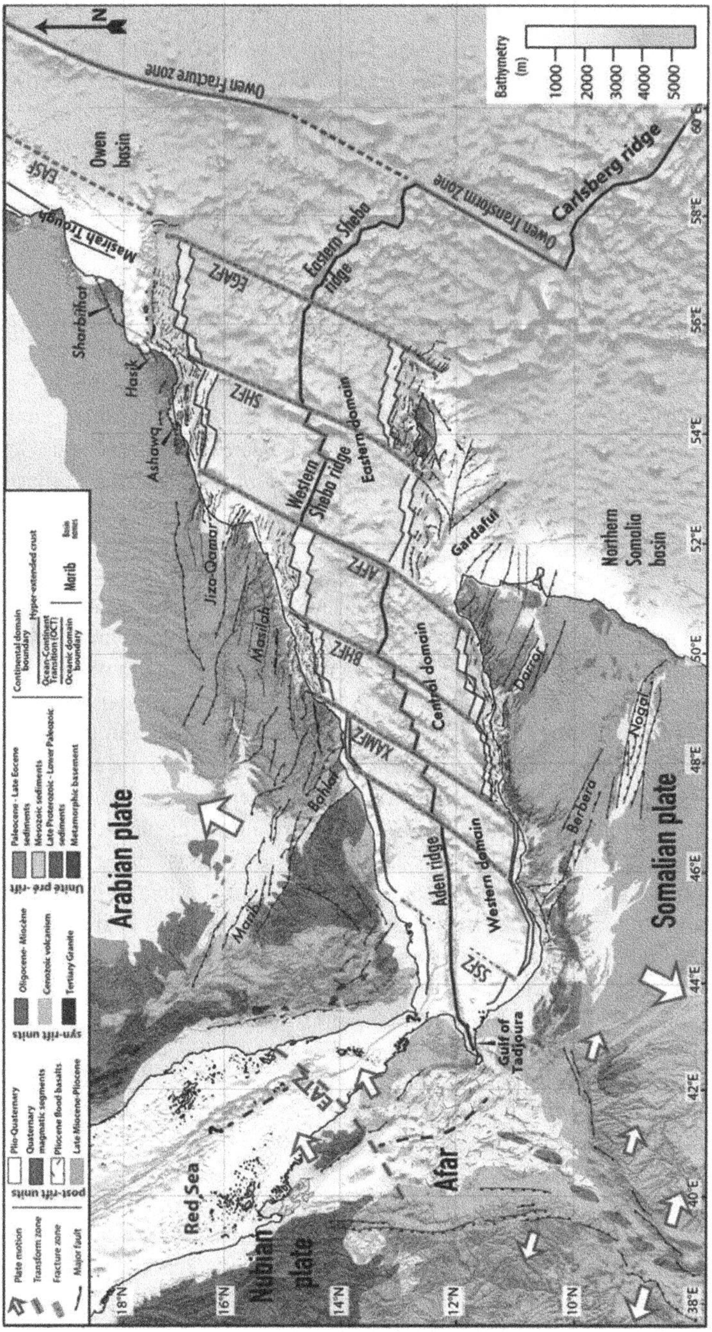

Figure 4.1. *Simplified geological map of the Gulf of Aden (source: adapted from Leroy et al. 2012)*

CONTINUATION OF CAPTION FOR FIGURE 4.1.– *The bathymetric and topographic data are from the GEBCO global database (GEBCO_2014 Grid). The Mesozoic basin names are noted onshore in black (Leroy et al. 2012). AFFZ: Alula Fartak Fracture Zone. AFTF: Alula Fartak Transform Fault. b: basin. BHFZ: Bosaso Hami Fracture Zone. EGAFZ: Eastern Gulf of Aden Fracture Zone. SHFZ: Socotra-Hadbeen Fracture Zone. SSFZ: Shukra El Sheik Fracture Zone. XAMFZ: Xiis Al Mukalla Fracture Zone. The location of the Ocean-Continent Transition and the hyper-stretched continental crust zone are from Leroy et al. (2010) and Nonn et al. (2017, 2019).*

4.3. Continental rifting timing, initiation and depositional environments

In the different sedimentary basins, the Tertiary series outcrop, notably in Sayhut-al Masilah and Qamar in Yemen, Ashawq-Salalah and Sharbithat in Oman, and in the southern part, on Socotra Island and in Somalia (Berbera, Darror and Gardafui) (Figure 4.1). In the area of Aden City in the western Gulf of Aden, the Tertiary series are volcanic and outcrop as remarkable Seaward Dipping Reflectors (SDRs) sequences from at least the early Oligocene (34 Myr) to 16 Myr (Burdigalian) (Tard *et al.* 1991).

The detailed observation of sedimentary facies has allowed us to compare the depositional environments during the rifting process in the western and eastern parts of the northern margin of the Gulf of Aden (see Figure 4.2; (Leroy *et al.* 2012; Robinet *et al.* 2013)). Thus, it appears that the western domain underwent uplift and erosion from the beginning of the rifting (from the Aden City area to the Al Mukalla City area), while a more continuous sedimentation characterizes the eastern domain. During the Oligocene, tectonic subsidence of the entire domain promoted the preservation of syn-rift sedimentary series, showing a clear palaeogeographic differentiation: continental terrigenous and evaporitic systems in the central part of the Gulf, and carbonate platform systems and turbiditic basins in its eastern part. The increase in the rate of subsidence during rifting is recorded in the western part of the central Gulf by the development of evaporitic, lacustrine and marine facies, and in the east by the development of turbiditic basins (Mughsayl Formation). The Lower Miocene series constitute the maximum marine inundation in the central domain, and in the eastern domain the deep facies persist (Figure 4.2).

The continental margins of both domains underwent major uplift at the transition between syn-rift and post-rift stages.

Syn-rift series have the same age throughout the Gulf, whether volcanic in the western domain of the Gulf or sedimentary in the central and eastern part (Nonn *et al.* 2019). Continental rifting thus appears synchronous across the Gulf of Aden since 34 Myr. The uplift phases of the western domain that drive erosions and clastic inflows are assumed to be related to the dynamics of the Afar mantle plume.

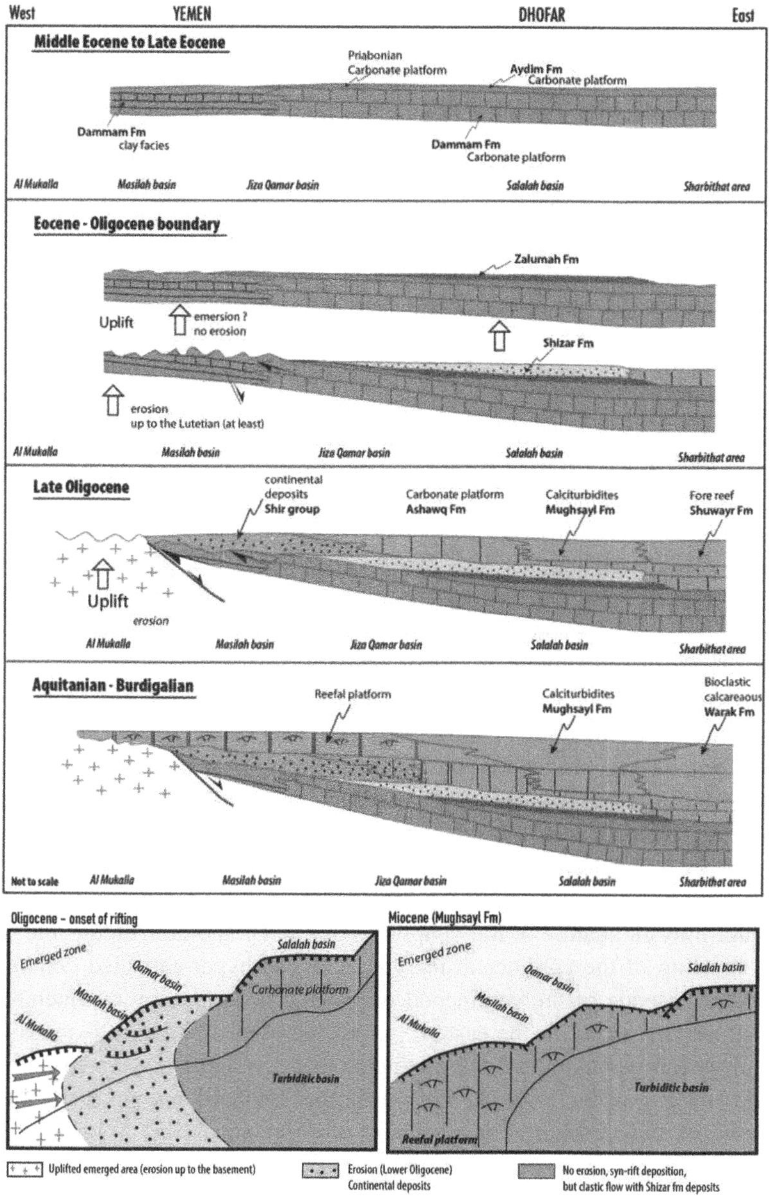

Figure 4.2. *Diagram of longitudinal evolution of sedimentary systems and eroding areas along the northern margin of the Gulf of Aden from the central region to the Salalah Basin in the east (between the city of Al Mukalla in Yemen in the west and to Sharbithat in the Sultanate of Oman in the east) (source: adapted from Leroy* et al. *2012)*

4.4. Structure of the conjugate margins

The Gulf of Aden is marked by the significant narrowness and asymmetry of its margins combined with a gradual decrease in magmatic intensity towards the east. The mapping of the different crustal domains can be seen in Figure 4.1.

The continental domain is structured by a succession of tilted blocks (see Figures 4.1 and 4.3). For the conjugate margins of the central Gulf of Aden, crustal thinning is dominated by north-verging normal faults (see Figure 4.3b). The northern margin is characterized by low fault displacements (100–500 m). In contrast, for the southern margin, the accommodation of extensive deformation occurs on faults with larger displacements (~1,000 to 4,000 m, see Figure 4.3a). For the margins of the eastern Gulf of Aden, thinning of the upper crust occurs along shallowly dipping faults (<30°), some of which were initiated at the onset of rifting on the western part of Socotra Island (Leroy *et al.* 2012). These faults are present throughout the continental domain, from proximal to distal domains (Figure 4.4).

Three crustal subdomains are identified (Figure 4.4) as follows: (1) The proximal domain: this domain corresponds to the outer parts of the rift where crustal thinning is very weak. The crustal thickness varies between ~35 and 40 km. (2) The necking domain: in this domain, crustal thickness decreases from ~35–40 km to 10 km. Crustal thinning is achieved by the interplay of high-displacement faults in the upper crust and shear zones in the lower crust. The necking domain is limited in its distal part by a remarkable point in the architecture of the margins, the Taper Break, which is located at the foot of major faults. It is the closest point to the coast where the crust is less than 10 km thick (Osmundsen and Redfield 2011). Major oceanward dipping faults border the necking domain and form scarps that participate in ~3–4 km of crustal thinning. They root into the mantle at the coupling point (CP) and are responsible for the crustal breakup of the continental margins. (3) The hyper-extended domain: this constitutes the edge of the continental domain. It is formed by continental crust less than ~10 km thick. In the eastern part of the Gulf of Aden, it extends over a width of 90 km cumulated on the conjugate margins and is asymmetric with the northern margin wider (~50–80 km) than the southern margin (~10–40 km). Consisting of a succession of small blocks of continental crust 2–14 km wide on the Omani margin, these blocks are decoupled by a system of major northward dipping detachments. Visible in seismic reflection by a reflector with a very high impedance contrast induced by the juxtaposition of continental crust and underlying exhumed mantle, it resembles the reflector observed in the

West Iberian margin (S reflector (Reston *et al.* 1996), or in the Porcupine basin, P reflector (Reston *et al.* 2001)).

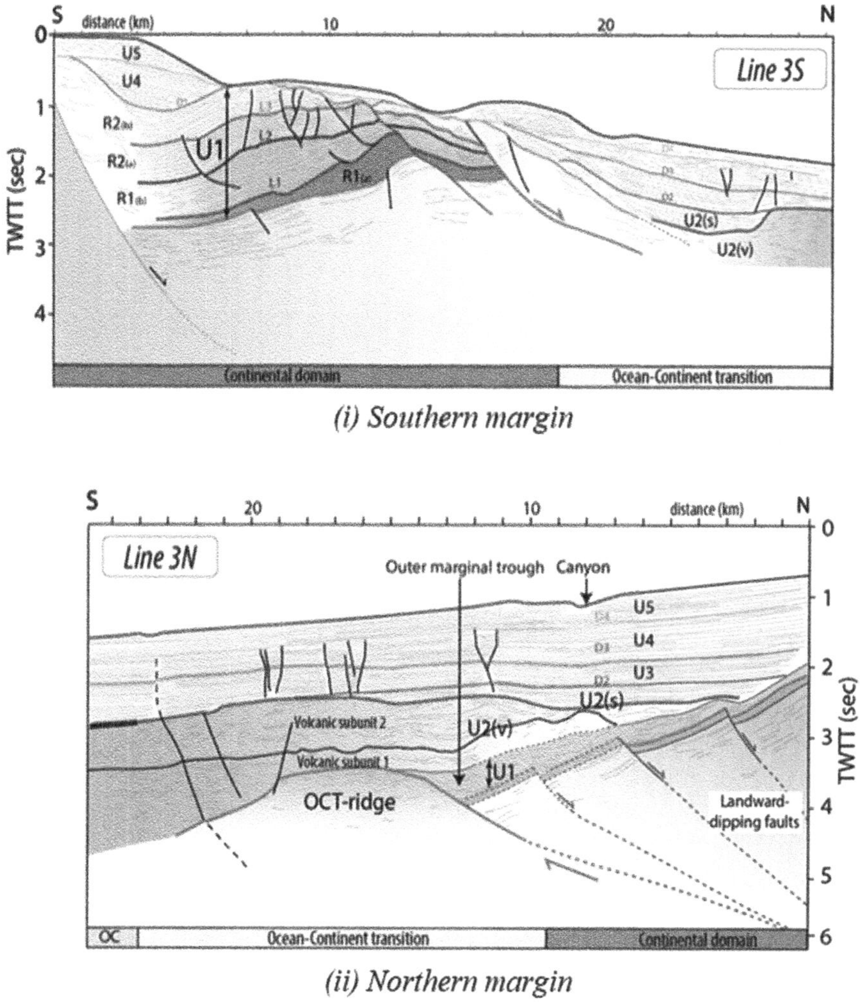

Figure 4.3. *Seismic interpretation of 3N and 3S conjugate sections in the Ahl Medo-Hami region of the central Gulf of Aden. The different seismic sequences are identified on the seismic profiles, as well as the different crustal domains (continental, oceanic and ocean-continent transition). (source: adapted from Nonn* et al. *2019)*

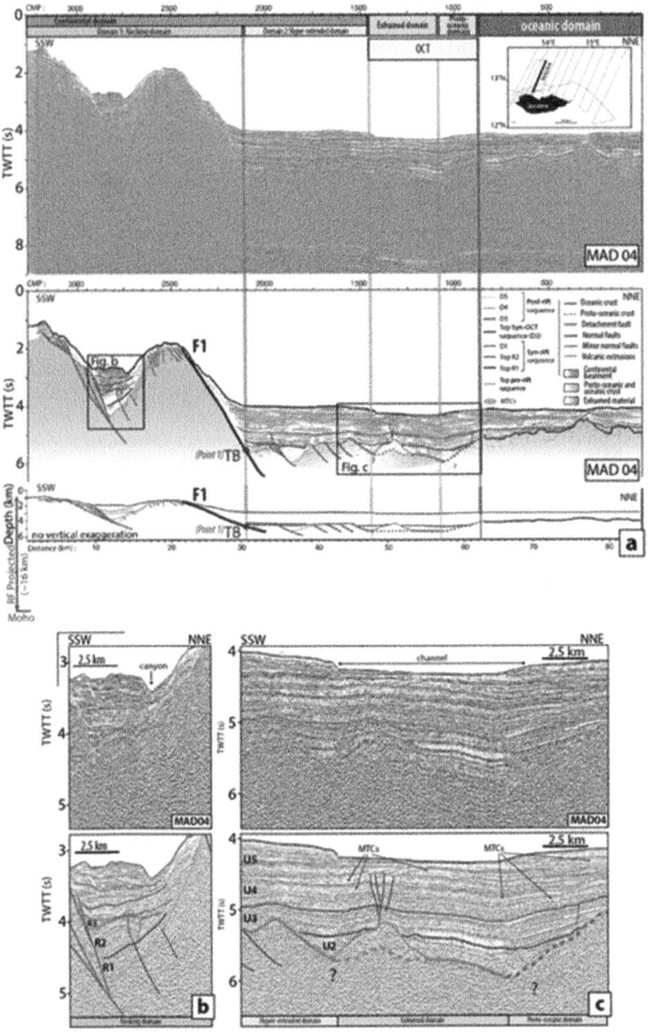

Figure 4.4a. *a) Seismic profile of the southern margin and its interpretation with the different sedimentary sequences identified on the conjugate margins and identification of the five crustal domains (continental, necking, hyper-extended, exhumed, proto-oceanic crust and steady-state oceanic domains). The depth of the Moho projected on the interpreted profile is from Ahmed et al. (2014). b) The interpretation of sedimentary sequences in the perched synrift basin. c) Interpretation of seismic bedrock structures and sedimentary units in the ocean-continent transition zone. MTC: Mass-transport deposits. OCT: Ocean-continent transition. TB: Taper Break. (source: adapted from Nonn* et al. *2017)*

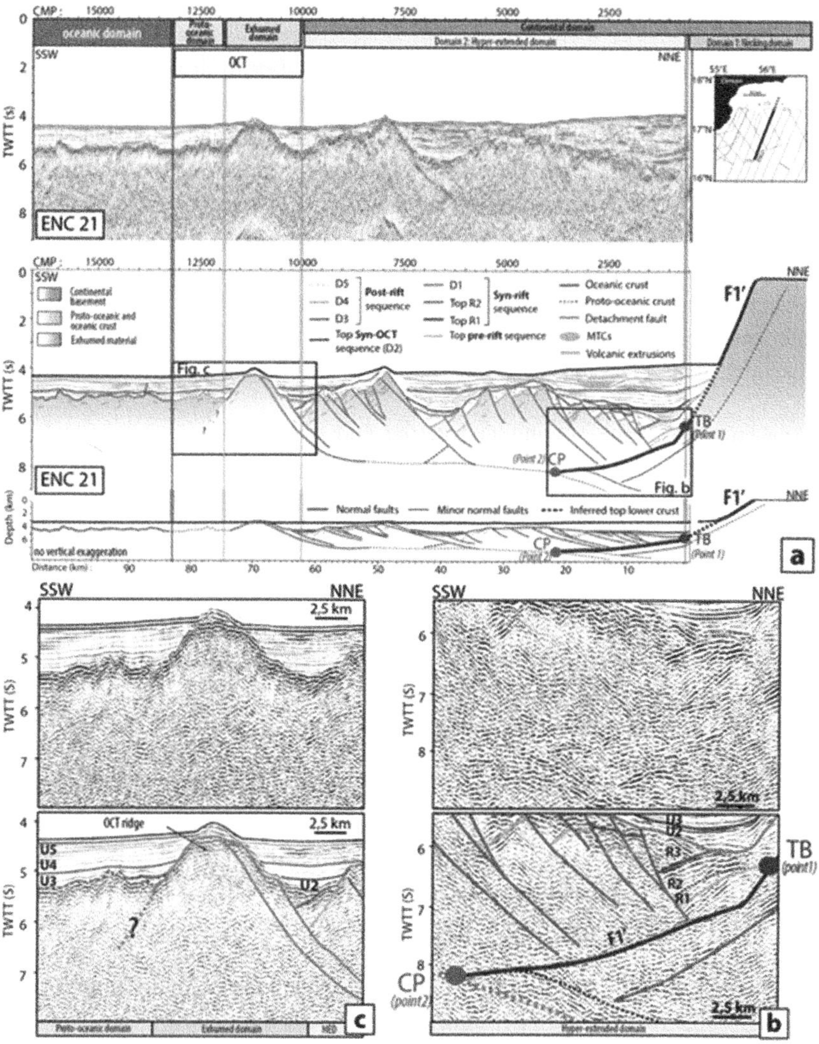

Figure 4.4b. *a) Seismic profile of the northern margin and its interpretation with the different sedimentary sequences identified on the conjugate margins and identification of the five crustal domains. b) Interpretation of seismic bedrock structures, location of the Taper Break (TB) and Coupling Point (CP) (see text for definitions). c) Ocean-Continent Transition (OCT) ridge and proto-oceanic crust. HED: Hyper-extended domain. MTC: Mass-transport Complex deposits. OCT: Ocean-continent transition. (source: adapted from Nonn* et al. *2017)*

4.5. Ocean-continent transition (OCT) and the birth of an oceanic ridge

Combined academic and industrial marine geophysical data provide a high-resolution magnetic coverage, associated with seismic profiles essential for locating the ocean-continent transition (Figure 4.1). The asymmetry of the margins and the observation of detachment surfaces in the hyper-extended continental domain of the northern margins support the hypothesis of the existence of north-verging exhumation systems (Figure 4.4). In our model (Figure 4.5), the northern margins in Yemen and Oman belong to the upper plate and are located in the hanging-wall of the main detachment. The margins located in the south (Somalia and Socotra Island) are the lower plates and are located in the footwall, in agreement with the model proposed by Lister *et al.* (1986).

Detachment systems are initiated at the end of the thinning phase and at the beginning of the exhumation phase, but the duration of their activity remains difficult to determine. For the central and eastern Gulf of Aden margins, exhumation faults are likely to have initiated the emplacement of the serpentinized subcontinental mantle in the OCT (Figures 4.4 and 4.5) (d'Acremont *et al.* 2006; Autin *et al.* 2010; Leroy *et al.* 2010, 2012; Watremez *et al.* 2011; Nonn *et al.* 2017, 2019).

In the central Gulf of Aden, the formation of the OCT is associated with significant volcanism related to the increasing westward influence of the Afar plume (Nonn *et al.* 2019).

The OCT in the eastern part of the Gulf is subdivided into two crustal subdomains: the exhumation domain consisting of serpentinized mantle and the proto-oceanic crustal domain, which is characterized here by serpentinized mantle strongly affected by volcanic intrusions and extrusions. The general morphology of the OCT is different on the upper and lower plate margins.

For margins located on the upper plate, the exhumation domain is characterized by an OCT ridge (also called a peridotite ridge (Sibuet *et al.* 2007; Davy *et al.* 2016)) that borders the hyper-extended domain. Several generations of detachment faults are identified; deformation delocalizes into the wall of the previous fault and causes the emplacement of a new exhumation fault (Sauter *et al.* 2013). Therefore, the duration of exhumation-initiating fault activity is time limited. For margins located on the lower plate, the OCT exhibits a basin-like morphology. The basement is probably locally composed of very thinned lower crust. In the eastern Gulf of Aden, the basement of the exhumed domain is affected by local volcanism. The significant increase in magmatic production occurs at the time of emplacement of the proto-oceanic domain and corresponds to the 5D magnetic anomaly (17.6 Myr; (Nonn *et al.* 2017)).

The oceanic domain is characterized by steady-state oceanic crust, emplaced around ~17 Myr (Leroy *et al.* 2012). The final lithospheric breakup occurs when active volcanism and asthenospheric dynamics take over lithospheric dynamics. Thus, oceanic crust formation occurs during a period of intense magmatic production that localizes deformation and creates the stable oceanic ridge (Jagoutz *et al.* 2007; Cannat *et al.* 2009; Bronner *et al.* 2011).

4.6. Opening of the Gulf of Aden

The interpretation of the data allows us to reconstruct the history of the opening of the Gulf of Aden without propagation within the African continent from the edge of this craton to the Afar region (Figure 4.1) since 34 Myr. This was followed by an oceanic opening that initiated around 17.6 Myr, with the emplacement of the proto-oceanic crust corresponding to the 5D magnetic anomaly identifiable from the edge of the Arabian-African craton in the east to the Shukra el Sheik fracture zone in the west (Figure 4.1).

The influence of the Afar mantle plume, as a lithospheric weakness (Bellahsen *et al.* 2003), is probably considerable in this ultrafast opening over the 700 km forming the Gulf of Aden. In the west, the influence of buoyancy forces (Afar mantle plume) are thus dominant, while in the east, plate boundary forces appear to drive extension throughout the rifting process (Bellahsen *et al.* 2013).

In the extreme east of the area, east of the Eastern Gulf of Aden Fracture Zone (EGAFZ), the oceanic ridge was initiated 20 Myr ago (Fournier *et al.* 2010) in an oceanic lithosphere of probably Late Cretaceous age, with a different rheology than the Afro-Arabian craton. This region would thus rather belong to the Carlsberg Ridge system than to the Gulf of Aden system, which starts west of the EGAFZ (Leroy *et al.* 2012) proper.

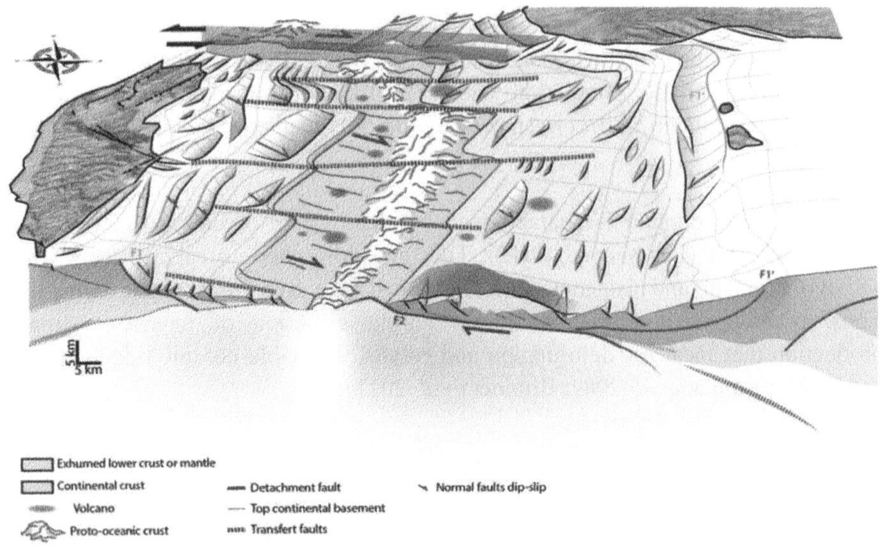

Figure 4.5. *3D block representing a bedrock scarp at the end of the ocean-continent transition period (Middle to Upper Burdigalian). This reconstruction takes into account the 3D morphology of the conjugate margins from the seismic interpretation of structures currently observed on seismic profiles (bedrock morphology, faults, nature of the crust). (source: adapted from Nonn* et al. *2017)*

4.7. References

d'Acremont, E., Leroy, S., Maia, M., Patriat, P., Beslier, M.-O., Bellahsen, N., Fournier, M., Gente, P. (2006). Structure and evolution of the eastern Gulf of Aden: Insights from magnetic and gravity data (Encens-Sheba MD117 cruise). *Geophysical Journal International*, 165, 786–803.

Ahmed, A., Leroy, S., Korostelev, F., Khanbari, K., Rolandone, F., Stuart, G., Obrebski, M. (2014). Crustal structure of the Gulf of Aden southern margin: Evidence from receiver functions on Socotra Island (Yemen). *Tectonophysics*, 637, 251–267.

Autin, J., Leroy, S., Beslier, M.-O., d'Acremont, E., Razin, P., Ribodetti, A., Bellahsen, N., Robin, C., Al-Toubi, K. (2010). Continental break-up history of a deep magma-poor margin based on seismic reflection data (northeastern Gulf of Aden margin, offshore Oman). *Geophysical Journal International*, 180, 501–519.

Bellahsen, N., Faccenna, C., Funiciello, F., Daniel, J.M., Jolivet, L. (2003). Why did Arabia separate from Africa? Insights from 3-D laboratory experiments. *Earth and Planetary Science Letters*, 216, 365–381.

Bellahsen, N., Husson, L., Autin, J., Leroy, S., d'Acremont, E. (2013). The effect of thermal weakening and buoyancy forces on rift localization: Field evidences from the Gulf of Aden oblique rifting. *Tectonophysics*, 607, 80–97.

Bronner, A., Sauter, D., Manatschal, G., Peron-Pinvidic, G., Munschy, M. (2011). Magmatic breakup as an explanation for magnetic anomalies at magma-poor rifted margins. *Nature Geoscience*, 4, 549–553.

Cannat, M., Manatschal, G., Sauter, D., Peron-Pinvidic, G. (2009). Assessing the conditions of continental breakup at magma-poor rifted margins: What can we learn from slow spreading mid-ocean ridges? *Comptes rendus geoscience*, 341, 406–427.

Davy, R.G., Minshull, T.A., Bayrakci, G., Bull, J.M., Klaeschen, D., Papenberg, C., Reston, T.J., Sawyer, D.S., Zelt, C.A. (2016). Continental hyperextension, mantle exhumation, and thin oceanic crust at the continent-ocean transition, West Iberia: New insights from wide-angle seismic. *Journal of Geophysical Research, Solid Earth*, 121(5), 177–3199. doi:10.1002/2016JB012825.

Fournier, M., Chamot-Rooke, N., Petit, C., Huchon, P., Al-Kathiri, A., Audin, L., Beslier, M.-O., d'Acremont, E., Fabbri, O., Fleury, J.-M., Khanbari, K., Lepvrier, C., Leroy, S., Maillot, B., Merkouriev, S. (2010). Arabia-Somalia plate kinematics, evolution of the Aden-Owen-Carlsberg triple junction, and opening of the Gulf of Aden. *Journal of Geophysical Research*, 115, B04102–24.

Jagoutz, O., Müntener, O., Manatschal, G., Rubatto, D., Peron-Pinvidic, G., Turrin, B.D., Villa, I.M. (2007). The rift-to-drift transition in the North Atlantic: A stuttering start of the MORB machine? *Geology*, 35, 1087.

Koptev, A., Gerya, T., Calais, E., Leroy, S., Burov, E. (2018). Afar triple junction triggered by plume-assisted bi-directional continental break-up. *Scientific Reports*, 8, 1–7.

Leroy, S., Lucazeau, F., d'Acremont, E., Watremez, L., Autin, J., Rouzo, S., Bellahsen, N., Tiberi, C., Ebinger, C., Beslier, M.-O., Perrot, J., Razin, P., Rolandone, F., Sloan, H., Stuart, G., Al-Lazki, A., Al-Toubi, K., Bache, F., Bonneville, A., Goutorbe, B., Huchon, P., Unternehr, P., Khanbari, K. (2010). Contrasted styles of rifting in the eastern Gulf of Aden: A combined wide-angle, multichannel seismic, and heat flow survey. *Geochemistry, Geophysics, Geosystems*, 11.

Leroy, S., Razin, P., Autin, J., Bache, F., d'Acremont, E., Watremez, L., Robinet, J., Baurion, C., Denèle, Y., Bellahsen, N., Lucazeau, F., Rolandone, F., Rouzo, S., Kiel, J.S., Robin, C., Guillocheau, F., Tiberi, C., Basuyau, C., Beslier, M.-O., Ebinger, C., Stuart, G., Ahmed, A., Khanbari, K., Ganad, I., Clarens, P., Unternehr, P., Toubi, K., Lazki, A. (2012). From rifting to oceanic spreading in the Gulf of Aden: A synthesis. *Arabian Journal of Geosciences*, 5, 859–901.

Lister, G.S., Etheridge, M.A., Symonds, P.A. (1986). Detachment faulting and the evolution of passive continental margins. *Geology*, 14, 246.

Nonn, C., Leroy, S., Khanbari, K., Ahmed, A. (2017). Tectono-sedimentary evolution of the eastern Gulf of Aden conjugate passive margins: Narrowness and asymmetry in oblique rifting context. *Tectonophysics*, 721, 322–348.

Nonn, C., Leroy, S., Lescanne, M., Castilla, R. (2019). Central Gulf of Aden conjugate margins (Yemen-Somalia): Tectono-sedimentary and magmatism evolution in hybrid-type margins. *Marine and Petroleum Geology*, 105, 100–123.

Osmundsen, P.T. and Redfield, T.F. (2011). Crustal taper and topography at passive continental margins. *Terra Nova*, 23, 349–361.

Reston, T.J., Krawczyk, C.M., Klaeschen, D. (1996). The S reflector west of Galicia (Spain): Evidence from prestack depth migration for detachment faulting during continental breakup. *Journal of Geophysical Research: Solid Earth*, 101, 8075–8091.

Reston, T.J., Pennell, J., Stubenrauch, A., Walker, I., Perez-Gussinye, M. (2001). Detachment faulting, mantle serpentinization, and serpentinite-mud volcanism beneath the Porcupine Basin, southwest of Ireland. *Geology*, 29, 587.

Robinet, J., Razin, P., Serra-Kiel, J., Gallardo-Garcia, A., Leroy, S., Roger, J., Grelaud, C. (2013). Tectonophysics. *Tectonophysics*, 607, 1–16.

Sauter, D., Cannat, M., Rouméjon, S., Andreani, M., Birot, D., Bronner, A., Brunelli, D., Carlut, J., Delacour, A., Guyader, V., MacLeod, C.J., Manatschal, G., Mendel, V., Ménez, B., Pasini, V., Ruellan, E., Searle, R. (2013). Continuous exhumation of mantle-derived rocks at the southwest Indian Ridge for 11 million years. *Nature Geoscience*, 6, 314–320.

Sibuet, J.-C., Srivastava, S., Manatschal, G. (2007). Exhumed mantle-forming transitional crust in the Newfoundland-Iberia rift and associated magnetic anomalies. *Journal of Geophysical Research*, 112, B06105.

Tard, F., Masse, P., Walgenwitz, F. (1991). The volcanic passive margin in the vicinity of Aden, Yemen. *Bulletin – Centres des recherches exploration-production*, 15, 1–9.

Watremez, L., Leroy, S., Rouzo, S., d'Acremont, E., Unternehr, P., Ebinger, C., Lucazeau, F., Al-Lazki, A. (2011). The crustal structure of the north-eastern Gulf of Aden continental margin: Insights from wide-angle seismic data. *Geophysical Journal International*, 184, 575–594.

5

The South China Sea – Rifting and the Evolution of a Marginal Basin

Manuel PUBELLIER, Sung-Ping CHANG and Matthias DELESCLUSE

Laboratoire de Géologie de l'ENS, CNRS UMR 8538, PSL Research University, Paris, France

Marginal basins are basins opened at the border of continental margins. Due to their moderate size, they allow more accurate observations of rifting and continental breakup processes than large oceans. Because their conjugate margins are close enough, precise correlations can be done, and boundary forces (e.g. free edge) are easier to identify. Among the most studied marginal basin, the South China Sea (SCS), which is not a back-arc basin, is a good example to illustrate how the initial conditions that shaped the geometry of the margins are important. The evolution of the rifting, spreading and cessation of rifting of this basin seem to be controlled by the effect of several structural parameters. Among these are the intrinsic structure of the continental crust inherited from the earlier structures, the thermal state and crustal thickness, and above all the V-shaped geometry of the basins, which reflects along-strike variation of the shear strain with time, which in turn has severe implications on the subsidence history.

For a color version of all the figures in this chapter, see www.iste.co.uk/peronpinvidic/margins2.zip.

Continental Rifted Margins 2,
coordinated by Gwenn PERON-PINVIDIC. © ISTE Ltd 2022.

The V-shape of the basin floor, which is the result of a propagating opening towards the SW, is also more pronounced than that of large oceans and allows us to pinpoint the diachronism of the breakup unconformity. The resulting southward migration of the zipper delayed the subsidence, which commonly follows the beginning of the spreading. This allowed the crustal extension to persist in areas where the spreading was already active, and also sustained the continental margin in a shallow marine or continental environment for long time. Older granites, which pierced the continental crust, seem to condition the wavelength of the crustal boudinage, allowing low-angle normal faults to reactivate pre-existing features, such as the granite-sediment boundaries, limbs of broad folds or former crustal thrusts. The thermal state of the crust, which was previously in a supra-subduction position, may be the cause of extreme continental stretching during long periods of time, and requires a very weak lower or intermediate crust, allowing mantle–crust decoupling and a preferred mantle breakup at depth.

In the case of the SCS, the crust rheology and the proximity of the nearby subduction of the Proto South China Sea (PSCS) allow us to catch the sensitivity of the rifting to the boundary conditions. The spreading changed direction at 23 Ma when the subduction of the neighboring PSCS oceanic crust was almost complete, and stopped when the shortening of the PSCS finished with the Palawan–Borneo collision by 16 Ma. The last stage of evolution is principally marked by vertical motion and gravity tectonics in Borneo from the Late Miocene (circa 12 Ma), concomitant with a drop of the SCS basement resulting in an unusual rapid subsidence, marked by a general draping of pelagic sediments. The synthesis of observations on the SCS shows how a small rifted basin may be sensitive to both the structural heritage and modifications of the boundary conditions.

5.1. Introduction

Rifting and continental breakup processes have been extensively studied in the recent years. So far, the modeling of the rifting has come mainly from knowledge on the margins of large oceans (Heezen 1960; Hess 1962; McKenzie 1978), and therefore, the margins are wide apart and the rifting features are often old and obliterated by later erosion, or covered by thick sedimentary shelf sediments if not passive margin volcanic series. In addition, the geodynamic context at the time of the margin rifting is only indirectly known from geological studies, even though a large amount of information comes with the exploration of the rifted basins (Ziegler 1988; Peron-Pindivic *et al.* 2007). The stretched continental crust tends to open marginal seas (Zhou 2014) by the separation of crustal slivers (Taylor *et al.* 1995; van der Broek and Gaina 2010).

On the other hand, small marginal basins display close conjugate margins and allow accurate reconstructions based on good datings. Many of these basins are back-arc basins (type 1 in Figure 5.1) and result from the arc splitting above a subduction zone. Some, however, have opened inside the downgoing plate (type 2 in Figure 5.1). The South China Sea is one of them and the only visible one at present. It is a well-studied basin floored with an oceanic crust and a magma-poor passive continental margin (Taylor and Hayes 1983; Ding *et al.* 2012). It also has limited ocean floor spreading, which allows relatively easy margin correlations over a short (400 km in the North and 100 km in the South) distance (Holloway 1982; Cullen *et al.* 2010; Morley 2016). The extension principally occurred during the Cenozoic and is thus relatively recent in geological time, allowing us to observe the tectonic context in terms of slab configuration at depth (Rangin *et al.* 1999; Wu and Suppe 2017), the former margin morpho-structures (Pubellier *et al.* 2016) and accurate dating of the surrounding rocks (Lunt 2019).

The South China Sea is also part of a series of basins which opened at the rim of a composite micro-continent surrounded by a Mesozoic volcanic arc, and its structural evolution is linked with that of the neighboring basins (Karig 1975). The aspect of opening of a basin to the detriment of another has been suggested for the reconstructions of older ocean basins (Stampfli and Borel 2002; Figure 5.1). The South China Sea offers the opportunity to observe the extension of a continental margin coeval with the shortening of an older one, the Proto South China Sea, which has now been entirely consumed (Hall 2002). This system is now fossilized and was followed by the opening of the Sulu Sea situated next to it (Figure 5.2; Rangin *et al.* 1990). Therefore, the SCS basin provides a unique setting to study the geological context of the extension of the continental margin with respect to the boundary conditions. It may also be compared with basins from similar contexts (van den Broek and Gaina 2020).

5.2. Geodynamic setting and particularities

Marginal basins are developed at the boundaries of continents and their extension is generally related to an ongoing subduction further away oceanward. Most of the documented examples are indeed back-arc basins (marginal basin type 1; Figure 5.1), which developed as a result of extension in the upper plate due to the slab roll-back of the downgoing plate. The South China Sea, being far from the

Sunda subduction zone, is not directly one of these cases. Instead, it directly opened in the midst of the China continental margin situated to the north (marginal basin type 2; Figure 5.1), and therefore, the southern margin is devoid of any leading volcanic arc.

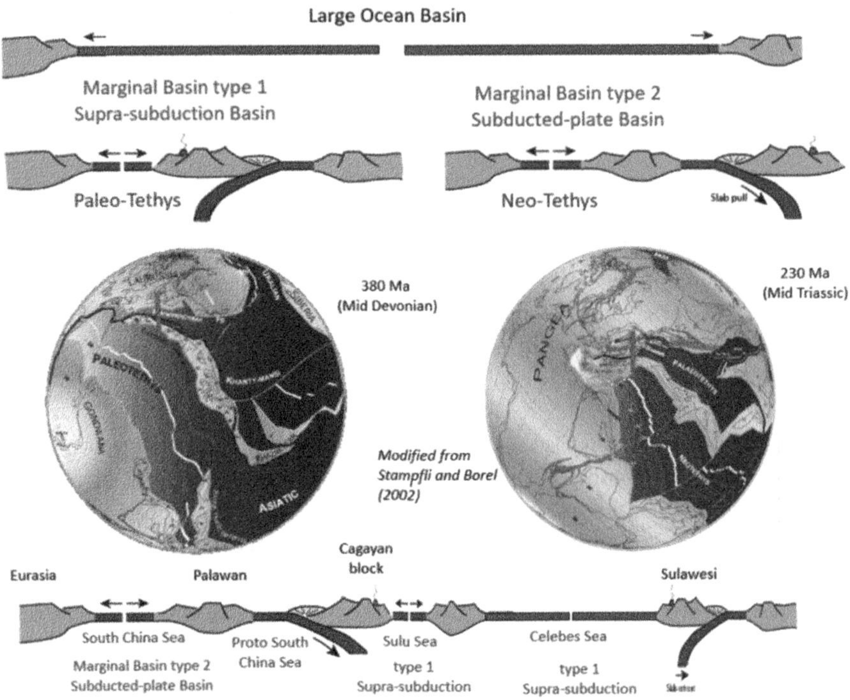

Figure 5.1. *Top: two distant margins separated by a large ocean will have lost the inherited structures that persisted until the late stages of the rifting, because of the erosion of the landmasses. There are two opposite settings resulting from the subduction-related extension in marginal basins. The left side is the classical back-arc situation and the right side is the rarely found case where extension develops within the downgoing plate. Middle: examples taken from reconstructions (Stampfli and Borel 2002) for the Paleotethys and the Neotethys. Bottom: a section of the SCS as a type 2 and the neighboring basin (PSCS) of type 1 before its shortening*

The South China Sea, surrounded by the China margin to the NW, the Indochina margin to the W, the NW Borneo wedge to the S and the SE, the Philippine islands to the E and Taiwan Island to the Northeast, is a rhomb-shaped basin which is diagonally transected by an NE–SW ocean floor (Taylor and Hayes 1983; Ru and Pigott 1986; Figure 5.2). Its evolution has undergone the far-field effect of the Indian-Asia collision zone and the disputed effect of the motion of Indochina with respect to the South China Block (Tapponnier *et al.* 1986; Peltzer and Tapponnier 1988; Cullen *et al.* 2010; Morley 2016), but was also under the influence of the subductions of the Proto Pacific, the Philippine Sea Plate and the Indian Ocean (Hall 2002; Pubellier and Morley 2014; Wu *et al.* 2016). The 1,000 km-large, stretched margin (Figures 5.2 and 5.3) is spectacular, with a morphology that resembles a "Basin and Range" of the Western United States in the offshore (Franke *et al.* 2011, 2014). The crust is thin from the shelf break of the South China and Vietnam margins (Nissen *et al.* 1995 and Hayes 1980) towards the E with a relatively flat Moho and a sharp transition to the oceanic crust (Figures 5.2 and 5.3).

The rifting started slowly after the mid-Cretaceous Yanshanian orogeny (Nanni *et al.* 2017) by Late Cretaceous and Paleocene. It was severe during the Eocene and Oligocene (Savva *et al.* 2014; Larsen *et al*, 2018) showing tilted blocks and often several decollement levels, and was globally followed by Late Oligocene to Middle Miocene spreading (Briais *et al.* 1983; Barckhausen *et al.* 2014). However, because of the propagation of opening towards the SW, continental extension continued during the spreading.

The South China Sea is also known for its southern margin having been partially underthrusted beneath the NW Borneo and Palawan wedge (Hamilton 1979; Cullen *et al.* 2010) creating a depression – the Sabah–Palawan Trough – and a forebulge along which reefal constructions still exist now (Steuer *et al.* 2014). The resulting characteristics create an example which offers a variety of morpho-structures created or reactivated gradually younger towards the SW (Franke *et al.* 2014), allowing cross sections which represent the margin frozen at different stages of evolution from the NE to the SW. The oceanic floor is well known and developed after a proven breakup which started at 32 Ma, as known from magnetic anomalies (Briais *et al.* 1983; Barkhausen *et al.* 2014), and IODP drilling (Li *et al.* 2015; Larsen *et al.* 2018). Older oceanic crust may exist South of Taiwan (Hsu and Sibuet 2011). However, volcanism was active with the oceanic crust even after the cessation of spreading.

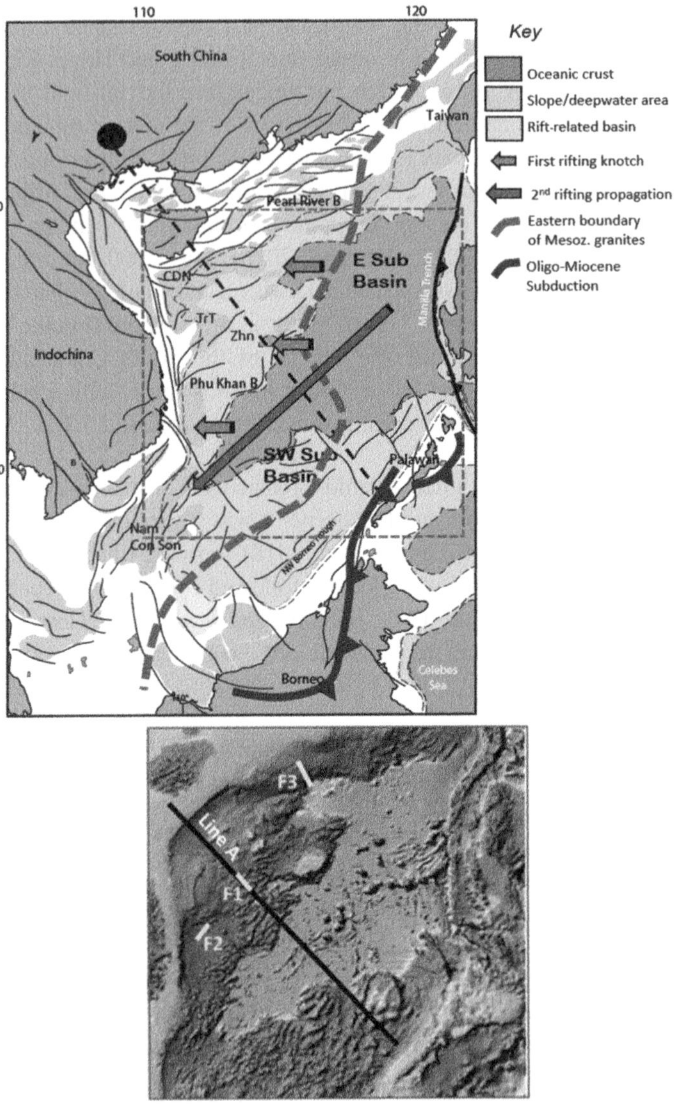

Figure 5.2. *Top: the intensely rifted continental crust of the South China Sea (light green), oceanic crust (light blue) and thickly sedimented crust (yellow) (source: adapted from Pubellier and Morley 2014). Note the very wide area of rifted crust. Small arrows represent the second-order "V"-shaped indentations on the margins. The large arrow indicates the main propagator of the SW sub-basin. The black circle indicates the section shown in Figure 5.4. Bottom: bathymetric morphology of the basin and location of profiles. B; Basin, CDN; Qiondongnan, TrT; Tri Ton Ridge, Zhn; Zhongshanan Basin*

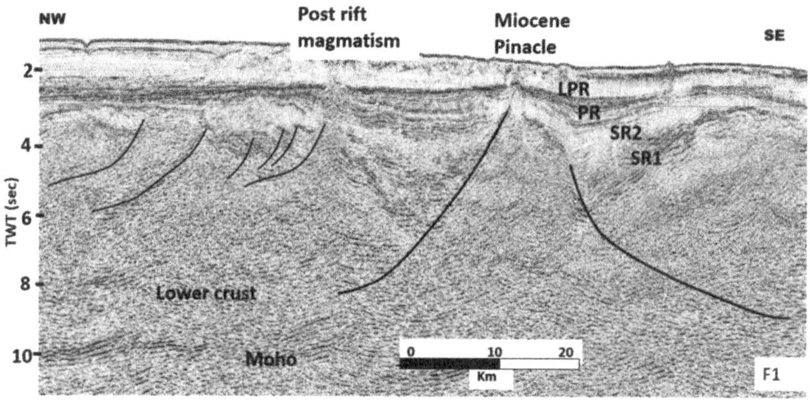

Figure 5.3. *Typical features of the South China Sea on the stretched part of the Northern margin between Tri Ton High and the Zhongshanan Basin (see the location of TrT and Zhn, respectively, in Figure 5.2). SR1: Syn-rift 1 (50-32 Ma), SR2: Syn-rift 2 (32-23Ma), PR: post-rift (23-16), LPR: late post-rift (15-0Ma). Note the existence of two levels of rooting of the low-angle normal faults (LANF), the Late Miocene and Pliocene magmatism, the presence of carbonate pinnacles (more prominent on the southern margin), erosional surfaces at the top of basement of the large tilted blocs, and prominent and rather flat Moho around 10 sec TWT*

5.3. Propagators and implications on the rifting and the timing of breakup

Most of the small-sized marginal basins have a clear V-shape, due to either the diachronism of the breakoff or an along-strike variation of the spreading velocity. If this does exist in large oceans like the North Atlantic or former oceanic domains (Figure 5.1), it is spectacular in small marginal basins and implies along-strike variations of the continental rifting rate and a migration of the breakup point through time (Courtillot 1982; Ranero and Peréz-Gussinyé 2006; Le Pourhier *et al.* 2018). Classical and striking examples include the Gulf of Aden and the Woodlark basin offshore eastern New Guinea Island (Taylor *et al.* 1995), but other less obvious cases are the Celebes/Makassar Basin in Indonesia, the Pyrenees (Gillard *et al.* 2017) or the Porcupine basin in offshore Ireland (Bulois *et al.* 2018). The South China Sea is a text-book example of such geometry, which opens from the NE to the SW and terminates by a remarkable propagator tip (Figures 5.2 and 5.4) that penetrated into the Sundaland crust (Taylor and Hayes 1983; Huchon *et al.* 2001).

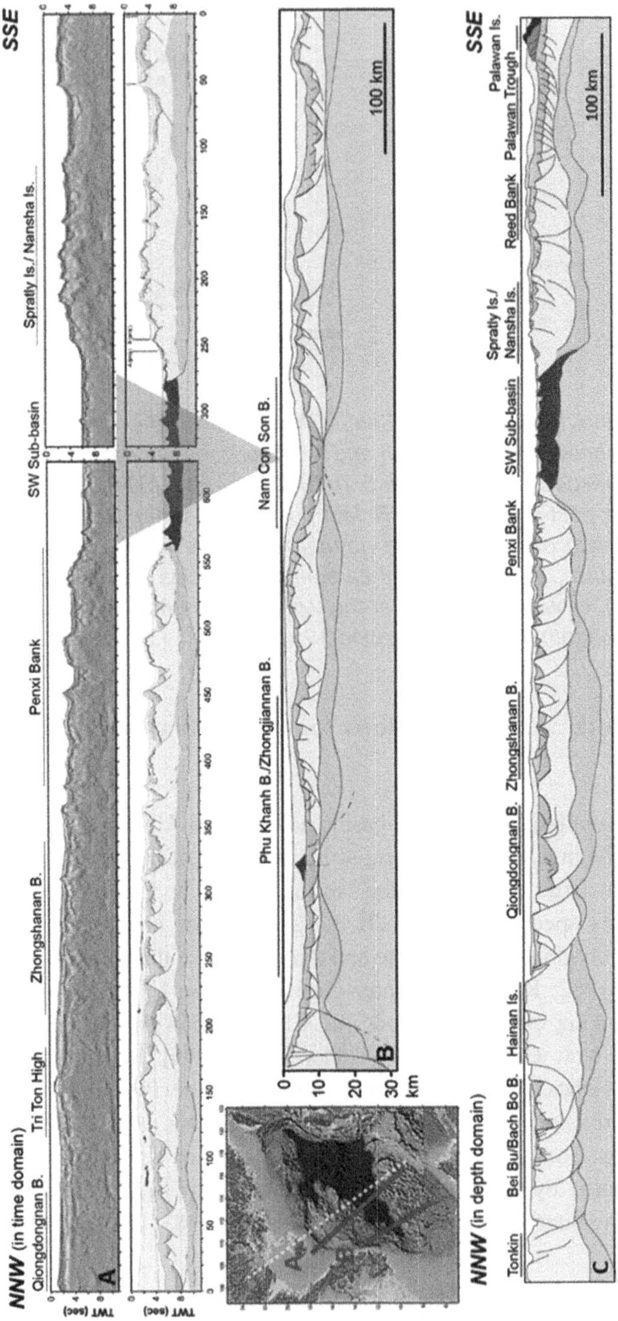

Figure 5.4. *Sketches of two seismic sections in the SW sub-basin of the South China Sea*

CONTINUATION OF CAPTION FOR FIGURE 5.4.– *Showing the considerable extension of the crust. Line A intersects the proximal regions affected by the first rifting as well as the second (post-23 Ma), and B is located in the southern part of the W Sub-basin where mostly the second rifting is preserved (source: top section adapted from Chang et al. 2022). Bottom, section C: schematic section of the whole SCS margin from Vietnam to Plawan (source: adapted from Pubellier et al. 2016).*

However, a first fast propagator initiated with the first rifting, and stalled in the Eastern Sub basin when the breakup took place (Le Pourhier *et al.* 2018), with the oceanic crust present in the eastern Xisha Trough (Briais *et al.* 1993). It is illustrated by an E–W V-shaped indentation of the continental crust. The bathymetry also shows second-order cracks in the margin (Figure 5.2). A southward remarkable ridge jump occurred at 23 Ma, followed by a second propagation towards the SW from 23 to 16 Ma. It terminates at the junction of the Nam Con Son Basin (Huchon *et al.* 2001; Sibuet *et al.* 2016; Chang *et al.* 2022; Figure 5.4). The phase of stalled propagation has been simulated by numerical modeling by introducing a slight component of compression in the direction of the propagation, possibly the load of the Indochina Peninsula (Le Pourhier *et al.* 2018). The time discrepancies of the opening have implications on the breakup configuration, the orientation of the normal faults and the fact that the margins remained sustained at shallow depth for a long period of time, even after breakup.

5.3.1. *Extreme thinning and detachment faulting away from narrow ocean–continent transition zones: far field and deep processes versus structural inheritance*

Some major characteristics of the South China Sea can be observed on every profile (Figure 5.3). The crust is thin from the shelf break to the OCT, there are two major levels of décollement with low-angle normal faults rooting in the intermediate crust – possibly the root of the flat Mesozoic granites – and the Lower crust, two main rift sequences, and a sharp post-rift unconformity, which marks the base of a homogeneous blanket of hemipelagic sediments. Both the continental margins and the oceanic crust are pierced by young Late Miocene and Pliocene magmatic bodies (Figure 5.3). All of these aspects are related to the rapid evolution of the crust, the inheritance of the early configuration of the margin and its thermal conditions.

The simple idea that the SW Sub-Basin experienced distributed extension of the continental lithosphere while the Eastern Sub-Basin oceanic accretion propagation stalled (Le Pourhier *et al.* 2018), is in agreement with an observed crustal thickness in the SW sub-basin that is rather constant (15 km, Pichot *et al.* 2014) (Figure 5.4). The crustal breakup is also known to be controlled by oceanic propagation dynamics rather than by a progressive cylindrical extensional process (Li *et al.* 2014b; Le Pourhier *et al.* 2018), resulting in a narrow OCT and a rapid onset of oceanic accretion without the emplacement of large exhumed mantle areas (Larsen *et al.* 2018; Nirrengarten *et al.* 2020). However, many basins in the South China Sea underwent extreme thinning (Qiongdongnan Basin, Xisha Basin), faulting along large detachment (e.g. basin C in Liang *et al.* (2019) and the ZS Trough in Huang *et al.* (2021)) and even exhumed serpentinized mantle (Savva *et al.* 2013; McIntosh *et al.* 2014) in the Phu Khanh Basin away from the OCTs. Figure 5.2 shows that most of these basins are located at the transition between the ESB and SWSB, in the area where the propagator remained stalled for 10 Myrs (32–23 Ma). Some are oriented E–W (Xisha, ZNSB), and can be easily interpreted as propagation attempts through the Mesozoic granitic plutons, while some others are oriented SW–NE (ZST and basin C). In the case of basin C (Liang *et al.* 2019), detachment faulting localized on a reactivated Yanshanian structure when the direction of opening in the SCS changed from N–S to NW–SE. If far-field forces and deep processes (oceanic propagation) explain the general properties of the conjugate margins (narrow OCT, wide rift), the reactivation of pre-existing structures and heterogeneous rheologies explain sparse failed necking zones, and were active in the vicinity of the propagator head.

5.4. A wide rift with boudinage crust resulting from contrasted crustal rheologies

5.4.1. *Hyperextended continental crust*

The continental crust of the SCS has been stretched and thinned considerably (Figures 5.2, 5.3 and 5.5) (Taylor and Hayes 1983; Clift and Lin 2001; Franke *et al.* 2014; Huismans and Beaumont 2011; Chang *et al.* 2022), with a propagation towards the OCT (Ranero and Peréz-Gussinyé 2006; Ding and Li 2016; Pérez-Gussinyé *et al.* 2020). A large number of industrial seismic reflection lines and academic refraction lines (Pichot *et al.* 2014; Wei *et al.* 2015) allow us to pinpoint the Moho in many places of the SCS (Franke *et al.* 2011, 2014; Gao *et al.* 2015). One line crosses the entire basin and shows a flat Moho underlying a 12–15 km thick crust (Pichot *et al.* 2014; dashed line in Figure 5.4). However, intense stretching has led the upper crust to disappear at several localities, such as near the OCT offshore Palawan (Franke *et al.* 2011; Savva *et al.* 2014; Clerc *et al.* 2018), in the Phu Khan basin, where the mantle reaches the base of the sediments (Figure 5.5) causing its partial melting and generating large volcanic edifices to construct (Cullen *et al.* 2010; Savva *et al.* 2013), or in the Tainan (Taxinan) Basin, where the lower crust may be exhumed (Gao *et al.* 2015; Yang *et al.* 2018).

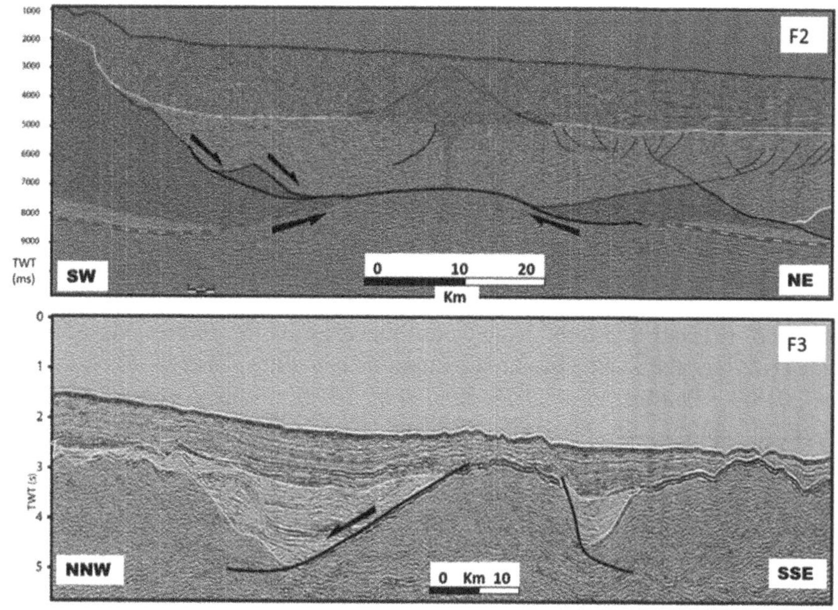

Figure 5.5. *Top: schematic section of the central part of the Phu Khanh basin illustrating hyper-extended crust where most of the lower crust has disappeared and the upper crust may be split by a mantle dome located beneath the sediments. The buried volcano sits on top of the syn-rift and marks the peak of the stretching late Early Miocene, suggesting adiabatic melting of the upper mantle. Other volcanoes (right) are located near the base of the syn-rift. Bottom: low-angle normal faults (dashed lines) activated along basement morpho-structures (source: adapted from Savva* et al. *2013)*

Two episodes of continental rifting have been described: an early one during the Paleocene–Eocene times and a second one located close to the OCT and more developed in the SW sub-basin (Taylor and Hayes 1983; Schlüter *et al.* 1996; Cullen *et al.* 2010; Franke *et al.* 2014). Numerical modeling has proposed a yet unchallenged model according to which a very weak (poorly viscous) lower crust and high thermal conditions at the Moho (Pérez-Gussinyé *et al.* 2020) may have caused a decoupling of the upper mantle during rifting, allowing the upper crust to be intensely stretched (Clift and Lin 2001; Huismans and Beaumont 2011; Shi *et al.* 2017; Brune *et al.* 2014), suggesting a very weak lower or intermediate crust (Li *et al.* 2014b), resulting in a preferred lithospheric mantle extension at depth (Clift and Lin 2001). It has been proposed that the crustal weakness may result from the supra-subduction origin of the margin during the Early Cretaceous (Xiao *et al.* 2019). The hyper-stretched margins were sustained at a shallow depth, allowing the continuous shallow marine or sub-continental lacustrine or fluvial environment to be preserved over the entire time of the rifting, from early Paleogene to the late Early Oligocene in the East Sub-Basin (Larsen *et al.* 2018), and to the early Late Miocene in the Southwest Sub-Basin (Franke *et al.* 2014).

5.4.2. *Reactivation of the Mesozoic Granite Belt*

It is frequent to observe that early continental fabrics are used during extension (Morley 2007; Festa *et al.* 2020). In the case of the SCS, the propagation of the early E–W rift and sea floor halted near the edge of the former Mesozoic arc, characterized by gneissic bodies or large granitic batholiths (Li *et al.* 2014; Pubellier *et al.* 2016; Le Pourhier *et al.* 2018; Xiao *et al.* 2019), resulting from a long-lasting subduction during the Mesozoic (Li and Li 2007). There, several grabens with extremely thinned crust are observed adjacent to the OCT or close to it (see Figure 5.5, bottom). Along the OCT of the northern margin, secondary small-sized grabens, which also have a triangular "V" shape, suggest a tear shear of the margin near the OCT. They are prominent on the distal part of the Pearl River Mouth basin on the NW margin, like the Liwan Trough (Ye *et al.* 2018; Larsen *et al.* 2018; Wang *et al*, 2018), the Enping Sag (Ye *et al.* 2018) or the flanks of the Parecel granites (Chang *et al.* 2022), as well as near the OCT of the Spratlys granite (Liang *et al.* 2019). Inside the thinned margin, the extension has been concentrated on the edge of the contact between the granites and the host rocks (Savva *et al.* 2014), with large low-angle normal faults (LANF). The batholiths have been reactivated and the short limbs of the broad folds of the Early Cretaceous Yanshanian orogeny have been re-used by LANF. As a result, the morphology of some parts of the sea bottom marks rounded shapes which correspond to the granites, and strips which are correlated with the folds. The folds are visible and dated offshore (Yan *et al.* 2014) and onshore in the Pearl River delta (Chan *et al.* 2010; Nanni *et al.* 2017). The abundance of granite in the composition of the margin of the South China Sea may also account for its thermal state. The SCS is well known for its intense magmatism, which followed the rifting (Chang *et al.* 2022; Figure 5.3).

5.5. Impact of the subduction and collision of the neighboring basin

The SCS, contrarily to most marginal basins, is not a back-arc basin. Instead, it is attached to the continental margin of the South China block. The subduction of the Indian Ocean is far to the south, and several basins have accommodated the upper plate extension, namely the Celebes Sea, the Banda Sea and the Damar basin. The nearest subduction is therefore that of the PSCS, which has disappeared since the Middle Miocene. Yet, some authors argue that there may be a subduction beneath the SCS on the grounds of the existence of a cold lithosphere underneath (Wu and Suppe

2017), but there is so far no evidence of any associated arc or subduction wedge. It is therefore commonly accepted that at least one part of the PSCS subducted to the south, where a volcanic arc and a large accretionary wedge exist in NW Borneo; the other part being an older slab of the PSCS or a relic of the Tethyan slab.

Therefore, it is likely that there is a relation between the subduction of the PSCS from the Eocene to the Mid Miocene and the rifting of the SCS from the Eocene to the Mid Miocene (Rangin *et al.* 1990; Sapin *et al.* 2011; Hall 2013; Chang *et al.* 2022). Sea floor spreading started about 32 Ma ago (Briais *et al.* 1993), and noticeably suffered a break and a change of spreading direction at 23 Ma (Briais *et al.* 1993; Sibuet *et al.* 2016), before it ceased at 16 Ma (Xie *et al.* 2014; Franke *et al.* 2014). Reconstructions of the South China Sea by the time of breakup around 23 Ma (Figure 5.6) suggest en echelon system of narrow grabens compatible with N–S opening. These basins would have become coalescent from then on, thus allowing the propagation of the second sub-basin oriented NE–SW (Chang *et al.* 2022). This scenario of evolution is to be paralleled with that of the NW Borneo wedge.

The wedge involves the sedimentary series of the PSCS from Eocene to Mid Miocene (Tongkul 2003; Sapin *et al.* 2011), which had been stacked under a sliver of Early Cretaceous oceanic crust, now present as ophiolite in Sabah and Palawan. The over-thrusting of the ophiolite and its dismembering occurred in Oligocene in terms of increasing ophiolitic sediments (van Hattum *et al.* 2013), with blocks of the ophiolite being engulfed in an Oligocene mélange. The post-tectonic series, often referred to as post-MMU (Figure 5.7), began at 16 Ma and stable sedimentation with subsidence resumed during the Late Miocene, with mini-basins limited by shale diapirs or "circular basins" (Clennell 1996; Morley 2009; Clark 2017). It is possible that the concomitant closure of a large basin to the NW of the SCS and a narrow continental crust in the south was responsible for the late anticlockwise rotation of Borneo (Fuller *et al.* 1999; McCabe 1985; Advokaat *et al.* 2018).

The striking correlation between these events (Figure 5.7) suggests that the rifting and spreading of the SCS with their internal events are therein triggered by the trench pull of the PSCS slab. The beginning of the subduction of the margin toe would have perturbed the subduction starting from the early Oligocene, and obducted the oceanic crust onto the wedge by Late Oligocene, often considered as the age of early deformation in the NW Borneo wedge. The peak of the collision in Borneo would correspond to the rearrangement of the spreading indicated by the rotation of the spreading directions (Sibuet *et al.* 2016) at 23 Ma, and the final over-thrusting marked the end of the collision and took place in the early Late Miocene at 16 Ma. Not only did the South margin of the SCS (Palawan block) stop entering the wedge, but the continental Cagayan ridge was also thrust on the margin by the Mid Miocene (Rangin *et al.* 1989). The exceptional setting of the SCS with the opening of V-shaped oceanic basin to the detriment of a long-lasting rifting within a thinned crust allows us to pinpoint the timing of the extension and its decay. Meanwhile, the subduction and complete closure of the neighboring PSCS are finely controlled by the stratigraphy. This paired succession of events permits us to apprehend the boundary forces and their modifications.

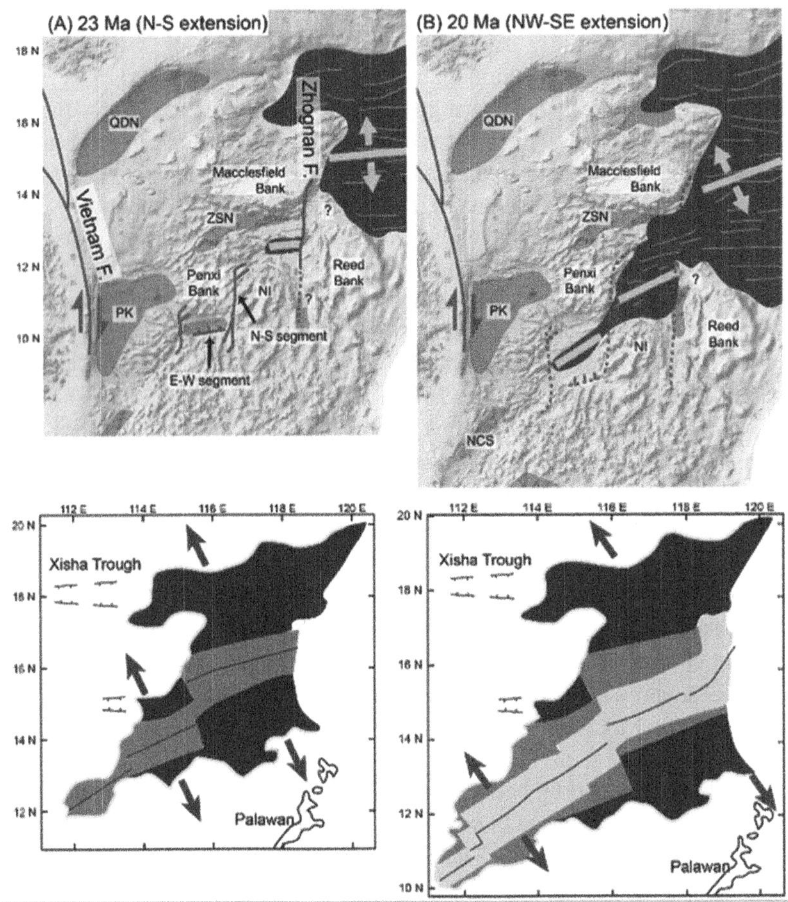

Figure 5.6. *Top: correlation of the ridge jump accompanied by a reorganization of the spreading directions (source: adapted from Chang* et al. *2022). The propagation of the first stage (Eocene–Oligocene up to 32 Ma) evolves to a series of en echelon pull apart basins, which develop during the intermediate period before the new spreading center connects them, allowing a rapid propagation of the second (post 23 Ma) system. Bottom: early interpretation which documented the two stages evolving with an intermediate short-lived stage around 24–23 Ma (source: Sibuet* et al. *2016)*

5.6. Post rifting evolution; uplift of margins and delayed subsidence of the basin

From then on, there was no significant shortening except by the Latest Miocene, when the Philippine arc docked against the Eurasian margin (Aurelio *et al.* 2012; Pubellier *et al.* 1996), and the movements became dominantly vertical. The relaxation in the system is synchronous with the opening of the Sulu Sea (Silver *et al.* 1989), on the back side of the Borneo–Palawan mountain range, as well as to the detachment of the Proto South China Sea (Sapin *et al.* 2011; Hall 2013). It is interesting to note that the subsidence of the basin was significantly delayed, especially in the North and NW of the South China Sea (Xie *et al.* 2006; Shi *et al.* 2017), possibly due to a thermal cooling episode with local magmatic event (Figure 5.8). The breakup took place at 32 Ma in the E Sub-basin (Briais *et al.* 1993; Li *et al.* 2014), and the subsidence was observed after the time of the MMU (Mid Miocene Unconformity) around 16 Ma. During this time, extension continued in the North of the South China Sea to a lesser extent, but was very active in the SW sub-basin. The Mid Miocene corresponds to the end of the rifting south of the propagator in the south.

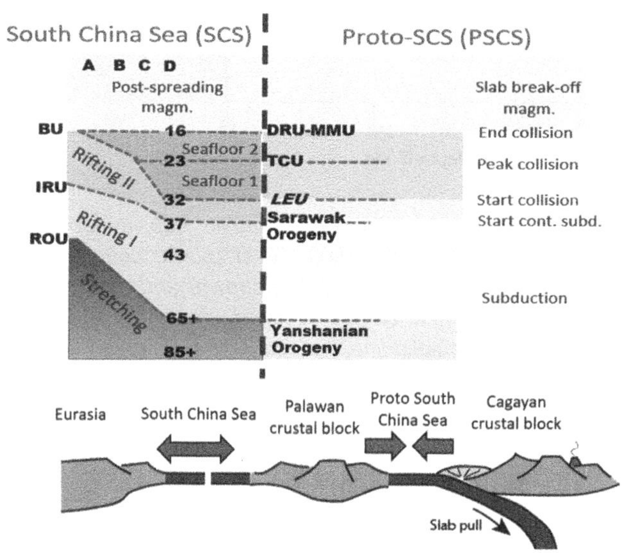

Figure 5.7. *Comparison of the tectonic evolution in the passive SCS margin and the active Borneo–Palawan Margin. ROU: rift-onset unconformity, IRU: intra-rift unconformity, BU: breakup unconformity, DRU: deep regional unconformity, MMU: Middle Miocene unconformity, and TCU: top Croker unconformity, LEU: Late Eocene Unconformity. A = Nsc/L; NamCon Son basin to Luconia Basin, B = PK/Si; Phu Khanh, C = CDN/RB, D = PRM/RB (source: schematic cross section adapted from Pubellier and Meresse 2013)*

Therefore, extension persisted in a smaller amount long after the breakup in the north of the South China Sea. This may provide an explanation for the intense post-rifting magmatism observed on the sea floor during the Late Neogene. The Late Neogene, mostly starting from the Late Miocene, is also marked by important vertical movements in Northern Borneo, which are often attributed to slab breakoff.

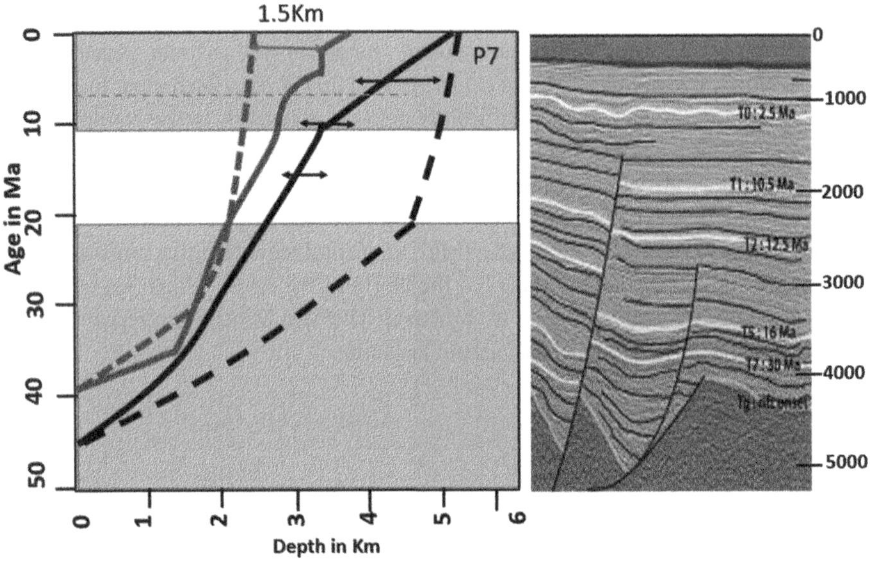

Figure 5.8. *Observed and predicted tectonic subsidence histories of 10 representative synthetic drills in the Pearl River (black; source: adapted from the compilation in Xie* et al. *2006, but only P7 is shown) and the Qiongdongnan Basin (red; source: adapted from Shi et al. 2017). The dashed lines represent the predicted tectonic subsidence, and the solid line represents the observed tectonic subsidence. An example of the seismic facies of the Pearl River Mouth basin (PRMB) is given for reference (on the right), with depth in meters*

5.7. Conclusion

The South China Sea is considered as an example of a medium-sized marginal basin, which may be considered as a model for identifying specific characteristics. The evolution of the rifting, spreading and cessation of rifting in the South China Sea seems to be controlled by the concomitant effect of several parameters. Among these, the location of the rifting appears to be along-strike the trend of the Yanshanian orogeny which existed previously. Then, the granites, which pierce the continental crust, seem to condition the wavelength of the crustal boudinage. The low-angle normal faults are typically rooted at both a shallow crustal level and a deep crustal level rooting into the lower crust. They reactivate pre-existing features, such as the crystalline basement-sediment boundaries or the broad folds of the Mesozoic orogens. The margins have been extremely stretched during a long period of time, suggesting a very weak lower or intermediate crust, allowing mantle–crust decoupling and a preferred mantle breakup at depth. The rifting geometry by the time of the breakup may be reconstructed and shows that the geometry of the continent–ocean boundary is in places marked by en echelon pull-apart basins, indicating N–S opening with a wrench component along N–S strike-slip segments of faults. The early stages of the spreading are sensitive to the changes of boundary forces, and a reorientation of the spreading is observed at 23 Ma. The time of the spreading shift and the end of the second rifting are concomitant with the age of the peak of the collision in Borneo and Palawan, when the neighboring Proto South China Sea was completely consumed by subduction. It is noticeable that contrarily to large oceans, the subsidence of the basin was delayed significantly, apparently awaiting the complete halt of the propagator in the south. Therefore, extension persisted in a smaller amount long after the breakup in the north of the South China Sea. This may provide an explanation for the intense post-rifting magmatism observed on the sea floor during the Late Neogene. Horizontal movements ceased by the Late Miocene after completion of the rifting at the tip of the SW sub-basin propagator, and cessation of the collision in Borneo and Palawan. From then on, vertical movements prevailed, with rapid subsidence in the South China Sea and strong uplift in Borneo, the latter often being attributed to slab breakoff.

It is therefore because of its recent age, the sharp angle of the propagator and the velocities of the plate involved that the South China allows us to highlight parameters which are difficult to see in large oceans. Studies on visible non-cylindrical opening of large oceans, such as the Atlantic with a slow propagator from the Early Cretaceous to the Eocene, or the early stages of the opening of the Indian Ocean along the NW shelf of Australia from Triassic to Early Cretaceous, could benefit from the knowledge acquired in smaller more recent marginal basins, such as the South China Sea.

5.8. Acknowledgements

This review paper benefitted from many years of studies in the South China Sea and adjacent basins in the SE Asian region. There was tight cooperation between the CNRS, Ecole Normale Supérieure/PSL University, Total, BGR in Hannover, Guangzhou Marine Geological Survey (GMGS) and geological surveys around the South China Sea, coordinated by the Commission for the Geological Map of the

World (CGMW). We are thankful for the comments of M. Nirrengarten, G. Peron-Pindivic and Zhen Sun on the first version of this manuscript.

FURTHER READING. – The above descriptions are abbreviated and often simplified. If interested in reading and learning further, the reader is referred to the following list of publications and references.

– General: (Huchon *et al.* 2001; Karig 1975; Pubellier and Meresse 2013; Taylor and Hayes 1983; van den Broek and Gaina 2020).

5.9. References

Advokaat, E.L., Marshall, N.T., Li, S., Spakman, W., Krijgsman, W., van Hinsbergen, D.J.J. (2018). Cenozoic rotation history of Borneo and Sundaland. SE Asia revealed by paleomagnetism, seismic tomography, and kinematic reconstruction. *Tectonics*, 37(8), 2486–2512.

Aurelio, M.A., Peña, R.E., Taguibao, K.J.L. (2012). Sculpting the Philippine archipelago since the cretaceous through rifting, oceanic spreading, subduction, obduction, collision and strike-slip faulting: Contribution to and strike-slip faulting: Contribution to IGMA5000. *Journal of Asian Earth Sciences* [Online]. Available at: http://dx.doi.org/10.1016/j.jseaes.2012.10.007.

Barckhausen, U., Engels, M., Franke, D., Ladage, S., Pubellier, M. (2014). Evolution of the South China Sea: Revised ages for breakup and seafloor spreading. *Marine and Petroleum Geology*, 58, 599–611.

Briais, A., Patriat, P., Tapponnier, P. (1993). Updated interpretation of magnetic anomalies and seafloor spreading stages in the South China Sea: Implications for the tertiary tectonics of Southeast Asia. *Journal of Geophysical Research: Solid Earth*, 98(B4), 6299–6328.

van den Broek, J.M. and Gaina, C. (2020). Microcontinents and continental fragments associated with subduction systems. *Tectonics*, 39, e2020TC006063 [Online]. Available at: https://doi.org/10.1029/2020TC006063.

Brune, S., Heine, C., Pérez-Gussinyé, M., Sobolev, S.V. (2014). Rift migration explains continental margin asymmetry and crustal hyper-extension. *Nature Communications*, 5, 4014.

Bulois, C., Pubellier, M., Chamot-Rooke, N., Watremez, L. (2018). From orogenic collapse to rifting: A case study of the northern Porcupine Basin, offshore Ireland. *Journal of Structural Geology*, 114, 139–162.

Chan, L.S., Shen, W., Pubellier, M. (2010). Polyphase rifting of greater Pearl River Delta region (South China): Evidence for possible rapid changes in regional stress configuration. *Journal of Structural Geology*, 32(6). 746–754.

Chang, S.-P., Pubellier, M., Delescluse, M., Yan, Q., Nirrengarten, M., Mohn, G., Chamot-Rooke, N., Yao, L. (2022). Crustal achitecture and evolution of the southwestern South China Sea in press: Implications on breakup. *Marine and Petroleum Geology*, 136.

Clark, J. (2017). Malaysia: Neogene tectonics of northern Borneo: A simple model to explain complex structures within Miocene-recent deltaic-deepwater sediments both onshore and offshore. *The Seapex Exploration Conference*, April 27th.

Clennell, B. (1996). Far-field and gravity tectonics in Miocene basins of Sabah. In *Rifted Margins: Ductile Deformation, Boudinage, Continentward-Dipping Normal Faults and the Role of the Weak Lower Crust*, Clerc, C., Ringenbach, J.-C., Jolivet, L., Ballard, J.-F. (eds). Gondwana Research, Malaysia.

Clerc, C., Ringenbach, J.-C., Jolivet, L., Ballard, J.-F. (2018). Rifted margins: Ductile deformation, boudinage, continentward-dipping normal faults and the role of the weak lower crust. *Gondwana Research*, 53, 20–40.

Clift, P. and Lin, J. (2001). Preferential mantle lithospheric extension under the South China margin. *Marine and Petroleum Geology*, 18(8), 929–945.

Courtillot, V. (1982). Propagating rifts and continental breakup. *Tectonics*, 1(3), 239–250.

Cullen, A., Reemst, P., Henstra G., Gozzard, S., Ray A. (2010). Rifting of the South China Sea: New perspectives. *Petroleum Geoscience*, 16, 73–282.

Ding, W., and Li, J. (2016). Propagated rifting in the Southwest Sub-basin, South China Sea: Insights from analogue modelling. *Journal of Geodynamics*, 100, 71–86.

Ding, W., Schnabel, M., Franke, D., Aiguo, R., Zhenli, W. (2012). Crustal structure across the northwestern margin of South China Sea: Evidence for magma-poor rifting from a wide-angle seismic profile. *Acta Geologica Sinica – English Edition*, 86(4), 854–866.

Festa, A., Balestro, G., Borghi, A., De Caroli, S., Succo, A. (2020). The role of structural inheritance in continental break-up and exhumation of alpine tethyan mantle (Canavese Zone, Western Alps). *Geoscience Frontiers*, 11(1), 167–188.

Franke, D., Barckhausen, U., Baristeas, N., Engels, M., Ladage, S., Lutz, R., Schnabel, M. (2011). The continent-ocean transition at the southeastern margin of the South China Sea. *Marine and Petroleum Geology*, 28(6), 1187–1204 [Online]. Available at: https://doi.org/10.1016/j.marpetgeo.2011.01.004.

Franke, D., Savva, D., Pubellier, M., Steuer, S., Mouly, B., Auxietre, J.-L., Chamot-Rooke, N. (2014). The final rifting evolution in the South China Sea. *Marine and Petroleum Geology*, 58, 704–720 [Online]. Available at: https://doi.org/10.1016/j.marpetgeo.2013.11.020.

Fuller, M., Ali, J.R., Moss, S.J., Frost, G.M., Richter, B., Mahfi, A. (1999). Paleomagnetism of Borneo. *Journal of Asian Earth Sciences*, 17(1–2), 3–24.

Gao, J., Wu, S., McIntosh, K., Mi, L., Yao, B., Chen, Z., Jia, L. (2015). The continent–ocean transition at the mid-northern margin of the South China Sea. *Tectonophysics*, 654, 1–19.

Gillard, M., Sauter, D., Tugend, J., Tomasi, S., Epin, M.E., Manatschal, G. (2017). Birth of an oceanic spreading center at a magma-poor rift system. *Scientific Reports*, 7(1), 1–6.

Hall, R. (2002). Cenozoic geological and plate tectonic evolution of SE Asia and the SW Pacific: Computer-based reconstructions, model and animations. *Journal of Asian Earth Sciences*, 20, 353–431.

Hall, R. (2013). Contraction and extension in northern Borneo driven by subduction rollback. *Journal of Asian Earth Sciences*, 76, 399e411.

Hamilton, W.B. (1979). Tectonics of the Indonesian region. Professional Paper 1078, U.S. Government Printing Office, Washington.

van Hattum, M.W.A., Hall, R., Pickard, A.L., Nichols, G.J. (2013). Provenance and geochronology of Cenozoic sandstones of northern Borneo. *Journal of Asian Earth Sciences*, 76, 266–282.

Heezen, B.C. (1960). The rift in the ocean floor. *Scientific American*, 203, 98–110.

Hess, H.H. (1962). History of ocean basins. *Petrologic Studies: A Volume to Honor A.F. Buddington*, Engel, A.E.J., James, H.L., Leonard, B.F. (eds). The Geological Society of American, Waverly Press, USA.

Holloway, N.H. (1982). North Palawan Block, Philippines – Its relation to Asian mainland and role in evolution of South China Sea. *AAPG Bulletin-American Association of Petroleum Geologists*, 66(9), 1355–1383.

Hsu, S.-K. and Sibuet, J.C. (2011). Continent–ocean transition of the northern South China sea and off southwestern Taiwan March 2004. *Marine Geophysical Researches*, 25(1), 1–4.

Huang, H., Klingelhoefer, F., Qiu, X., Li, Y., Wang, P. (2021). Seismic imaging of an intracrustal deformation in the northwestern margin of the South China Sea: The role of a Ductile Layer in the Crust. *Tectonics*, 40(2021).

Huchon, P., Nguyen, T.N.H., Chamot-Rooke, N. (2001). Propagation of continental break-up in the southwestern South China Sea. In *Non-Volcanic Rifting of Continental Margins: A Comparison of Evidence from Land and Sea*. Wilson, R.C.L., Whitmarsh, R.B., Taylor, B., Froitzem, N. (eds). Geological Society of London, Special Publications.

Huismans, R. and Beaumont, C. (2011). Depth-dependent extension, two-stage breakup and cratonic underplating at rifted margins. *Nature*, 473, 74–79.

Karig, D.E. (1975). Basin genesis in the Philippine Sea. In *Repts. DSDP*, Karig, D.E., Ingle Jr., J.C. *et al.* (eds). U.S. Government Printing Office, Washington, DC.

Larsen, H.C., Mohn, G., Nirrengarten, M., Sun, Z., Stock, J., Jian, Z., Klaus, A., Alvarez-Zarikian, C.A., Boaga, J., Bowden, S.A. *et al.* (2018). Rapid transition from continental breakup to igneous oceanic crust in the South China Sea. *Nature Geoscience*, 11(10), 782–789.

Le Pourhiet, L., Chamot-Rooke, N., Delescluse, M., May, D.A., Watremez, L., Pubellier, M. (2018). Continental break-up of the South China Sea stalled by far-field compression. *Nature Geoscience*, 11(8), 605–609.

Li, Z. and Li, X. (2007). Formation of the 1300-km wide intracontinenta lorogeny and postorogenic magmatic province in Mesozoic South China: A flat-slab subduction model. *Geology*, 35(2), 179–182.

Li, C.-F., Xu, X., Lin, J., Sun, Z., Zhu, J., Yao, Y., Zhang, G.-L. (2014a). Ages and magnetic structures of the South China Sea constrained by deep tow magnetic surveys and IODP Expedition 349. *Geochemistry, Geophysics, Geosystems*, 15(12), 4958–4983.

Li, L., Clift, P.D., Stephenson, R., Nguyen, H.T. (2014b). Non-uniform hyper-extension in advance of seafloor spreading on the Vietnam continental margin and the SW South China Sea. *Basin Research*, 26(1), 106–134.

Li, J., Zhang, Y., Dong, S., Johnston, S.T. (2014c). Cretaceous tectonic evolution of South China: A preliminary synthesis. *Earth-Science Reviews*, 134, 98–136.

Li, C.-F., Li, J., Ding, W., Franke, D., Yao, Y., Shi, H., Pang, X., Cao, Y., Lin, J., Kulhanek, D.K. *et al.* (2015). Seismic stratigraphy of the central South China Sea basin and implications for neotectonics. *Journal of Geophysical Research: Solid Earth*, 120(3), 1377–1399.

Liang, Y., Delescluse, M., Qiu, Y., Pubellier, M., Chamot-Rooke, N., Wang, J., Nie, X., Watremez, L., Chang, S.P., Pichot, T., Savva, D., Meresse, F. (2019). Décollements, detachments, and rafts in the extended crust of dangerous ground, South China Sea: The role of inherited contacts. *Tectonics, American Geophysical Union (AGU)*, 38(6), 1863–1883. 10.1029/2018TC005418ff.

Lunt, P. (2019). A new view of integrating stratigraphic and tectonic analysis in South China Sea and North Borneo basins. *Journal of Asian Earth Sciences*, 177, 220–239.

McIntosh, K., Lavier, L., van Avendonk, H., Lester, R., Eakin, D., Liu, C.-S. (2014). Crustal structure and inferred rifting processes in the northeast South China Sea. *Marine and Petroleum Geology*, 58, 612–626.

McKenzie, D. (1978). Some remarks on the development of sedimentary basins. *Earth and Planetary Science Letters*, 40(1), 25–32.

Morley, C.K. (2007). Variations in Late Cenozoic recent strike-slip and oblique extensional geometries, within Indochina: The influence of pre-existing fabrics. *Journal of Structural Geology*, 29(1), 36–58.

Morley, C.K. (2009). Growth of folds in a deep-water setting. *Geosphere*, 5(2), 59–89.

Morley, C.K. (2016). Major unconformities/termination of extension events and associated surfaces in the South China Seas: Review and implications for tectonic development. *Journal of Asian Earth Sciences*, 120, 62–86.

Nanni, U., Pubellier, M., Chan, L.S., Sewell, R.J. (2017). Rifting and reactivation of a Cretaceous structural belt at the northern margin of the South China Sea. *Journal of Asian Earth Sciences*, 136, 110–123.

Nirrengarten, M., Mohn, G., Kusznir, N.J., Sapin, F., Despinois, F., Pubellier, M., Chang, S.P., Larsen, H.C., Ringenbach, F.J.C. (2020). Extension modes and breakup processes of the southeast China-Northwest Palawan conjugate rifted margins. *Marine and Petroleum Geology*, 113, 104123.

Nissen, S.S., Hayes, D.E., Bochu, Y., Zeng, W., Chen, Y., Nu, X. (1995). Gravity, heat flow, and seismic constraints on the processes of crustal extension: Northern margin of the South China Sea. *Journal of Geophysical Research: Solid Earth*, 100(B11), 22447–22483.

Peltzer, G. and Tapponnier, P. (1988). Formation and evolution of strike-slip faults, rifts, and basins during the India-Asia collision: An experimental approach. *Journal of Geophysical Research*, 93, 15085–15117.

Pérez-Gussinyé, M., Andrés-Martínez, M., Araujo, M., Xin, Y., Armitage, J., Morgan, J.P. (2020). Lithospheric strength and rift migration controls on syn-rift stratigraphy and breakup unconformities at rifted margins: Examples from numerical models, the Atlantic and South China Sea margins. *Tectonics*, 39(12), 1–37.

Peron-Pinvidic, G., Manatschal, G., Minshul, T.A., Sawyer, D.S. (2007). Tectonosedimentary evolution of the deep Iberia-Newfoundland margins: Evidence for a complex breakup history. *Tectonics*, 26(2).

Pichot, T., Delescluse, M., Chamot-Rooke, N., Pubellier, M., Qiu, Y., Meresse, F., Auxiètre, J.-L. (2014). Deep crustal structure of the conjugate margins of the SW South China Sea from wide-angle refraction seismic data. *Marine and Petroleum Geology*, 1–17 [Online]. Available at: https://doi.org/10.1016/j.marpetgeo.2013.10.008.

Pubellier, M. and Meresse, F. (2013). Phanerozoic growth of Asia: Geodynamic processes and evolution. *Journal of Asian Earth Sciences*, 72, 118–128.

Pubellier, M. and Morley, C.K. (2014). The basins of Sundaland (SE Asia): Evolution and boundary conditions. *Marine and Petroleum Geology*, 58, 555–578.

Pubellier, M., Quebral, R., Aurelio, M., Rangin, C. (1996). Docking and post-docking tectonics in the southern Philippines. In *Tectonic Evolution of SE Asia*, Hall, R., Blundell, D. (eds). Geological Society, Special Publication, London.

Pubellier, M., Savva, D., Sapin, F., Aurelio, M. (2016). Structural map of the South China Sea at scale 1:3M. *Commission for the Geological Map of the World*. doi:10.14682/2017 STRUCTUSCS.

Ranero, C. and Peréz-Gussinyé, M. (2006). A new model for the development of the asymmetry of conjugate margins during continental rifting. *Geophysical Research Abstracts*, 8.

Rangin, C., Silver, E.A., Berner, U., Bertrand, P., Betzler, C., Brass, G.W., Hsu, V., Huang, Z.F., Jarrard, R.D., Lewis, S., Linsley, B.K., Merril, D.L., Müller, C., Nederbragt, A.J., Nichols, G., Pubellier, M., Sajona, F.G., Scherer, R.P., Sheu, D.D., Shibuya, H., Shyu, J.P., Smith, R.K., Smith, T.E., Solidum, R.U., Spadea, P., Breyman, M.V. (1989). Forages dans les bassins marginaux du SE asiatique : résultats préliminaires du leg 124 (Ocean Drilling Program).

Rangin, C., Bellon, H., Benard, F., Letouzey, J., Müller, C., Tahir, S. (1990). Neogene arccontinent collision in Sabah, N. Borneo (Malaysia). *Tectonophysics*, 183, 305–319.

Rangin, C., Spakman, W., Pubellier, M., Bijwaard, H. (1999). Geological and tomographic constraints on the subduction of the SE Asia marginal basins. *Bulletin de la Société Géologique de France*, 170(6), 775–788.

Ru, K. and Pigott, J.D. (1986). Episodic rifting and subsidence in the South China Sea. *American Association of Petroleum Geologists Bulletin (United States)*, 70(9), 1136–1155.

Sapin, F., Pubellier, M., Lahfid, A., Janots, D., Aubourg, C., Ringenbach, J. (2011). Onshore record of the subduction of a crustal salient: Example of the NW Borneo Wedge. *Terra Nova*, 100(0), 1–9.

Savva, D. (2013). Variabilité des processus d'extension continentale en Mer de Chine du Sud. Thesis, Pierre and Marie Curie University, Paris.

Savva, D., Meresse, F., Pubellier, M., Chamot-Rooke, N., Lavier, L., Wong Po, K., Franke, D., Steuer, S., Auxietre, J.L., Lamy, G. (2013). Seismic evidence of hyper-stretched crust and mantle exhumation offshore Vietnam. *Tectonophysics*, 608.

Savva, D., Pubellier, M., Franke, D., Chamot-Rooke, N., Meresse, F., Steuer, S., Auxietre, J.L. (2014). Different expressions of rifting on the South China Sea margins. *Marine and Petroleum Geology*, 58, 579–598.

Schlüter, H.U., Hinz, K., Block, M. (1996). Tectono-stratigraphic terranes and detachment faulting of the South China Sea and Sulu Sea. *Marine Geology*, 130(1–2), 39–78.

Shi, X., Jiang, H., Yang, J., Yang, X., Xu, H. (2017). Models of the rapid post-rift subsidence in the eastern Qiongdongnan Basin, South China Sea: Implications for the development of the deep thermal anomaly. *Basin Research*, 29(3), 340–362.

Sibuet, J.C., Yeh, Y.C., Lee, C.S. (2016). Geodynamics of the South China Sea. *Tectonophysics*, 692, 98–119.

Silver, E., Rangin, C., Von Breymann, B.U., Bertrand, P., Betzler, C., Brass, G.W., Huang, Z., Jarrard, R., Lewis, S., Linsley, B., Merril, D., Müller, C., Nederbragt, A., Nichols, G., Pubellier, M., Sajona, F.G., Scherer, R.P., Sheu, D.D., Shibuya, H., Shyu, J.P., Smith, R., Smith, T., Solidum, R.U., Spadea, P., Tannan, D.D. (1989). Origins of marginal basins. *Nature*, 338, 380.

Song, T., Li, C.-F., Wu, S., Yiao, Y., Gao, J. (2019). Extensional styles of the conjugate rifted margins of the South China Sea. *Journal of Asian Earth Sciences*, 177, 117–128.

Stampfli, G.M. and Borel, G.D. (2002). A plate tectonic model for the Paleozoic and Mesozoic constrained by dynamic plate boundaries and restored synthetic oceanic isochrones. *Earth and Planetary Science Letters*, 196(1–2), 17–33.

Steuer, S., Franke, D., Meresse, F., Savva, D., Pubellier, M., Auxietre, J.-L. (2014). Oligocene–Miocene carbonates and their role for constraining the rifting and collision history of the dangerous grounds, South China Sea. *Marine and Petroleum Geology*, (58), 644–657.

Tapponnier, P., Peltzer, G., Armijo, R. (1986). On the mechanisms of the collision between India and Asia. *Tectonics*, 19, 115–157.

Taylor, B. and Hayes, D. (1983). Origin and history of the South China Sea basin. *The Tectonic and Geologic Evolution of Southeast Asian Seas and Islands*, Part 2, Geophysical Monograph, 27.

Taylor, B., Goodliffe, A., Martinez, F., Hey, R. (1995). Continental rifting and initial sea-floor spreading in the Woodlark basin. *Nature*, 374(6522), 534–537 [Online]. Available at: https://doi.org/10.1038/374534a0.

Tongkul, F. (2003). The structural style of Lower Miocene sedimentary rocks, Kudat Peninsula, Sabah. *Geological Society of Malaysia, Bulletin*, 49, 119–124.

Wang, J., Pang, X., Liu, B., Zheng, J., Wang, H. (2018). The Baiyun and Liwan Sags: Two supradetachment basins on the passive continental margin of the northern South China Sea. *Marine and Petroleum Geology*, 95(April), 206–218.

Wei, X., Ruan, A., Zhao, M., Qiu, X., Wu, Z., Niu, X. (2015). Shear wave velocity structure of Reed Bank, southern continental margin of the South China Sea. *Tectonophysics*, 644, 151–160.

Wu, J. and Suppe, J. (2017). Proto-South China Sea plate tectonics using subducted slab constraints from tomography. *Journal of Earth Science*, 29(6), 1304–1318.

Wu, J., Suppe, J., Lu, R.Q., Kanda, R. (2016). Philippine Sea and East Asian plate tectonics since 52 Ma constrained by new subducted slab reconstruction methods. *Journal of Geophysical Research: Solid Earth*, 121(6), 4670–4741.

Xiao, M., Yao, Y.-J., Cai, Y., Qiu, H.-N., Xu, Y.-G., Xu, X., Jiang, Y.-D., Li, Y.-B., Xia, X.-P., Yu, Y.-J. (2019), Evidence of Early Cretaceous lower arc crust delamination and its role in the opening of the South China Sea. *Gondwana Research*, 76, 123–145.

Xie, X., Müller, R.D., Li, S., Gong, Z., Steinberger, B. (2006). Origin of anomalous subsidence along the Northern South China Sea margin and its relationship to dynamic topography. *Marine and Petroleum Geology*, 23(7), 745–765.

Xie, H., Zhou, D., Li, Y., Pang, X., Li, P., Chen, G., Li, F., Cao, J. (2014). Cenozoic tectonic subsidence in deepwater sags in the Pearl River Mouth Basin, northern South China Sea. *Tectonophysics*, 615–616, 182–198.

Yan, P., Wang, L., Wang, Y. (2014). Late Mesozoic compressional folds in Dongsha Waters, the northern margin of the South China Sea. *Tectonophysics*, 615–616, 213–223.

Yang, L., Ren, J., McIntosh, K., Pang, X., Chao, L., Zhao, Y. (2018). The structure and evolution of deepwater basins in the distal margin of the northern South China Sea and their implications for the formation of the continental margin. *Marine and Petroleum Geology*, 92, 234–254.

Ye, Q., Mei, L., Shi, H., Shu, Y., Camanni, G., Wu, J. (2018). A low-angle normal fault and basement structures within the Enping Sag, Pearl River Mouth Basin: Insights into late Mesozoic to early Cenozoic tectonic evolution of the South China Sea area. *Tectonophysics*, 731–732, 1–16.

Zhou, D. (2014). Marginal seas. In *Encyclopedia of Earth Sciences Series*, Harff, J., Meschede, M., Petersen, S., Thiede, J. (eds). Springer, Dordrecht.

Ziegler, P.A. (1988), Evolution of the Arctic-North Atlantic and the Western Tethys. *AAPG Memoir*, 43, 164–196.

6

Arctic Continental Margins

Carmen GAINA
Centre for Earth Evolution and Dynamics, Department of Geosciences, University of Oslo, Norway

6.1. Introduction

The Arctic region occupies continental and oceanic areas situated in the northern part of our planet but its extent has numerous definitions. The southern boundary of the Arctic region could be defined as: (1) the Arctic Circle (66° 33'N), which marks the latitude above which the sun does not set on the summer solstice and does not rise on the winter solstice; (2) the tree line or (3) the 10 degrees Celsius isotherm (or the area where the average temperature of the warmest month of the year is below 10°C). As the climate changes, the 10°C isotherm will migrate north and the treeline is also expected to move northward over time. A more inclusive definition is the "circum-Arctic" region, which includes the land masses and oceanic basins north of 60° N.

The Arctic Ocean is the smallest ocean in the world and is bounded by the wide continental shelves of North America, Greenland and Eurasia. This ocean has two main deep basins: the Eurasia Basin, situated north of the Eurasian continent, and the Amerasia Basin, to the north of Canada, Alaska and Siberia (Figure 6.1).

The formation of the Arctic's wide continental shelves and deep oceanic basins is the consequence of rifting that probably started in the Jurassic (ca. 150 million years ago) and of subsequent seafloor spreading, which resulted in the Cretaceous

For a color version of all the figures in this chapter, see www.iste.co.uk/peronpinvidic/margins2.zip.

Continental Rifted Margins 2,
coordinated by Gwenn PERON-PINVIDIC. © ISTE Ltd 2022.

Canada Basin, the Cenozoic Eurasia Basin and possibly other small, isolated basins (such as the Makarov Basin) of controversial age, scattered among the vast extended continental crust around the North Pole (Figure 6.1).

The Arctic region is slowly revealing its complex tectonic structure and history, as the geoscientific data collection has been increased fewfold in the last two decades due to political and legal reasons – like the Law of the Sea – and facilitated by the decrease in sea ice extent and thickness. Irregular-spaced geophysical data collected from ships, airplanes and submarines was complemented with satellite data that covers vast areas at equally spaced sampling rates. A plethora of regional compilations and models are now available (see Figure 6.1 and Gaina *et al.* (2014) for a review) allowing more detailed or accurate interpretation of the present-day structure of this remote region, and implicitly make progress into untangling its tectonic evolution.

6.2. Passive continental margins in the Arctic

The continental margins in the Arctic formed due to several rifting episodes, which were followed, in most cases, by breakup and seafloor spreading. However, the nature and formation mechanisms of these continental margins may depart from classically defined categories.

To exemplify this, conjugate profiles across the continental margins of the oldest basin, the Canada Basin, and across the youngest Arctic oceanic basin, the Eurasia Basin, are shown and discussed below.

6.2.1. *The Canada Basin continental margins*

After decades of research and data collection (for a review, see Coakley *et al.* 2016), the formation of the Canada Basin (Figure 6.1) remains puzzling. The most widely used model to explain its pseudo-triangular shape invokes a counter-clockwise motion of the northern part of Alaska, the North Slope terrane, relative to the northern Canada margin (e.g. Carey 1955; Grantz *et al.* 1998). This rotational opening may have led to the formation of oceanic crust identified by fan-shaped magnetic anomalies along a mid-ocean ridge that is visible in the gravity field (Figure 6.1b, c). The timing of the Canada Basin opening has not yet been resolved but is believed to have occurred in Early or Mid-Cretaceous (130 or 80 Ma).

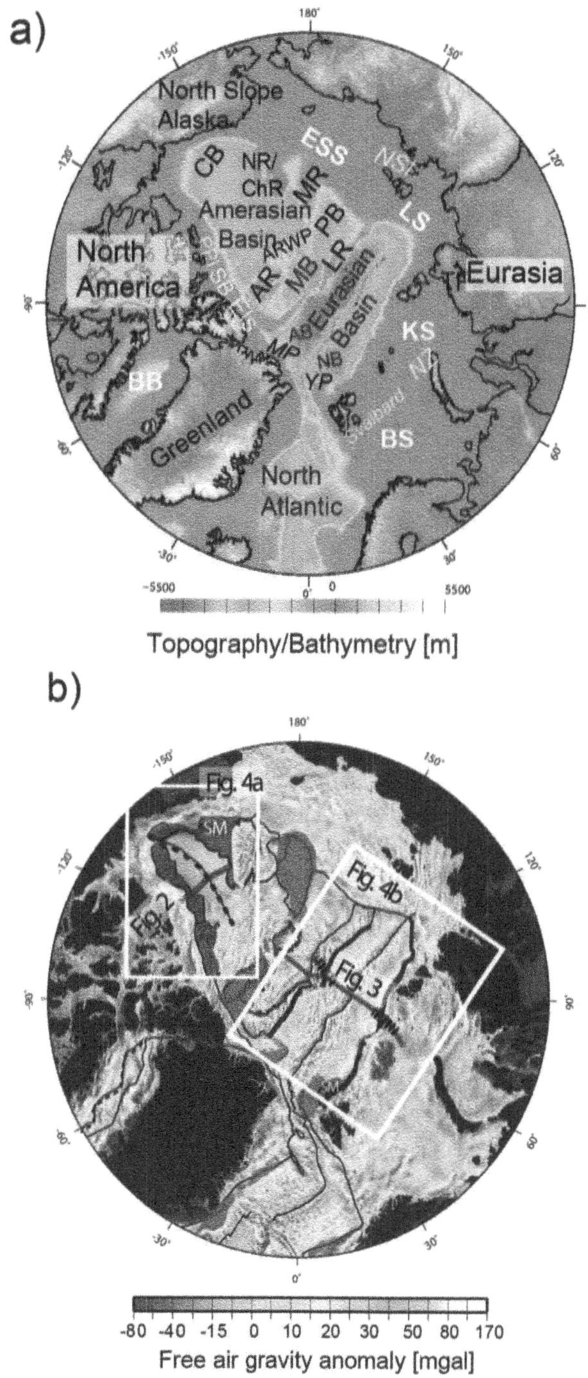
a)
180°
-150°
150°
North Slope
Alaska
-120°
120°
CB
NR/
ChR
MR
ESS
NSI
LS
Amerasian
Basin
PB
ARWP
North
America
MB
LR
AR
Eurasia
Eurasian
Basin
-90°
90°
KS
MP
NB
YP
BB
NZ
Greenland
BS
-60°
60°
North
Atlantic
-30°
30°
0°
-5500
0
5500
Topography/Bathymetry [m]
b)
180°
-150°
150°
Fig. 4a
SM
-120°
120°
Fig. 4b
Fig. 2
Fig. 3
-90°
90°
-60°
60°
-30°
30°
0°
-80 -40 -15 0 10 20 30 50 80 170
Free air gravity anomaly [mgal]

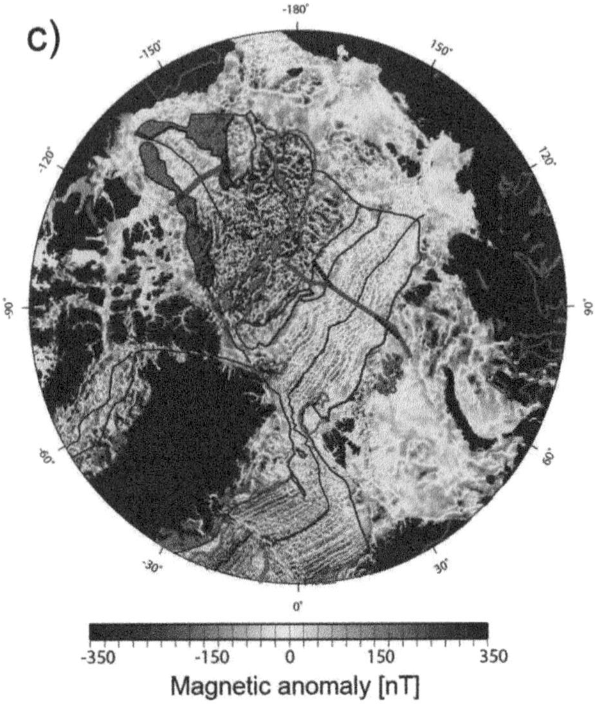

Figure 6.1 *a) Topography and bathymetry of the Circum-Arctic region (north of 68° N) (Amante and Eakins 2009). Abbreviations: AR, Alpha Ridge; ARWP, Alpha Ridge West Plateau; AB, Amundsen Basin; BB, Baffin Bay; BS, Barents Sea; CB, Canada Basin; EIs, Ellesmere Islands; ESS, East Siberian Shelf; KS, Kara Sea; LR, Lomonosov Ridge; LS, Laptev Sea; MB, Makarov Basin; MP, Morris Jessup Plateau; NB, Nansen Basin; NR/ChR, Northwind Ridge/Chukchi Ridge; NSI, New Siberian Islands; NZ, Novaya Zemlya; PB, Podvodnikov Basin; PPI, Prince Patrick Island; SB, Sverdrup Basin; YP, Yermak Plateau. b) Free air gravity (Andersen* et al. *2014). Thin blue lines show the locations of active mid-ocean ridges, and dashed black lines indicate extinct mid-ocean ridges (in the Canada Basin and Baffin Bay). Red polygons indicate magmatic provinces. The interpreted continent–ocean boundaries (COBs) and outlines of major tectonic features in the High Arctic are indicated by thick gray lines. Blue lines in the Eurasia Basin show the narrow COB locations. The yellow segments in both Eurasia and Canada basins indicate transform margins. SM and gray transparent polygons indicate regions of exhumed subcontinental mantle (in the Canada Basin, they are shown together with thinned continental crust). Magenta lines show the location of profiles shown in Figures 6.2 and 6.3. c) Magnetic anomaly map of Circum-Arctic region (Gaina et al. 2011). Note the linear magnetic anomalies in the North Atlantic, Eurasia Basin and partially in the Canada Basin. The Alpha-Mendeleev Ridge in the central High Arctic shows very large values of magnetic anomalies, characteristic of magmatic provinces*

The rotational model implies a large overlap between the Northwind Ridge/Chukchi Borderland (situated north of the Alaska margin) and the Canadian margin or intra-continental compression, of which no evidence exists. Recent Canadian and American geophysical data, in particular seismic velocities derived from sonobuoy measurements, revealed that extended continental crust and serpentinized mantle is flooring a considerable part of the Canada Basin, and only a narrow band along the extinct mid-ocean ridge floor has classical oceanic crust characteristics (Chian *et al.* 2016). In addition, more evidence converges towards a scenario in which the Canada Basin opened in an oblique, transtensional motion. Hutchinson *et al.* (2017) have used a range of geological and geophysical data sets and interpreted a few linear structures as evidence of strike slip or transtensional motion in the Canada Basin prior to sea-floor spreading. According to their interpretation, the Alaska northern margin is a passive extensional margin, and the Canadian northern margin and the conjugate Northwind Ridge are transform margins (Figure 6.1). Furthermore, Døssing *et al.* (2020) demonstrated that the subsequent seafloor spreading in the Canada Basin developed in a highly oblique pattern, requiring a translational motion of the Alaska/Chukotka tectonic block, replacing the decades-old rotational model of the Cretaceous High Arctic.

A joint interpretation of seismic refraction and reflection data, collected along a profile that crosses the Canada Basin from the Northwind Ridge to the northern Canadian margin (Prince Patrick Island), shows the architecture of the transform margin, with a relatively abruptly terminating extended continental crust, to the wider and more extended continental crust of the northern Canadian margin (Figure 6.2a). The Northwind Ridge seems to be underplated by a high-velocity lower crustal body (HVLC), which may explain the large positive magnetic anomaly along this margin. In the seismic reflection data, high-amplitude reflectors (red segments in Figure 6.2b) are identified within the upper basement, at the foot of the steep Northwind Ridge margin. Based on calculated seismic velocities, Chian *et al.* (2016) interpreted the transition between the extended continental crust and the oceanic crust as exhumed serpentinized mantle.

The Canada Basin continental margins are therefore composed of elements characteristic of transform, magmatic and amagmatic passive margins. The Northwind Ridge margin has a linear magnetic and gravity signature characteristic of transform margins. This margin may have been affected by volcanic activity but is also bordered by a region of exhumed mantle – therefore falling into the category of both magmatic and amagmatic passive margins. The Canadian northern margin indicates a very prominent linear trend, characteristic of transform margins, but is also bordered by a large area of extended continental crust and exhumed mantle, which is usually found in hyperextended continental margins.

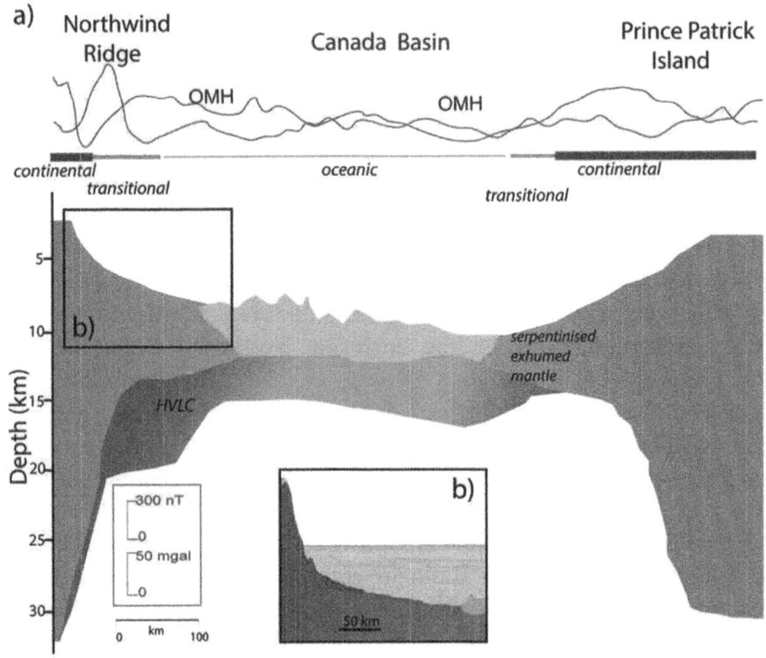

Figure 6.2. *Conjugate continental margins in the Canada Basin. a) Crustal structure of Canada Basin interpreted from seismic reflection and refraction data (modified from Chian* et al. *2016). HVLC is high-velocity lower crust. Red and magenta profiles show magnetic and gravity anomaly data, respectively (see Figure 6.1b, c for maps of this data) along the same profile. Interpretation of basement and sediment layers based on seismic profile LSL2009-31-32 published by the Geological Survey of Canada (Mosher* et al. *2016). Red lines indicate the location of high-reflectivity basement. OMH: outer magnetic high*

6.2.2. *The Eurasia Basin continental margins*

The Eurasia Basin is floored by Paleocene to Recent (ca. 57–0 Ma) old oceanic crust (e.g. Nikishin *et al.* 2018) that formed after the breakup between the northern Eurasian margin, and a narrow sliver of continental crust, the Lomonosov Ridge. The continental nature of the Lomonosov Ridge was established some time ago by noticing the morphological similarities with the Barents and Kara Sea margins and confirmed by continental rocks recovered from its flanks (Backman *et al.* 2008;

Knudsen *et al.* 2018). The Eurasia Basin is divided in two sub-basins: the Nansen Basin, which extends from the Barents and Kara Sea margins to the Gakkel mid-ocean ridge and the conjugate Amundsen Basin, which extends from the Gakkel Ridge to the Lomonosov Ridge margin (Figure 6.1).

The location and architecture of the transition from the continental margin towards the oceanic lithosphere is usually established by interpreting a range of geophysical data that concur in establishing variation in densities, magnetic properties and seismic velocities, which in turn are proxies for various crustal types. The Eurasia Basin is relatively well covered by magnetic data from airborne and ship tracks, as well as gravity data from ship and submarine tracks, complemented by satellite-derived data (see data compilation by Gaina *et al.* (2011)).

The acquisition of seismic reflection and refraction data was hampered by poor accessibility due to ice conditions but, in the last 10 years, a wealth of new data has been collected (e.g. Nikishin *et al.* 2018). As part of the Russian Federation's recent geophysical data campaign, a complete and continuous seismic reflection profile across the Eurasia Basin and its conjugate margins has been acquired (Nikishin *et al.* 2018). This has allowed us to image the continental margins and their transition to the oceanic domain in detail (Figure 6.3). Since no seismic refraction profile is available along this transect through the Eurasia Basin, a model of the conjugated rifted margins based on gravity inversion and sparse seismic reflection data is presented instead (Minakov *et al.* 2012).

According to this data and published interpretation (e.g. Berglar *et al.* 2016; Nikishin *et al.* 2018), the Nansen Basin continental margin is relatively abrupt, showing a narrow transition from continental to oceanic crust (Figure 6.3). Modern seismic reflection data shows that the continental basement blocks of Barents and Kara Sea northern margins are bounded by steep faults (e.g. Berglar *et al.* 2016; Nikishin *et al.* 2018) and the transition from extended continental crust to oceanic crust occurs within 50–60 km. No clear volcanic sequences are visible in the available seismic reflection data, but a positive magnetic anomaly landward from the identified COB (Figure 6.3a) may indicate the existence of volcanic material in the continental margin region (possibly as magmatic intrusive). On the conjugate Lomonosov Ridge margin, a gentle slope with faulted blocks bounded by normal faults is well visible in the selected seismic reflection data, although along other segments of this ridge a sharper transition to the oceanic domain is also observed.

New high-resolution aerogeophysical data combined with seismic reflection and sparse refraction data have been used to model the Lomonosov Ridge–Eurasian margin and to infer the age of breakup and seafloor spreading (Døssing *et al.* 2014).

This data highlights in greater detail NNW-SSE-oriented features bordering the Lomonosov Ridge margin situated north of Greenland and may indicate the direction of transtension prior to breakup. These features are bordered offshore by a prominent magnetic anomaly high, which has been interpreted as magnetic chron C25 (ca. 57 Ma, according to the Ogg (2012) timescale) by Brozena *et al.* (2003) and Dossing *et al.* (2014).

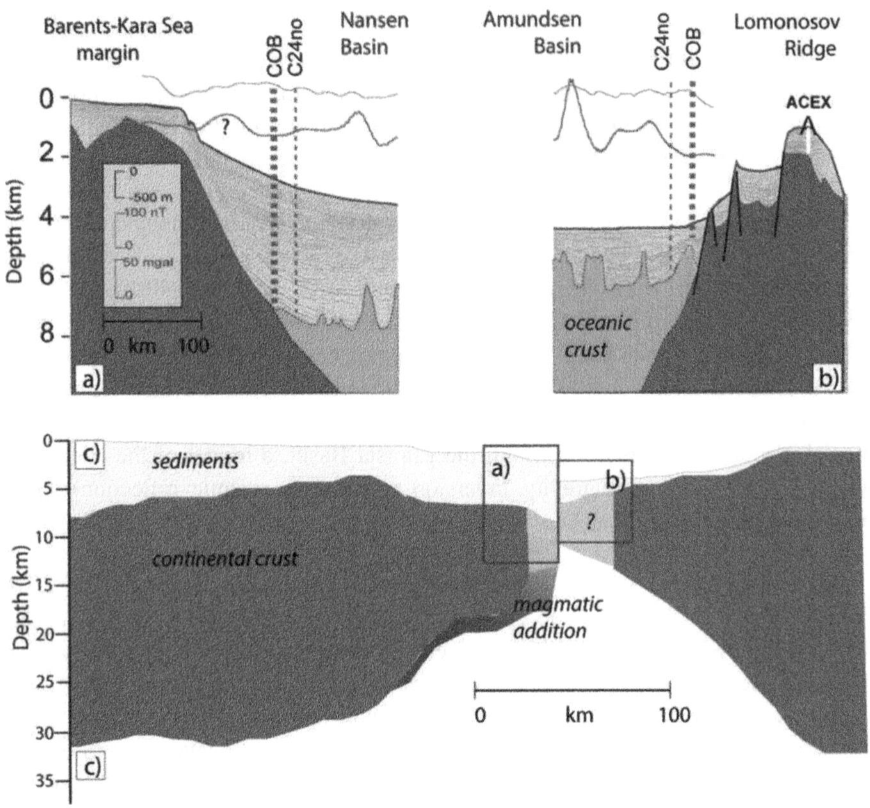

Figure 6.3. *Conjugate continental margins in the Eurasia Basin. a) and b) Interpretation of continental and oceanic basement and continent–ocean transition (COB) based on seismic reflection data along a continuous profile (magenta line in Eurasia Basin, Figure 6.1b) crossing the Eurasia Basin and its margins (modified from profile ARC-1407, Nikishin* et al. *(2018)). Red and magenta profiles show magnetic and gravity anomaly data, respectively (see Figure 6.1b, c for maps of this data) along the same profile. C24no is the oldest identified magnetic chron of ca. 54 million years. c) Modeled structure of the Eurasia Basin margins along conjugate profiles (dashed lines next to magenta profile in the Eurasia Basin, Figure 6.1b) modified from profile D, Minakov* et al. *(2012)*

On the conjugate Barents Sea margin, Lutz *et al.* (2018) interpreted rugged seafloor topography, observed on new seismic reflection profiles, as exhumed mantle, which was subsequently serpentinized and partially covered by volcanic material, a structure that may explain the origin of a weaker magnetic anomaly situated landward of the oldest oceanic crust of chron 24 age (ca. 54 Ma).

In summary, following the state-of-the-art geophysical data and latest interpretation, the Eurasia Basin continental margins are relatively steep and narrow, and their structure shows evidence of pre-breakup transtension and possibly strike-slip motion, but once the oceanic crust formation has been established, the direction of the basin opening was perpendicular on the passive margin. The conjugate margins of the western Eurasia Basin are asymmetric, with a wider region of faulted extended continental basement on the Lomonosov Ridge site, and a narrower Barents Sea margin. Recently, the interpretation of exhumed continental mantle offshore the Barents Sea continent–ocean boundary (Lutz *et al.* 2018) is challenging the conventional model where oceanic crust is directly juxtaposed to extended continental crust. The asymmetric margin structure requires more data to be validated, and a geodynamic explanation has to be found for the diachronous opening (chron 25 time on the Lomonosov Ridge side and chron 24 on the Barents Sea side) time of the Eurasia Basin.

6.3. Discussion and concluding remarks

The continental margins of the Arctic region that border the two main oceanic basins, the Canada and Eurasia basins, are segmented into individual parts, reflecting the complex tectonic history of this area. Rifting and seafloor spreading occurred in three main episodes: (1) Late Jurassic-Early/Mid Cretaceous, which led to the formation of the Canada Basin and possibly other small basins north of it (Figure 6.4a); (2) Mid-Late Cretaceous with rifting in the central Amerasia Basin and seafloor spreading in the Makarov Basin; and (3) Paleocene-Recent continental breakup and seafloor spreading in the Eurasia Basin (Figure 6.4b). The driving forces for these tectonic activities were more likely a "tug-of-war" between subduction of older oceanic crust in the Pacific domain and the extension between two main tectonic plates: the North America and Eurasia, which led to the formation of the North Atlantic. A third triggering mechanism could have been the emplacement of the High Arctic Large Igneous province (HALIP) in the Cretaceous Period, but its age and extent are controversial, and a causal relationship between it and rifting in Canada Basin is still debated.

The southward motion of the North Slope Alaskan terrane and adjacent Northwind Ridge/Chucki Borderland was probably the result of Pacific slab pull or rollback which created the necessary extension, forcing the Northwind Ridge/Chuckchi Borderland to slide along the northern Canadian margin and to thin the crust and exhume the subcontinental mantle between the northern Alaskan margin and Prince Patrick Island margin. An older oceanic basin, the South Anyui Ocean, which may have been part of the proto-Pacific Ocean, was subducted along the Siberian margin (South Anyui Suture (SAS), Figure 6.4), as evidenced by ophiolite outcrops along the Eastern Siberian margin and on the New Siberian Islands (e.g. Sokolov *et al.* 2002; Franke *et al.* 2008). As shown by numerical modeling, strike-slip motion weakens continental crust and facilitates rapid breakup (Brune *et al.* 2016). In the case of the Canada Basin, oceanic crust was produced after a fair amount of extension and mantle exhumation took place, and the decisive breakup was possibly linked to some magmatic activity, as indicated by deep magmatic underplating (Figure 6.2). Seafloor spreading in the Canada Basin came to a halt possibly due to a change in the Pacific subduction dynamics in the Northern Pacific.

The second rifting episode in the Arctic region affected the High Arctic region on the Amerasian part of the Lomonosov Ridge. Numerous shallow plateaus separate the Eurasia Basin and Lomonosov Ridge from the Canada Basin (Figure 6.1). Most of them are part of the Alpha-Mendeleev Ridge, an igneous province that may have formed in the Cretaceous (e.g. Dossing *et al.* 2017). The igneous material obstructs the original basement in this region, and therefore the nature of the Alpha-Mendeleev Ridge is still disputed with studies that consider it continental, oceanic or both (e.g. Oakey and Saltus 2016). Late Cretaceous rifting of this igneous plateau and subsequent seafloor spreading led to the formation of a small, enclosed basin, the Makarov Basin (Figures 6.1 and 6.4), situated between the continental Lomonosov Ridge and the Alpha Ridge West Plateau of unknown origin (Dossing *et al.* 2017). The deep Makarov Basin extends towards the East Siberian Shelf and towards northern Greenland with shallower basins (the Podvodnikov Basin and the Alpha Basin, respectively) but it is postulated that they are floored by extended continental crust (Nikishin *et al.* 2014). The margins of the oceanic Makarov Basin are relatively steep, and its formation was probably due to a combination of extension/transtension between the North American and Eurasian plates, and far-field stresses induced by northward Pacific plate subduction.

The last rifting and subsequent seafloor spreading episode in the High Arctic region started in the Paleocene when the Lomonosv Ridge was rifted from the northern Eurasian margin. In the young Eurasia Basin, the conjugate margins display rifted structures that may point to strike-slip or transtensional motion just before breakup (Minakov *et al.* 2012; Dossing *et al.* 2014) and narrow continent–ocean transitions (narrower in the north Kara and Barents Seas) (Nikishin *et al.* 2018). Patches of exhumed mantle have been also described in the SW Eurasia

Basin in a region close to the continent–ocean boundary. The patchy oldest magnetic anomaly observed in that area may have been sourced by later magmatic additions that partially covered the serpentinized mantle (Lutz *et al.* 2018), refuting its interpretation as oceanic crust of chron 25 age (ca. 57 Ma). On the other hand, geophysical data and gravity modeling of the conjugate margin, the Lomonosov Ridge (Døssing *et al.* 2014), suggest an earlier breakup age (before chron 25 time) of the NW Eurasia Basin. The two different ages of these conjugate margin segments may require plate boundary relocation after chron 25 time, a scenario proposed by Brozena *et al.* (2003) and Døssing *et al.* (2014). At chron 24 time (ca. 54 Ma), orthogonal seafloor spreading started to create oceanic crust in the Eurasia Basin (Figure 6.4b), a process that continues today along the slowest mid-ocean ridge on Earth: the Gakkel Ridge.

All basins in the Arctic Ocean formed as enclosed basins, surrounded by continental crust, some of it with ancient, cratonic nature. Lithospheric heterogeneities due to multiple periods or rifting, seafloor spreading and subsequent subduction and collisions have been exploited in the last three episodes of rifting and continental breakup (Figure 6.4a). Due to the confined region, and also proximity to opening poles of rotation, Arctic oceanic basins have fan-shaped or even circular shapes, illustrating the need to accommodate extension in limited spaces. Few continental margin segments in the Arctic display exhumed continental mantle which may indicate prolonged episodes of extension and non-magmatic margin development. As illustrated for both the Canada and Eurasia basins, breakup may have been preceded by short periods of strike-slip motions which, in most circumstances, facilitate rapid continental breakup (Brune *et al.* 2016). Magmatic additions, both at the base of the crust and/or as shallow intrusions, have also been described along Arctic margins. Short episodes of magmatic activity, some of it probably associated with the formation of the High Arctic Large Igneous Province (HALIP), may have also contributed to continental breakup.

Lastly, the Arctic region is the locus where the present-day plate boundary between Eurasia and North America, the ultra-slow Gakkel mid-ocean ridge, is propagating into continental crust in a diffuse manner, as illustrated by the multiple rifting episodes in the Laptev Sea and presented-day distributed seismicity (e.g. Nikishin *et al.* 2018).

Although in the last decade the joint effort of various Arctic countries and research teams from elsewhere has made a substantial contribution to the understanding of the Arctic tectonic history, additional geophysical data is needed in poorly-explored regions, such as the central Arctic. The Arctic region is still a relatively unexplored region and the lack of rock samples from deep drilling precludes precise identification of the age and composition of various tectonic features, the age of continental breakup and the timing of major tectonic events that marked the formation of polar oceanic basins.

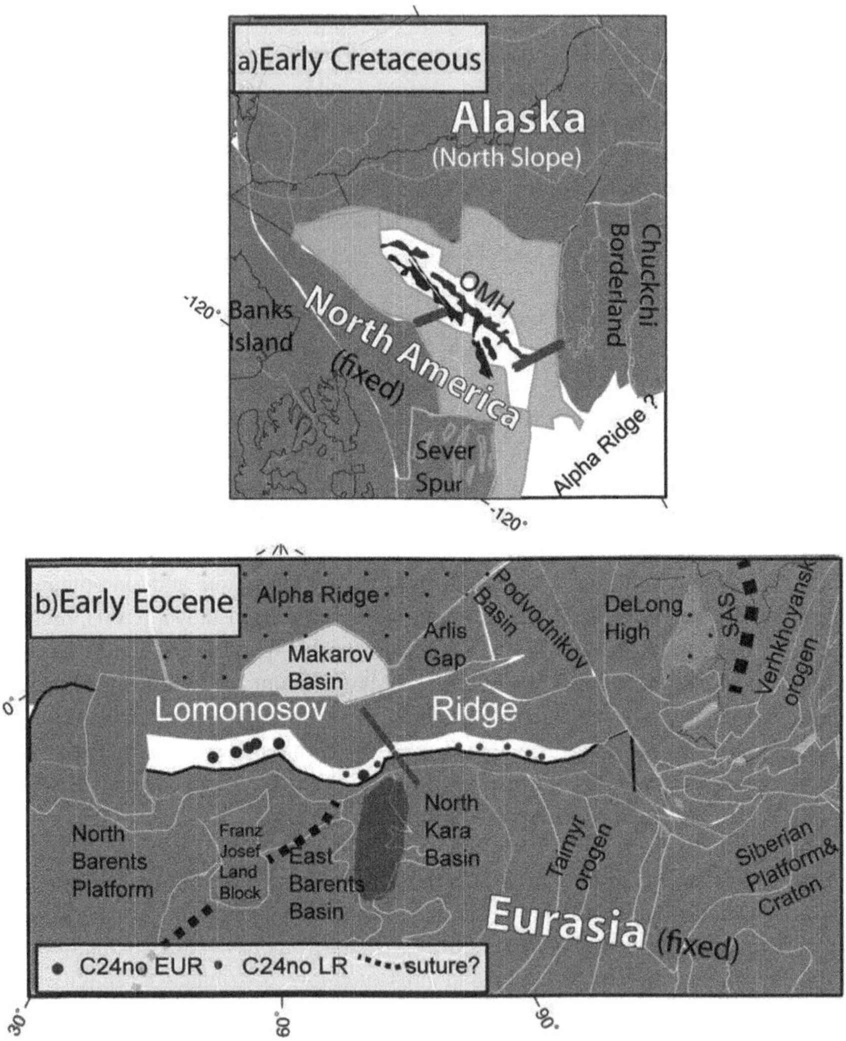

Figure 6.4. *Plate tectonic reconstructions for the time of first seafloor spreading in the Canada Basin (a) and Eurasia Basin (b) (modified from Døssing* et al. *(2020) and Nikishin* et al. *(2018), respectively)*

CONTINUATION OF CAPTION FOR FIGURE 6.4. – *Dark gray regions indicate continental lithosphere, light gray – thinned continental crust and/or exhumed mantle. The outlines of various tectonic continental blocks are shown by thin white lines. Blue polygons and circles represent evidence of seafloor spreading based on the interpretation of magnetic anomalies. Magenta segments indicate the approximate position of seismic profiles shown in Figures 6.2 and 6.3. OMH: outer magnetic high. SAS indicates the approximate location of the Cretaceous South Anyui Suture.*

FURTHER READING.– The above descriptions are abbreviated and often simplified. If interested in reading and learning further, the reader is referred to the following list of publications and references.

– General: (Brandes and Piepjohn 2021; Drachev *et al.* 2010; Drachev and Shkarubo 2018; Evangelatos and Mosher 2016; Franke and Hinz 2009; Ilhan and Coakley 2018; Jackson *et al.* 2010; Jakobsson *et al.* 2003; Kristoffersen 2011; Kristoffersen *et al.* 2020).

6.4. References

Amante, C. and Eakins, B.W. (2009). ETOPO1 1 Arc-Minute Global Relief Model: Procedures, data sources and analysis. NOAA Technical Memorandum NESDIS NGDC-24. National Geophysical Data Center, Boulder.

Andersen, O.B., Knudsen, P., Kenyon, S., Holmes, S. (2014). Global and arctic marine gravity field from recent satellite altimetry [DTU13]: Extended abstract. *Proceedings of the 76th EAGE Conference*, Amsterdam, Netherlands.

Backman, J., Jakobsson, M., Frank, M., Sangiorgi, F., Brinkhuis, H., Stickley, C., O'Regan, M., Lovlie, R., Palike, H., Spofforth, D., Gattacecca, J., Moran, K., King, J., Heil, C. (2008). Age model and core-seismic integration for the Cenozoic Arctic Coring Expedition sediments from the Lomonosov Ridge. *Paleoceanography*, 23, PA1S03, doi:10.1029/2007PA001476.

Berglar, K., Franke, D., Lutz, R., Schreckenberger, B., Damm, V. (2016). Initial opening of the Eurasian Basin, Arctic Ocean. *Frontiers in Earth Science*, 4, doi.org/10.3389/feart.2016.00091.

Brandes, C. and Piepjohn, K. (2021). Eurekan faults on northern Ellesmere Island, Arctic Canada: From Cenozoic strike-slip tectonics to recent seismicity. *Journal of Geodynamics*, 144, 101816, doi.org/10.1016/j.jog.2021.101816.

Brozena, J.M., Childers, V.A., Lawver, L.A., Gahagan, L.M., Forsberg, R., Faleide, J.I., Eldholm, O. (2003). New aerogeophysical study of the Eurasia Basin and Lomonosov Ridge: Implications for basin development. *Geology*, 31, 825–828.

Brune, S., Williams, S.E., Butterworth, N.P., Muller, R.D. (2016). Abrupt plate accelerations shape rifted continental margins. *Nature*, 536, 201.

Carey, S.W. (1955). The orocline concept in geotectonics – Part I. *Papers and Proceedings of the Royal Society of Tasmania*, 89, 255–288.

Chian, D., Jackson, H.R., Hutchinson, D.R., Shimeld, J.W., Oakey, G.N., Lebedeva-Ivanova, N., Li, Q., Saltus, R.W., Mosher, D.C. (2016). Distribution of crustal types in Canada Basin, Arctic Ocean. *Tectonophysics*, 691, 8–30, doi.org/10.1016/j.tecto.2016.01.038.

Coakley, B., Brumley, K., Lebedeva-Ivanova, N., Mosher, D. (2016). Exploring the geology of the central Arctic Ocean: Understanding the basin features in place and time. *Journal of the Geological Society*, 173, 967–987.

Døssing, A., Hansen, T.M., Olesen, A.V., Hopper, J.R., Funck, T. (2014). Gravity inversion predicts the nature of the Amundsen Basin and its continental borderlands near Greenland. *Earth and Planetary Science Letters*, 408, 132–145.

Døssing, A., Gaina, C., Brozena, J.M. (2017). Building and breaking a large igneous province: An example from the High Arctic. *Geophysical Research Letters*, 44, 6011–6019.

Døssing, A., Gaina, C., Jackson, R. (2020). Cretaceous ocean formation in the High Arctic. *Earth and Planetary Science Letters*, 551, doi.org/10.1016/j.epsl.2020.116552.

Drachev, S.S. and Shkarubo, S.I. (2018). Tectonics of the Laptev shelf, Siberian Arctic. *Geological Society, London, Special Publications*, 460(1), 263–283, doi:10.1144/sp460.15.

Drachev, S.S., Malyshev, N.A., Nikishin, A.M. (2010). Tectonic history and petroleum geology of the Russian Arctic Shelves: An overview. *Geological Society, London, Petroleum Geology Conference Series*, 7(1), 591–619, doi:10.1144/0070591.

Evangelatos, J. and Mosher, D.C. (2016). Seismic stratigraphy, structure and morphology of Makarov Basin and surrounding regions: Tectonic implications. *Marine Geology*, 374, 1–13, doi.org/10.1016/j.margeo.2016.01.013.

Franke, D. and Hinz, K. (2009). Geology of the shelves surrounding the new Siberian islands, Russian Arctic. *Stephan Mueller Special Publication Series*, 4, 35–44.

Franke, D., Reichert, C., Damm, V., Piepjohn, K. (2008). The south Anyui suture, Northeast Arctic Russia, revealed by offshore seismic data. *Norwegian Journal of Geology*, 88, 189–200.

Gaina, C., Werner, S.C., Saltus, R., Maus, S., Grp, C.-G. (2011). Circum-Arctic mapping project: New magnetic and gravity anomaly maps of the Arctic. *Geological Society, London, Memoirs*, 35, 39–48, doi: 10.1144/M35.3.

Gaina, C., Medvedev, S., Torsvik, T.H., Koulakov, I., Werner, S.C. (2014). 4D Arctic: A glimpse into the structure and evolution of the Arctic in the light of new geophysical maps, Plate tectonics and Tomographic models. *Surveys in Geophysics*, 35, 1095–1122.

Grantz, A., Clark, D.L., Phillips, R.L., Srivastava, S.P., Blome, C.D., Gray, L.B., Haga, H., Mamet, B.L., McIntyre, D.J., McNeil, D.H., Mickey, M.B., Mullen, M.W., Murchey, B.I., Ross, C.A., Stevens, C.H., Silberling, N.J., Wall, J.H., Willard, D.A. (1998). Phanerozoic stratigraphy of northwind ridge, magnetic anomalies in the Canada basin, and the geometry and timing of rifting in the Amerasia basin, Arctic Ocean. *Geological Society of America Bulletin*, 110, 801–820.

Hutchinson, D.R., Jackson, H.R., Houseknecht, D.W., Li, Q., Shimeld, J.W., Mosher, D.C., Chian, D., Saltus, R.W., Oakey, G.N. (2017). Significance of northeast-trending features in Canada Basin, Arctic Ocean. *Geochemistry Geophysics Geosystems*, 18, 4156–4178.

Ilhan, I. and Coakley, B.J. (2018). Meso–Cenozoic evolution of the southwestern Chukchi Borderland, Arctic Ocean. *Marine and Petroleum Geology*, 95, 100–109, doi:https://doi.org/10.1016/j.marpetgeo.2018.04.014.

Jackson, H.R., Dahl-Jensen, T., Group, L.W. (2010). Sedimentary and crustal structure from the Ellesmere Island and Greenland continental shelves onto the Lomonosov Ridge, Arctic Ocean. *Geophysical Journal International*, 182(1), 11–35, doi:10.1111/j.1365-246X.2010.04604.x.

Jakobsson, M., Grantz, A., Kristoffersen, Y., Macnab, R. (2003). Physiographic provinces of the Arctic Ocean seafloor. *GSA Bulletin*, 115(12), 1443–1455.

Knudsen, C., Hopper, J.R., Bierman, P.R., Bjerager, M., Funck, T., Green, P.F., Ineson, J.R., Japsen, P., Marcussen, C., Sherlock, S.C., Thomsen, T.B. (2018). Samples from the Lomonosov Ridge place new constraints on the geological evolution of the Arctic Ocean. In *Circum-Arcic Lithosphere Evolution*, Pease, V. and Coakley, B. (eds). Geological Society, London.

Kristoffersen, Y. (2011). Chapter 45: Geophysical exploration of the Arctic Ocean: The physical environment, survey techniques and brief summary of knowledge. *Geological Society, London, Memoirs*, 35(1), 685–702.

Kristoffersen, Y., Ohta, Y., Hall, J.K. (2020). On the the origin of the Yermak plateau in the Arctic Ocean north of Svalbard. *Norwegian Journal of Geology*, 100, doi.org/10.17850/njg100-1-5.

Lutz, R., Franke, D., Berglar, K., Heyde, I., Schreckenberger, B., Klitzke, P., Geissler, W.H. (2018). Evidence for mantle exhumation since the early evolution of the slow-spreading Gakkel Ridge, Arctic Ocean. *Journal of Geodynamics*, 118, 154–165.

Minakov, A., Faleide, J.I., Glebovsky, V.Y., Mjelde, R. (2012). Structure and evolution of the northern Barents-Kara Sea continental margin from integrated analysis of potential fields, bathymetry and sparse seismic data. *Geophysical Journal International*, 188, 79–102.

Mosher, D., Shimeld, J., Hutchinson, D., Jackson, R. (2016). Canadian UNCLOS extended continental shelf program seismic data holdings (2006–2011). Document, Geological Survey of Canada, Ottawa.

Nikishin, A.M., Malyshev, N.A., Petrov, E.I. (2014). *Geological Structure and History of the Arctic Ocean*. EAGE Publications, Houten.

Nikishin, A.M., Gaina, C., Petrov, E.I., Malyshev, N.A., Freiman, S.I. (2018). Eurasia Basin and Gakkel Ridge, Arctic Ocean: Crustal asymmetry, ultra-slow spreading and continental rifting revealed by new seismic data. *Tectonophysics*, 746, 64–82.

Oakey, G.N. and Saltus, R.W. (2016). Geophysical analysis of the Alpha-Mendeleev ridge complex: Characterization of the High Arctic Large Igneous Province. *Tectonophysics*, 691, 65–84.

Ogg, J.G. (2012). The geomagnetic polarity timescale. In *The Geologic Time Scale 2012*, Gradstein, F.M., Ogg, J.G., Schmitz, M., Ogg, G. (eds). Elsevier, Amsterdam.

Sokolov, S.D., Bondarenko, G.Y., Morozov, O.L., Shekhovtsov, V.A., Glotov, S.P., Ganelin, A.V., Kravchenko-Berezhnoy, I.R. (2002). South Anyui suture, northeast Arctic Russia: Facts and problems. *Geological Society of America Special Paper*, 360, 209–224.

7

Alpine Tethys Margins

Gianreto MANATSCHAL and Pauline CHENIN
Université de Strasbourg, CNRS, ITES UMR 7063, Strasbourg, France

7.1. Background and historical overview

7.1.1. *Preamble*

When writing a review on the Alpine Tethys margins, we must keep in mind that there are only a few places on Earth where rocks, sediments and structures have been examined in such detail and where debates have been so heated, such as in the Alps of Western Europe. Due to the short format of this chapter, it is neither possible to present a comprehensive historical review, nor to discuss all of the observations, data and concepts that were developed over more than one century. For readers looking for a detailed historical perspective, we suggest the paper by Şengör (2014). For those seeking details about specific aspects of the Alpine Tethys margins, we encourage the exploration of the rich literature available on these topics.

A major challenge we faced while writing this chapter was the integration of the various terminologies and concepts used to describe the Alpine Tethys margins into a coherent language accessible to readers from different geological backgrounds, including those who are not familiar with Alpine geology. Unfortunately, there is only so much we can do, and we are aware that some new concepts are not necessarily explained and discussed as much as they should. For the sake of

For a color version of all the figures in this chapter, see www.iste.co.uk/peronpinvidic/margins2.zip.

Continental Rifted Margins 2,
coordinated by Gwenn PERON-PINVIDIC.

accessibility, we referred to specific Alpine locations, nappes, formations and deformation phases as sparingly as possible.

We introduced some of the terminologies used to describe present-day “magma-poor” rifted margins. Indeed, although the Alpine Tethys may not be comparable to present-day oceans, neither in size nor in fate, the Alpine Tethys margins display several first-order features that can be recognized at many of the world’s “magma-poor” rifted margins.

Last but not least, even if we try to provide a snapshot of the state of the art on the Alpine Tethys geology based on observations and data, we stress that this view is neither fully impartial nor unique. Although we do not aim to review all of the studies and ideas that have been proposed for the Alpine Tethys here, we mention alternative views that conflict with those presented in this chapter. Some of the open debates result from the lack of data, which makes it impossible to favor one interpretation over the another at this stage. Other debates may be considered outdated because they were based on inappropriate assumptions or flawed interpretations, violated physical principles and/or ignored new or existing data. While the former debates are important and need to continue, the latter are hopefully going to disappear.

In this chapter, we aim to provide a snapshot of the state of the research on the Alpine Tethys rift system and discuss some of the new concepts and ideas that have been recently developed. Finally, we highlight the potential of the Alpine Tethys as a laboratory for future research.

7.1.2. *Geological and paleogeographical settings*

Remnants of the Alpine Tethys and related rifted margins are exposed as scattered fragments throughout the Alpine orogen of Western Europe (Figure 7.1b and 7.1c). Here, we use the term “Alpine Tethys” to refer to the (proto-)oceanic domain(s) separating the European–Iberian and African–Adriatic (micro)continents (Figure 7.1a). By proto-oceanic, we mean embryonic oceanic crust formed by an immature slow-spreading system that is linked to the exhumation of subcontinental lithospheric mantle to the seafloor and associated with unequal magmatic activity.

The Alpine Tethys evolution is part of an *incomplete Wilson Cycle* (Chenin *et al.* 2018a) that initiated with the late- to post-orogenic collapse of the Variscan orogen, evolved through different rift phases into a proto-oceanic domain, which was subsequently closed via Ampferer-type subduction (i.e. closure of hyperextended continental basins with minor volumes of oceanic crust, and whose subduction is largely limited to the convective upper mantle; McCarthy *et al.* 2020). The resulting phases of continental collision gave rise to the present-day Alpine orogen exposed in Western Europe (Figure 7.1c).

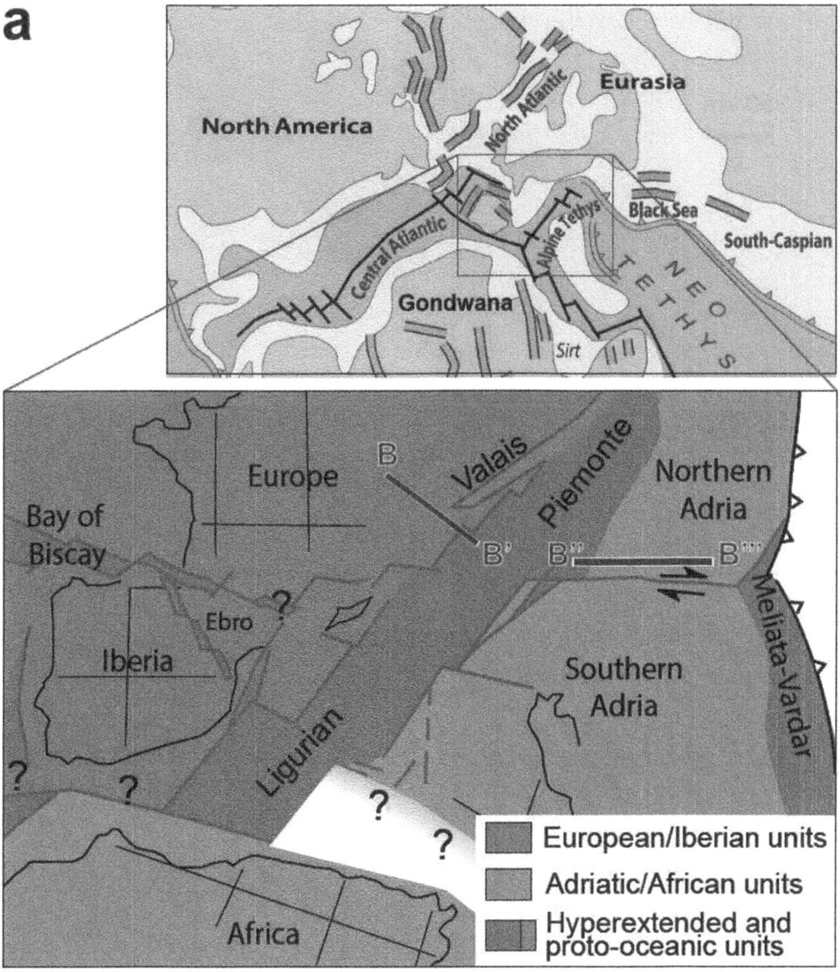

Figure 7.1a. *Schematic paleogeographic maps of the Alpine Tethys region during the Early Cretaceous. The B–B' transect across the former European margin is schematized on the left-hand side of Figure 7.3b; the B"–B"' is the Grischun transect is schematized on the right-hand side of Figure 7.3b and is described in this chapter. Note that these two transects are not conjugates (source: top panel adapted from Frizon de Lamotte* et al. *2015; bottom panel adapted from Manatschal and Müntener 2009)*

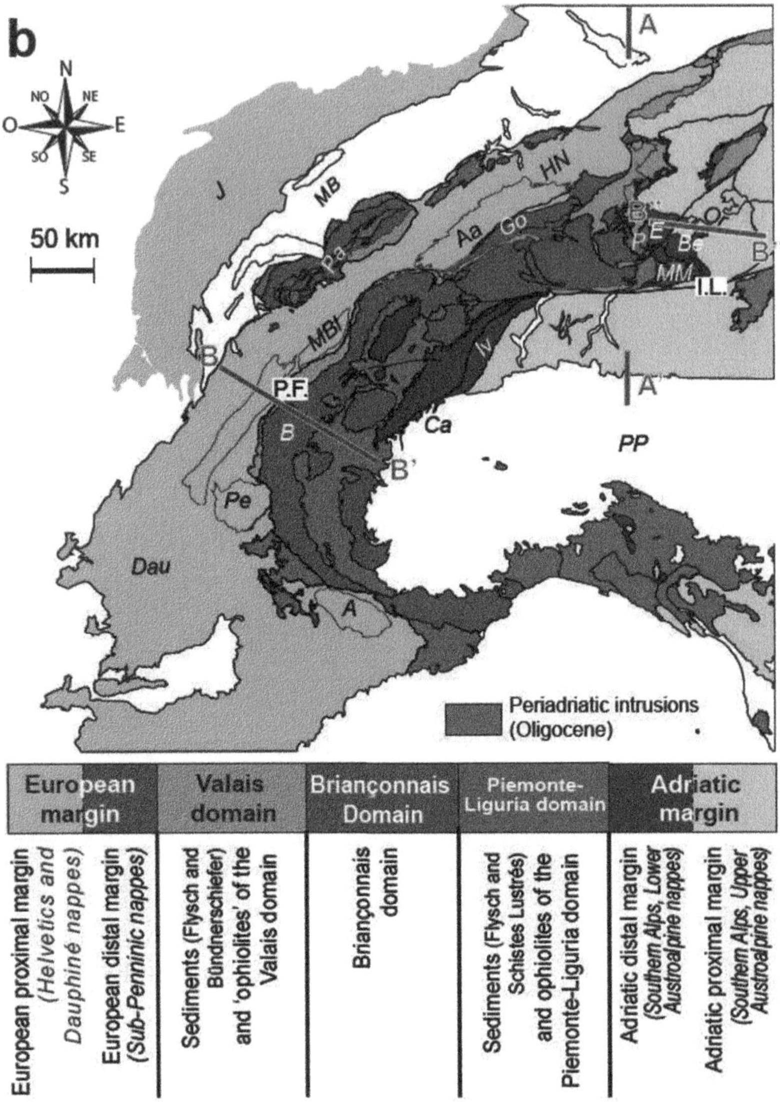

Figure 7.1b. *Simplified tectonic map of the Alpine orogen straddling France, Switzerland and Italy. A: Argentera massif, Aa: Aar massif, B: Briançonnais nappes, Be: Bernina Ca: Cannavese nappes, Dau: Dauphiné nappes, E: Err nappe, Go: Gothard massif, HN: Helevetic nappes, Iv: Ivrea, J: Jura, MB: Molasse basin, MBl: Mont Blanc massif, MM: Malenco-Margna nappes, Or: Ortler nappe, P: Platta nappe, Pa: Prealps, Pe: Pelvoux massif, PP: Po Plain (source: adapted from Schmid* et al. *2004)*

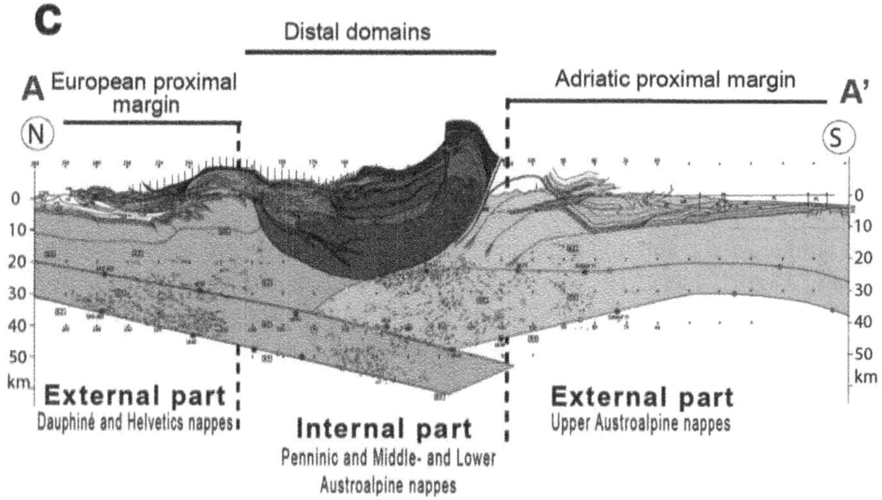

Figure 7.1c. *Interpretation of the NFP 20 seismic section across the Central Alps in terms of paleogeographic units. From a large-scale perspective, collisional orogens can be regarded as composed of two buttresses (the "external domains") formed by the former proximal margins on either side of an "internal domain" composed of the former distal margins and the sedimentary cover that was scraped off during the closure of the Alpine Tethys (source: adapted from Schmid* et al. *1996)*

The Alpine Tethys formed during the Jurassic, contemporaneously with the subduction of the Paleo- and/or Neo-Tethys ocean(s) to the east and with the opening of the Central Atlantic Ocean to the west (Figure 7.1a). From the Late Sinemurian to the Pliensbachian (at 185 ± 5 Ma), the Alpine Tethys was a segmented rift system composed of individual sub-basins, named the *Liguria*, *Piemonte* and *Valais* basins (Figure 7.1a). These individual rift segments merged at 165 ± 5 Ma into one proto-oceanic basin that linked the Central Atlantic Ocean to the Tethys realm (Baumgartner 2013). Whether the Alpine Tethys developed into a mature, steady-state spreading system remains debated. However, no evidence for a mature Penrose crust has been found and described from the Alpine Tethys domain so far. The duration of active seafloor spreading, whenever there was any, has implications for the maximum size of the Alpine Tethys, and hence for the subsequent subduction and collisional history recorded in the Alps.

Aside from the Alpine Tethys, other basins such as the *Meliata-Vardar*, the *Pyrenean rift system* and the *Bay of Biscay* (Figure 7.1a) formed between the Paleo- and/or Neo-Tethys and Atlantic domains from the Late Triassic to the Early

Cretaceous. These (proto-)oceanic domains developed successively westwards, resulting in a complex plate-kinematic framework that will be discussed later.

The Alpine Tethys margins are exposed over an area that extends from the Carpathians to the Apennines and further south in the Tell region of northern Algeria. In this chapter, we focus on the former northwestern Adriatic margin, today exposed in the Austroalpine and south Penninic nappes of southeastern Switzerland and northern Italy, and on the former European margin exposed in the western and central Alps (Figure 7.1b and 7.1c). The northeastern termination of the Alpine Tethys has been less well investigated, and the southwestern part has been heavily overprinted by the later subduction and formation of back arc basins (Gulf of Lyon and Tyrrhenian Sea).

In this context, we stress the difficulty of proposing robust paleogeographic and kinematic reconstructions of the Alpine Tethys rift system. Major debates arise from the restoration of the Iberian plate (Vissers and Meijer 2012; Barnett-Moore *et al.* 2016; Nirrengarten *et al.* 2018) due to the lack of undisputed magnetic anomalies and the existence of intracontinental deformation preceding breakup (Nirrengarten *et al.* 2017). The restorations of the Alpine Tethys is also disputed due to the heavy Alpine overprint of several terranes (Handy *et al.* 2010). Here, we use the interpretation of Decarlis *et al.* (2018), who suggested a segmentation of the Alpine Tethys into a Ligurian and a Piemonte sub-basin separated by a Jurassic transform system (Figure 7.1a). Based on this assumption, the Grischun transect discussed in this chapter (B''–B''' section in Figures 7.1a and 7.1b) was part of the Piemonte segment of the Alpine Tethys and was conjugate to the southeast European margin currently exposed in Sardinia/Calabria. However, this hypothesis needs to be confirmed by future work.

7.1.3. *Historical overview*

As documented by Şengör (2014), the Alps of Western Europe are presumably the place on Earth that most inspired new concepts on the formation and reactivation of oceanic domains and orogens dear to the upholders of Plate Tectonics (Figure 7.2), starting with the pioneering works of Suess (1883; Figure 7.2a), Steinmann (1905; Figure 7.2b) and Argand (1916; Figure 7.2c). This may look surprising today because the Alpine Tethys may have never developed a steady-state seafloor spreading system and never went to mature, Pacific-type subduction (i.e. long-lasting subduction of oceanic lithosphere), the two most important processes of Plate Tectonics. It may, however, explain why Alpine geologists did not discover Plate Tectonics (Trümpy 2001) and why Steinman's trinity (Steinmann 1905; Figure 7.2b), Ampferer and Hammer's (1911) idea of "subduction" and Decandia and Elter's (1969) idea of exhumed mantle were not able to convince the protagonists of Plate Tectonics.

At present, no petrological data was able to demonstrate the existence of a depleted oceanic mantle lithosphere (Picazo *et al.* 2016) in the Alpine Tethys domain. Age constraints from ophiolites suggest a short-lived accretion (Li *et al.* 2013; Tribuzio *et al.* 2016), which is not consistent with a wide Alpine Tethys domain either. Moreover, the lack of arc-related magma (McCarthy *et al.* 2018, 2020), which is expected during the subduction of a wide oceanic domain, tips the scale against the existence of a mature, Pacific-type subduction. Yet, many authors have attempted to apply the Plate Tectonic "dogmas" in the Alps in the 1970s and subsequent decades, namely: (1) continental margins are composed of tilted blocks of continental crust; (2) the ocean–continent boundary (OCB) is sharp as "Penrose-type oceanic crust" starts abruptly where continental crust is thinned to zero; (3) oceanic domains are wide and (4) ocean closure is achieved via long-lasting subduction of oceanic lithosphere (Pacific-type subduction). For these authors, whatever did not fit into the Plate Tectonic template (e.g. the existence of exhumed serpentinized mantle directly overlain by deep-water sediments, the lack of a magmatic arc in the Alpine orogen) was interpreted as overprinted, destroyed or subducted.

In the Alps, the Plate Tectonics revolution prompted geologists to reinterpret stratigraphic contacts between mantle and sediments, or between mid-crustal rocks and sediments, which had been nicely described as pre-compressional prior to the 1970s (e.g. Steinmann 1905; Decandia and Elter 1969; Dietrich 1970; Trümpy 1975), mélanges or Alpine tectonic contacts (e.g. Hsü and Briegel 1991; Dürr 1992; Hsü 1995). Thus, while the integration of the Plate Tectonics theory was a huge leap forward for the understanding of oceans and subduction zones, it was a leap backward for the interpretation of field observations in the Alps.

It was only in the late 1980s and early 1990s, when serpentinized mantle rocks were drilled offshore Iberia (Boillot *et al.* 1987) and detachment systems were described in southwest United States (Wernicke 1985), that geologists started to reconsider former field observations of exhumed mantle (Florineth and Froitzheim 1994; Lagabrielle 1994) and detachment faults (Froitzheim and Eberli 1990; Froitzheim and Manatschal 1996). These concepts were reintroduced in the interpretations of Alpine Tethys remnants (Lemoine, *et al.* 1987; Froitzheim and Eberli 1990; Lagabrielle and Cannat 1990) and applied to present-day "magma-poor" rifted margins (Whitmarsh *et al.* 2001).

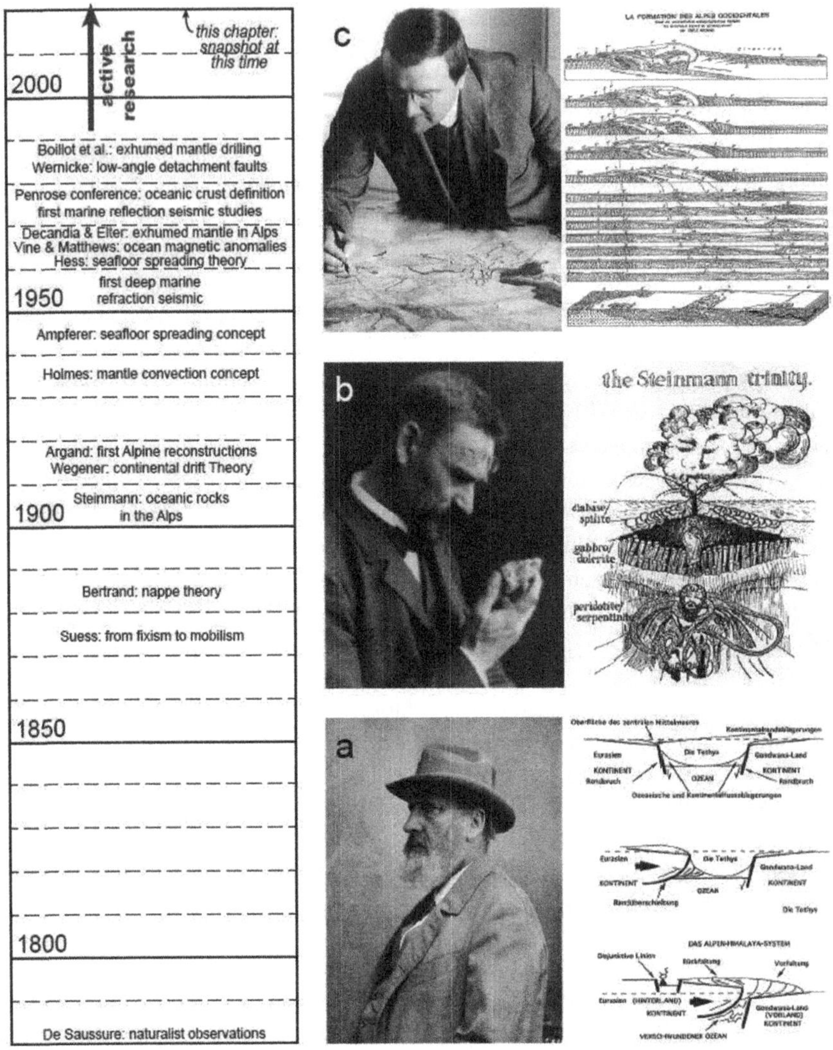

Figure 7.2. *Major players of the "Plate Tectonic revolution" in the Alps: a) Eduard Suess was the first to recognize the existence of a former basin in the Alps, which he named Tethys, and to propose a mobilistic explanation to the Alpine orogen; b) Gustave Steinmann was the first to recognize the serpentinized peridotite–"diabase"–radiolarite association mapped throughout the Alps as characteristic of a deep ocean floor; c) Emile Argand is presumably the author of the first tectonic and paleogeographic restoration ever, which he started in the Alps*

The significant progress in understanding rifted margins achieved during the last decades is intimately linked to the development of new, high-resolution seismic imaging methods and numerical modeling, and their comparison with renewed field studies (Lavier and Manatschal 2006). Seismic imaging of entire rifted margins and attempts to model the entire rifting process enabled us to integrate the local-scale and fragmented observations derived from the Alps into a modern rifted margin template. This is especially true for the Grischun transect (right-hand side of Figure 7.3b) that is described later.

7.2. From dispersed outcrops to the 2D architecture of the Alpine Tethys rift system

The Alpine Tethys margins have been accreted into the Alpine orogenic wedge during a multi-stage compressional event. When we talk about the Tethyan margins in the Alps, we thus refer to "remnants" of the former margins, now caught in the Alpine nappe stack. These remnants can be considered as pieces of a puzzle, which can be used to reconstruct the rift system template. Reconstructing a rift system based on isolated outcrops dispersed in an orogen is not easy, and there is not a unique solution. In this section, we explain how the Alpine Tethys margin template was built. In particular, we describe: (1) the datasets and how they were used to build the so-called "building blocks" (BB); (2) how the initial spatial framework was restored and how the temporal dimension was integrated in the restoration and (3) the reliability of the proposed template and what data are missing to improve and better constrain the restoration of the Alpine Tethys.

7.2.1. *Datasets and building blocks (BB)*

The Alps of Western Europe provide access to entire crustal sections, as well as upper- and lower-mantle outcrops. Extensive petrological and thermochronological data from crustal rocks and ophiolites provide insights into the conditions and timing of crustal thinning, mantle exhumation and continental breakup. There is also an important stratigraphic record, from which the tectonic and isostatic evolution of rift domains can be deduced, from the onset of rifting to the final breakup. Thus, even though the former Alpine Tethys margins are dismembered, deformed and/or partially lost due to the Alpine subduction and orogeny, they are presumably among the best-documented rifted margins in the world.

To restore the Alpine Tethys rifted margins, geologists defined what we call here "building blocks" (BB). BB are typically kilometer-scale outcrops, where primary rift relationships are preserved between basement rocks, fault structures and syn-rift stratigraphic levels. BB therefore provide information about the local: (1) nature of the basement (continental crust, magmatic crust or exhumed mantle); (2) characteristics of depositional environments (e.g. stratigraphic thickness, accommodation space, paleobathymetry and hence, basement thickness by applying isostatic models; e.g. Tugend *et al.* (2014)); and (3) geometry and age of the associated structures. In the Alps, six types of BB have been described so far. In the following, we refer to them as BB1, BB2, ... BB6 and exemplify each one of these BB with one or several type localities (Figure 7.3a).

BB1 (Figure 7.3a) displays classical fault-bounded tilted blocks framed by high-angle normal faults. The normal faults affect a thick continental basement, usually covered with continuous pre-rift sedimentary layers. The syn-tectonic deposits thicken towards the bounding faults. Crustal thinning is typically low (ß values < 1.5), and the bathymetry remains shallow to moderately deep. The type locality of BB1 is the Il Motto block in the Ortler Nappe, exposed near Livigno in northern Italy. Other examples are the Generoso Basin (Bernoulli 1964) and the Bourg d'Oisans basins (Lemoine *et al.* 1986). BB1 is the classical structure of rifted margins. It has been extensively recognized and described at both intracontinental rifts and proximal rifted margins since the 1960s.

BB2 (Figure 7.3a) shows crustal-scale shear zones responsible for major crustal thinning by juxtaposing upper and lower crustal levels via the attenuation of ductile mid-crustal layers. The best-preserved examples are the Eita shear zone and Grosina detachment fault exposed in the Grosina/Campo nappes (Mohn *et al.* 2012, 2014). The link between basement structures and the stratigraphic record is not constrained in the Grosina/Campo nappes, but studies on the Mont Blanc (western Alps; Ribes *et al.* 2020) and on the Cusio-Biellese zones (southern Alps; Beltrando *et al.* 2015; Decarlis *et al.* 2018) suggest that these zones show a particular isostatic evolution resulting in uplift, and sometimes even erosion (Chenin *et al.* 2020, 2019).

BB3 (Figure 7.3b) displays a major unconformity, sometimes named "the Briançonnais unconformity". This prominent structure was recognized throughout the Alps in the so-called Briançonnais domain. It was first identified by Lory (1860, 1866) and Kilian (1891) based on its characteristic stratigraphic position, which marks the missing Liassic and early Dogger sediments between Triassic platform carbonates and late Middle to Upper Jurassic pelagic carbonates. Haug (1909), Debelmas (1955) and Lemoine (1961) defined this unconformity as a major characteristic of the former distal European margin of the Alpine Tethys rift system. The type locality of BB3 is the Fond Froid – Lac de l'Ascension area in the

Briançonnais nappes, near Briançon in southeastern France (Claudel and Dumont 1999).

BB4 (Figure 7.3a) shows an exhumation system composed of successive in-sequence detachment faults. The detachment faults exhume deep crustal levels and can accommodate large ß-values. The exhumed basement is locally overlain with allochthonous fragments of upper crust or pre-rift sediments. These structures are sealed by syn- or post-tectonic, deep-water sediments. The type locality of BB3 is the Err-detachment system exposed in the Lower Austroalpine Err nappe, north of the Julier Pass (e.g. Froitzheim and Eberli 1990; Manatschal and Nievergelt 1997; Masini *et al.* 2012; Epin and Manatschal 2018). Another example is the Bernina detachment system (Mohn *et al.* 2012).

BB5 (Figure 7.3a) displays tectonically exhumed serpentinized mantle capped by extensional detachment faults and overlain by deep-water sediments. The nature of the mantle is subcontinental (Mantle-type 1 from Picazo *et al.* (2016)). The type locality is the Tasna ocean–continent transition (OCT) exposed north of Scuol in the Lower Engadin valley (Florineth and Froitzheim 1994; Manatschal *et al.* 2006; Ribes *et al.* 2019b). Other examples include the Upper Platta unit (Epin and Manatschal 2018), Northern Lanzo (Kaczmarek and Müntener 2008) or the External Ligurides (Marroni *et al.* 2002).

BB6 (Figure 7.3a) shows syn-tectonic magmatic additions related to mantle exhumation, as well as high- and low-angle faults affecting the infiltrated subcontinental mantle (Mantle-type 2 from Picazo *et al.* (2016)). The magma is related to incipient magma production corresponding to OCT-basalts (Amann *et al.* 2020). The type locality is the Lower Platta Unit exposed in the Sursees valley in Grischun. Other examples include the Chenaillet Ophiolite (Manatschal *et al.* 2011), Southern Lanzo (Müntener and Piccardo 2003) or the Internal Ligurides (Tribuzio *et al.* 2004).

All observations derived from field studies in the Alps are not considered in terms of BB, and yet they are important for the restoration of the former Alpine Tethys rifted margins. Examples include the discovery of a fossil Moho in the Malenco unit (Hermann *et al.* 1997) and lower crustal granulites linked to the intrusion of Permian gabbros in the Ivrea zone (Quick *et al.* 1992; Barboza *et al.* 1999) and in the Sondalo region (Petri *et al.* 2016). These outcrops enabled us to understand the pre-rift crustal structure, determine the exhumation history and assess the pressure–temperature-time (PTt) evolution of these rocks during rifting.

7.2.2. *Building a spatial and temporal template for the Alpine Tethys rifting*

There are two ways to reconstruct the Alpine Tethys margins. We can either restore the Alpine nappe stack back to its initial position, or we can use a "modern" margin section as a template. In the past, both approaches have been used. Argand (1916) was one of the first to restore the Alpine nappe stack using a mobilistic view, and describe paleogeographic domains.

Based on the first reflection seismic images in the 1970s, in which a series of "tilted blocks" (BB1) were imaged in the proximal margin (Montadert *et al.* 1979), it was assumed that the entire margin was formed by such structures. The continental margin was supposed to terminate sharply, where a classical, 7-km-thick, "Penrose-type" oceanic crust starts (Montadert *et al.* 1979). This template dictated all margin restorations in the 1970s and early 1980s. This method clearly ignored the observations made in the Penninic units of the Alps (Dietrich 1970; Trümpy 1975) and in the Apennines (Decandia and Elter 1969). One of the first to understand that margins may not only be built by BB1 were Lemoine *et al.* (1987). Froitzheim and Eberli (1990) were the first to evidence the existence of rift-related extensional detachment faults (BB4) based on field examples.

From the 1990s onwards, the structures related to the former Adriatic margin in the Grischun transect have been decrypted. Eberli (1988) and Froitzheim and Eberli (1990) demonstrated that the proximal and distal margins were controlled by different fault systems, namely pre-Sinemurian high-angle faults in the proximal margin and post-Sinemurian detachment faults in the distal margin. Lemoine *et al.* (1987) and Froitzheim and Manatschal (1996) proposed that the Alpine Tethys margins had to be asymmetric based on the mapping of extensional detachment faults (BB4). Although Decandia and Elter (1969) had already mentioned the existence of exhumed mantle, Florineth and Froitzheim (1994) and Manatschal and Nievergelt (1997) emphasized field evidence for the structures related to mantle exhumation in the Tasna and Platta nappes (BB5). Many studies had investigated the petrology of magmatic and mantle rocks (Piccardo *et al.* 1990; Rampone *et al.* 1995; 1997); however, the tectonic link between mantle exhumation along detachment faults and syn-magmatic activity was only evidenced by Manatschal and Müntener (2009) and Manatschal *et al.* (2011) for the Chenaillet ophiolite, and by Desmurs *et al.* (2001) and Epin *et al.* (2019) for the Lower Platta unit (BB6).

While many studies dealt with the peculiar nature of the Alpine ophiolites, the mechanisms of crustal thinning were poorly studied prior to the start of the 21st century. The first models suggested that crustal thinning was controlled by simple shear ('Wernicke-type', 1985) and detachment systems (e.g. Lemoine *et al.* 1987; Froitzheim and Manatschal 1996). It was only the work of Mohn *et al.* (2012) that revealed how and when the crust may have thinned. This study enabled us to identify BB2. However, to understand and integrate BB3 into the evolution of the Alpine Tethys margins required the use of numerical modeling (Chenin *et al.* 2019). Thus, it took more than two decades to find the structures and mechanisms that could explain the differing evolution of the proximal and distal margins from Sinemurian–Pliensbachian time onward (i.e. 185 ± 5 Ma), which were first described by Eberli (1988) and Froitzheim and Eberli (1990).

The discovery, mapping and characterization of the BB in the Alps went in parallel with the research at present-day rifted margins, in particular along the Iberia-Newfoundland margins (see Manatschal 2004; Manatschal *et al.* 2007) and the Pyrenean domain (Jammes *et al.* 2009; Lagabrielle *et al.* 2010). Geologists attempted to identify and describe BB all over the Alps; however, it was only in the Grischun transect that the spatial template of the different blocks could be restored by kinematic inversion of the Alpine nappe stack, because it was in the hanging wall of the Alpine subduction, and thus escaped pervasive deformation (Mohn *et al.* 2011). The key was the work done in the Austroalpine and South-Pennic nappes in Grischun by Froitzheim *et al.* (1994) and Manatschal and Nievergelt (1997), and more recently by Epin *et al.* (2017). Indeed, in the Grischun transect, five BB out of six have been identified in the nappe stack, each one in a different nappe. The restoration of the nappe stack back to the pre-Alpine stage (for details, see Epin *et al.* 2017) enabled us to put the building blocks back to a relative position, with the BB1 in the Ortler nappe belonging to the proximal margin, the BB2 in the Grosina-Campo nappe belonging to the necking zone, the BB4 in the Err-Bernina nappes belonging to the hyperextended domain, the BB5 in the Upper Platta and the BB6 in the Lower Platta unit to the proximal and distal ocean–continent transition (Figure 7.3b). The restoration of this thrust stack enabled us to assess the distribution of these BB in the margin, but not to establish the detailed structure and dimension of the margin.

For the European margins of the Alpine Tethys, a comprehensive restoration like that proposed for the Grischun margin is not possible. As identified today, the only preserved domains in this margin are the proximal parts (BB1), the distal Briançonnais domain (BB3), remnants of BB5 (Tasna) and BB6 (Chenaillet). These BB cannot be restored to their initial position from the kinematic inversion of the nappe stack only. Indeed, most of the former distal parts of the European margin were destroyed during the Alpine subduction. Moreover, the change in kinematics from N–S- to E–W-directed convergence and the polyphase orogenic overprint (e.g. Apennines) render the restoration of the nappe stack difficult (Rosenbaum *et al.* 2002).

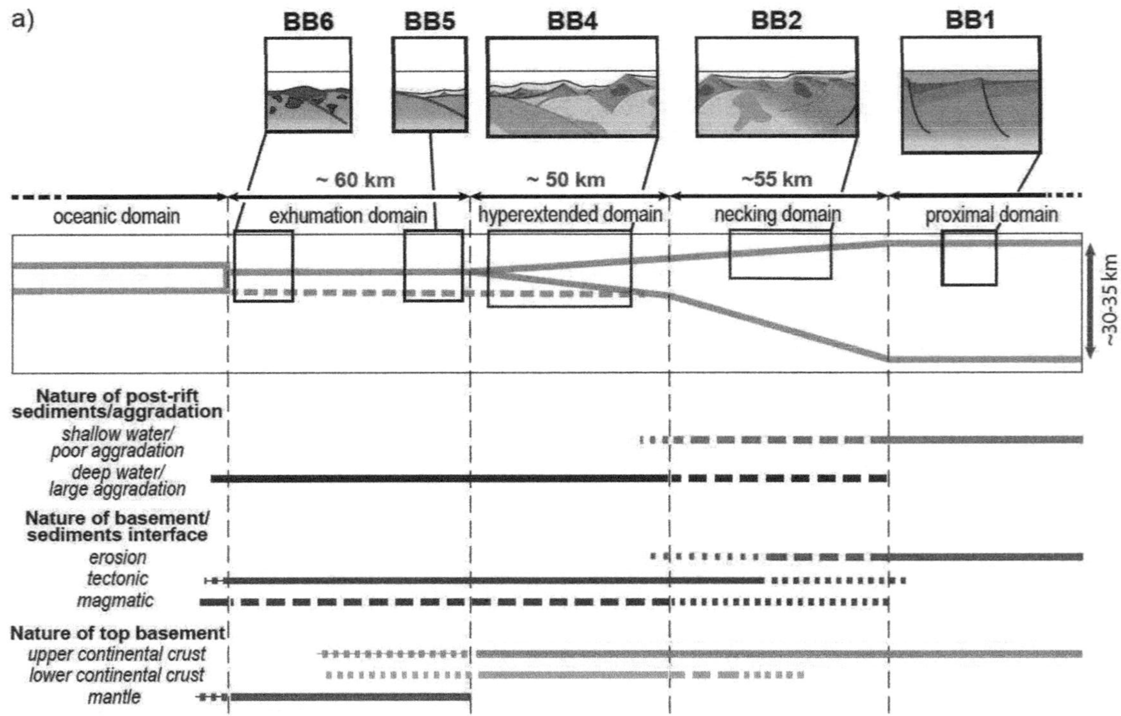

Figure 7.3a. *Schematic section of a magma-poor, lower plate rifted margin and location of the building blocks (BB) characteristic of each domain (see text for details). The natures of top basement, post-rift sediments and their interface vary depending on the domain (source: adapted from Tugend* et al. *2014)*

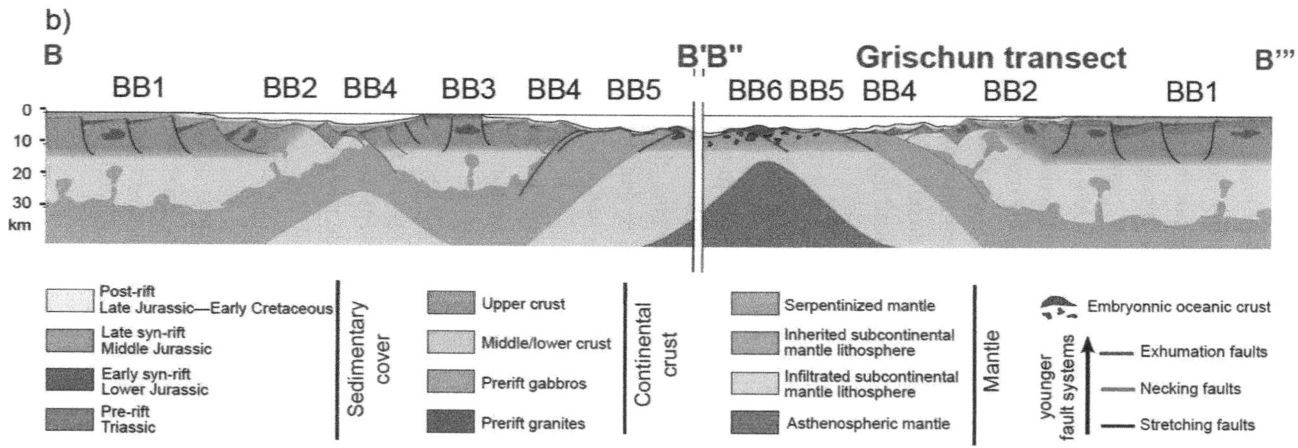

Figure 7.3b. *Synthetic reconstruction of a composite transect across the Alpine Tethys rift system (see Figure 7.1 for the location of B–B' and B''–B''' transects)*

In order to integrate time into this spatial reconstruction, we can rely on two types of data: on the one hand, petrological and thermochronological data provide insight into the PTt path of rocks; on the other hand, sedimentary deposits act as a tape recorder for the evolving deformation and depositional environments. Whenever timelines can be identified throughout the margin, they give insight into the migration of deformation through time along the margin (e.g. Ribes *et al.* 2019a). As each BB represents a "frozen" structural template with specific structural, stratigraphic, thermochronological and petrological characteristics, the comparison between the different BB characteristics provides insights into the temporal evolution of the entire rift system. This methodology enabled Masini *et al.* (2013) and Ribes *et al.* (2019a) not only to demonstrate the migration of rift activity from the future proximal margins towards the future distal margins, but also to point out the onset of rifting in the Alpine Tethys domain, the moment of necking and the timing of first magma production in the exhumed domain.

7.2.3. *What is missing?*

In order to reconstruct the Alpine Tethys margins more accurately, we miss the lateral continuity in dip and strike, as well as the dimensions of margins and their subsidence history. However, when we assume that the anatomy and isostatic evolution of former margins are comparable to that of present-day rift margins, we can propose restorations that are somehow more realistic.

For instance, Tugend *et al.* (2014) estimated the total accommodation space and crustal thinning based on the nature and thickness of sediments and the nature of the underlying top basement (Figure 7.3a). We can get these pieces of information from each BB. The biggest limitation in working in a fossil margin remains, however, the determination of the width/scale of the former margin. Chenin *et al.* (2017) measured characteristic width/scales of present-day rifted margin domains. If we assume that these width/scales are comparable to that of the Alpine margins, we can restore a generic section and replace local high resolution (BB) into a lower resolution framework (Figure 7.3a).

BB3, which is characteristic of an upper-plate margin (e.g. Haupert *et al.* 2016), is missing in the Grischun section. This, together with the polarity of faults, the location of the fault breakaways and the transport direction along the detachment faults leaves little doubt about the Grischun section being a lower-plate margin (Epin and Manatschal 2018).

Based on all of these studies and using the new approaches developed in modern margins, better and more realistic templates have been built for the Alpine Tethys margins (Figure 7.3b). Moreover, upper and lower plates were recognized for each segment of the Alpine Tethys rift system (Decarlis *et al.* 2017).

Except for the Grischun section, which was reconstructed based on the restoration of Alpine structures (for explanations of the restoration, see Epin *et al.* (2017)), all other sections through the former Alpine Tethys margins were reconstructed based on a margin template. Thus, the Grischun section is at present, unique, and can be considered the best documented fossil margin worldwide. Additional work is necessary to improve the details of the other restored sections. Geologists are also currently missing more reliable kinematic models, which would enable them to refine the spatio-temporal evolution of the Alpine Tethys system.

7.3. Evolution of the Alpine Tethys rift system

Petrological studies and dating of rocks from all pre-rift lithospheric levels in the Alps enabled us to establish PTt datasets throughout the former Alpine Tethys margins. Their combination with stratigraphic and structural data, as well as information from magmatic rocks and hydrothermal fluids from each of the BB enabled us to get insights into the polyphase rifting history of the Alpine Tethys rift system. The Alpine Tethys rifting (Figure 7.4) initiated with (1) distributed stretching during the earliest Jurassic (200–185 Ma), leading to (2) necking at 185 ± 5 Ma, (3) hyperextension from ca. 180 to 165 Ma and finally, (4) subcontinental mantle exhumation and onset of magmatic activity. Details about each of these characteristic steps are given below:

During the stretching phase from 200 to 185 Ma (e.g. the "first phase of rifting" in Froitzheim and Eberli (1990); Figure 7.4a), extension was distributed over a wide area (from the Cevennes in France to the Trento-plateau in Northern Italy) with the development of many rift basins bounded by listric normal faults that soled out in the middle crust (e.g. Bertotti 1991). The total amount of extension accommodated during this event may have been minor and, even though some basins may have accommodated several kilometers of thick syn-tectonic sedimentary deposits, the overall crustal thinning remained moderate and widely distributed.

During the necking phase at 185 ± 5Ma (i.e. the transition between the "first" and "second phases of rifting" in Froitzheim and Eberli (1990); Figure 7.4b), extension localized abruptly into the future distal margin. At this stage, there is evidence for local uplift and major unconformities formed in the future distal parts of the Alpine Tethys margins and at the transition between the future proximal and distal parts of margins. Based on numerical modeling, we assume that the necking of the upper lithospheric mantle (usually the strongest layer within the lithosphere) occurred prior to significant crustal thinning (Chenin *et al.* 2018b). The flexural rebound related to the rupture of the upper mantle associated with the buoyancy of the overlying little-thinned continental crust may explain the uplift of the future distal part and the formation of the so-called necking unconformity (Chenin *et al.* 2019).

During the hyperextension phase between 180 and 165 Ma (Figure 7.4c), which is well documented in the Lower Austroalpine units and in the Pre-Piemontais units, the residual thin crust was dismembered by extensional detachment faults that propagated in-sequence towards the future site of breakup. This phase is the only one where deformation was markedly asymmetrical and resulted in the differentiation between an upper- and lower-plate margin. At this stage, the entire margin was subsiding. While the proximal parts, which were tectonically inactive, subsided thermally, the distal parts subsided because of tectono-thermal processes.

The successive in-sequence detachment faults that developed during the hyperextension phase rooted in the mantle and eventually led to the exhumation of mantle to the seafloor (Figure 7.4d). There is widespread evidence for this so-called "exhumation phase" in the Alps. Recent work showed that the first exhumed mantle is the subcontinental lithospheric mantle (Mantle-type 1 from Picazo *et al.* (2016)). In more advanced stages of exhumation, deeper parts of the subcontinental mantle can be exhumed (Mantle-type 2 from Picazo *et al.* (2016)). This type of mantle shows two important features: i) it was depleted during the Permian (e.g. Rampone *et al.* 1998; Müntener *et al.* 2004; McCarthy and Müntener 2015) and/or showed inherited depleted heterogeneities (Tribuzio *et al.* 2004; Sanfilippo *et al.* 2019); and ii) it was infiltrated during the subsequent rifting (e.g. Rampone *et al.* 1997; Rampone, *et al.* 2020; Piccardo *et al.* 2004, 2007; Müntener *et al.* 2010). Mantle exhumation was associated with minor magmatic activity, sometimes interpreted as embryonic seafloor spreading (Jagoutz *et al.* 2007; Manatschal *et al.* 2011; Lagabrielle *et al.* 2015; Epin *et al.* 2019). The magma-poor nature of the Alpine Tethys may have originated from Permian depletion (Chenin *et al.* 2017), which was linked to a massive magmatic event throughout W-Europe following the Variscan orogenic collapse (see Bonin *et al.* 1998; Petri *et al.* 2017 and the references therein). The Jurassic rift-related infiltration may have been protracted, with a magmatic plumbing system that was built from the base of the lithosphere to the top. It is assumed to have initiated with asthenospheric melt infiltration at ca. 185 Ma and to have reached the surface with first the magma extrusions (basalts) and shallow intrusions (gabbros) at about 165 Ma. Recent studies on the first basalts (Amann *et al.* 2020; Renna *et al.* 2018) showed a complex magmatic evolution during the mantle exhumation phase, partially controlled by the inherited mantle lithosphere composition. At present, there is no evidence that the Alpine Tethys developed a mature, steady-state spreading system (Picazo *et al.* 2016).

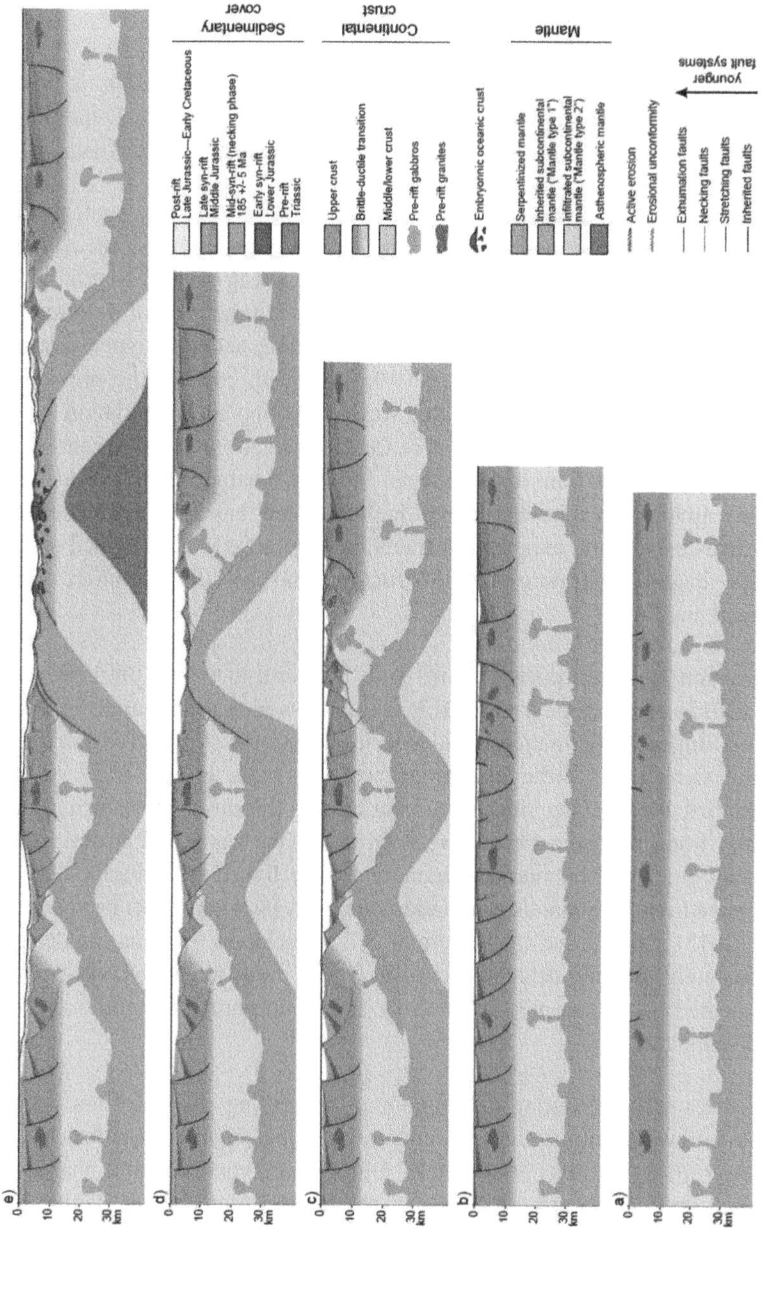

Figure 7.4. *Main phases of the Alpine Tethys rifting: a) initial stage; b) stretching phase; c) necking phase; d) hyperextension phase; and e) mantle exhumation phase and embryonic seafloor spreading. See text for details*

7.4. Continuous deformation vs. succession of distinct deformation phases

Whether rift systems are formed through progressive deformation or via a succession of distinct deformation events is a longstanding debate. Trends have recently been identified from stratigraphic, fluid, magmatic and structural data.

The stratigraphic tape recorder shows evidence for an abrupt switch from distributed to localized deformation at the Sinemurian/Pliensbachian (185 ± 5 Ma) based mainly on ammonite stratigraphy (Lemoine *et al.* 1986; Eberli 1988; Froitzheim and Eberli 1990). More recent studies show a migration of deformation that can be described by four successive tracts (stretching, necking, hyperextension and exhumation tracts; see Ribes *et al.* (2019a)). Indeed, each of the previously described phases is related to a characteristic way of creating accommodation space, that is, space below sea level that is available for sedimentation. The stretching tract is linked to the accumulation of sediments showing characteristic syn-tectonic wedge-shaped architectures in distributed fault-bounded basins. The subsequent necking, hyperextension and exhumation tracts have been little investigated so far, but may show diagnostic features of the structural and isostatic evolution of the Alpine Tethyan margins.

Little attention has been paid to fluids and their link to rifting until recently. Pinto *et al.* (2015) and Incerpi *et al.* (2017, 2020) showed that the nature of fluids was intimately linked to the hydrothermal activity during the Alpine Tethys rifting, and that fluid signature is different during each rift phase. Incerpi *et al.* (2020) demonstrated that the necking phase was marked by the onset of Si-rich fluids, which resulted from the interaction between fresh surface water and crustal rocks. Hyperextension and onset of mantle exhumation were linked with marine-derived fluids that reacted with the mantle that underlaid a thin (< 6-km-thick) brittle curst. Pinto *et al.* (2015) showed that hydrothermal cells developed from marine-derived fluids that channelized along detachment faults. These fluids interacted with mantle rocks, as indicated by the enrichment in V, Cr and Ni in fault rocks and overlying syn-tectonic sediments.

The structures accommodating deformation during rifting changed from mainly listric, concave upward normal faults to concave downward extraction/detachment faults (Mohn *et al.* 2012; Petri *et al.* 2019). During the hyperextension phase, the dominant fault systems were in-sequence extensional detachment faults that migrated towards the future ocean and delaminated allochthonous blocks from the hanging-wall (upper-plate) block (e.g. Epin and Manatschal 2018). During mantle exhumation and the onset of magmatic activity, the system showed an interplay between exhumation and high-angle faulting (Manatschal *et al.* 2011). This deformation phase, which is observed in the Platta nappe (Epin *et al.* 2019), may approach that observed at slow-spreading mid-ocean ridges (Sauter *et al.* 2013; Cannat *et al.* 2019).

Magmatic systems show an evolution from subcontinental mantle infiltration prior to the mantle exhumation phase, to magma emplacement as small dykes, gabbroic bodies and sporadic volcanic flows during mantle exhumation and finally, to post-tectonic alkaline magmatic activity. Although the age of the first mantle infiltration is difficult to determine, it is believed to have occured at ca. 180 Ma (Picazo *et al.* 2016). Classical MORB magmatism forming basalt and gabbro is documented between 166 and 158 Ma at the Chenaillet ophiolite (Li *et al.* 2013) and 165–161 Ma for the Ligurian ophiolites (Tribuzio *et al.* 2016). More recently, Epin *et al.* (2019) and Amann *et al.* (2020) evidenced post-rift alkaline magmatic activity manifested as sills and dykes within Cretaceous sediments in the Lower Platta unit.

These observations suggest that, although extension may have been continuous in the Alpine Tethys Realm, different rift phases can be clearly identified and linked to characteristic stratigraphic system tracts, fluid systems, extension styles and magmatic systems.

7.5. Three-dimensional aspect of the Alpine Tethys rift system

Our description of the Alpine Tethys rift system and of its evolution is mainly based on the central Piemonte segment of the Alpine Tethys (Figure 7.1a). The corresponding data derives from the Austroalpine and Penninic units in the Central and Western Alps. However, this 2D view underscores the 3D complexity of the Alpine Tethys, which can be regarded from different scale perspectives.

From a large-scale perspective, the Alpine Tethys is only one out of at least three rift systems, referred to as *Meliata/Vardar/E-Mediterranean*, *Alpine Tethys* and *Iberia/Pyrenees/Bay of Biscay* (Figure 7.1a). These three rift systems accommodated extension and transtension between the European, African and North American plates from Late Triassic to Late Cretaceous time. The partitioning of extension between these rift systems remains little constrained, since no undisputed data exist on the width, rate or the detailed kinematic movements of any of these systems. Our preferred view is that the Alpine Tethys rifting initiated during the earliest Jurassic and stopped in the Late Jurassic time as a failed, approximately 300- to 400-km-wide proto-oceanic domain (Li *et al.* 2013). The width of each margin might have been of the order of 165 km, which is the average width of magma-poor rifted margins according to Chenin *et al.* (2017). Note, however, that the width of the Alpine Tethys rift system may have largely varied laterally, as a result of its complex architecture initially segmented into several independent rift basins. Moreover, most studies that attempt to reckon the width of the Alpine rift system are either limited to one rift basin (e.g. the Ligurian, Piemonte or Valais sub-basin) or do not clearly state what that width represents (does it include (parts of) the conjugate margins, or is it limited to the exhumed mantle and/or "oceanic" domain?). Finally, as a basin width is defined in a dip direction, and not in a strike direction, the kinematic direction of opening needs to be defined.

At the scale of the Alpine Tethys, we propose here a subdivision into three segments, the *Ligurian*, *Piemonte* and *Valais* basins (Figure 7.1a). Thermochronological and stratigraphic data suggest that these basins may have started to individualize during the Sinemurian/Pliensbachian and may have merged into a unique rift system, so-called Alpine Tethys, during the Callovian/Bathonian (Baumgartner 2013). However, many questions remain open, especially about the southern Ligurian domain. To date, it remains unclear whether this basin had a Triassic precursor basin and was linked to the eastern Mediterranean, as suggested by its thick and locally deep Triassic sequences (Speranza *et al.* 2012).

Another debate is the origin and age of the Valais branch of the Alpine Tethys. While Frisch (1979), Stampfli *et al.* (2002) and Handy *et al.* (2010) suggested a Cretaceous origin and a link to the Iberian-Biscay system, recent studies by Galster *et al.* (2012), Tavani *et al.* (2018) and Ribes *et al.* (2019b) clearly show that this correlation is neither supported by kinematic data nor by geological observations. They suggest that the Valais domain formed in Jurassic time, simultaneous with the Piemonte domain. Thus, there are no reasons for this debate to persist. The detailed interactions between the Valais and Piemonte domains are yet to be clarified. Here, we advocate for a segmented Alpine Tethys rift system with relay ramps, similar to what has been proposed for the Cretaceous basins in the Pyrenean domain (Lescoutre and Manatschal 2020). The formation of continental ribbons such as the Briançonnais, separating two V-shaped hyperextended domains that were set up as relay ramps and merged during breakup, may be a reasonable explanation. This hypothesis is yet to be confirmed by field data.

Regardless of the actual architecture of what the Alpine Tethys looked like, it was clearly not cylindrical. The relationships between the Ligurian and Piemonte segments of the rift system appear, in contrast, simpler and seem to coincide with the Peri-Adriatic strike slip/transfer system. Neither the southwestern nor the northeastern terminations of the Alpine Tethys are well understood to date.

At the scale of individual segments through the former Alpine Tethys rift system, each transect is composed of conjugate upper- and lower-plate margins. BB3 is the typical fingerprint of an upper-plate margin (e.g. Haupert *et al.* 2016). Decarlis *et al.* (2017b) showed that, along the Adriatic margin, a flip from an upper- to a lower-plate margin occurred across the Peri-Adriatic system. This interpretation is fundamental for the kinematic restoration and the definition of rift segments in the Alpine Tethys. Following this interpretation, we suggest that the Austroalpine/Carpatian segment is conjugate to the Calabria/Sardinia-Western Alps segment and that the Grischun section is conjugate to a section at present in Calabria/Sardinia. However, more research is needed to confirm these correlations.

The complex paleogeography of the former Alpine Tethys rift system is not very different from that observed at many Atlantic margins. The paleogeography of the Alpine Tethys rift system has important implications on how and where sediments were transported in the deep margins (e.g. Ribes *et al.* 2019a). The segmented paleogeography of the Alpine Tethys explains why the Alpine terminology is so complex and the paleogeographic domains cannot be correlated along the Alps in a simple way. Thus, most of the controversies based on correlations between Alpine units and domains may be solved, or at least better defined, when the complex pre-Alpine geography is better understood.

7.6. The Alpine Tethys remnants in the Alps: a laboratory to investigate and calibrate rift systems

7.6.1. *The role of rift-inheritance in building collisional orogens*

The Alpine Tethys has been described as a trace of a lost ocean in an orogen (Popier 1991). New research results show, however, that it is much more the remnants of margins that are archived in the Alpine orogen than the remnants of a true ocean. This questions how rift structures control the architecture of the orogen upon closure, and to what extent the fate of an orogen depends on its former life as a rift system and/or an ocean. This question has not yet been addressed in many studies. It appears that the width of the oceanic domain is important and can determine the subsequent orogenic evolution (Sengör 1991; Chenin *et al.* 2017; Chenin *et al.* 2018; McCarthy *et al.* 2018, 2020). For narrow and magma-poor oceanic domains such as the Alpine Tethys, the closure and subsequent collision are essentially mechanical processes. Thus, the rift architecture, in particular the location of decoupling levels, the distribution of buttresses and the paleogeographical complexity of the rift system involved, fundamentally controls the architecture of the orogen. Future studies are necessary to progress and better integrate rift inheritance in orogenic studies. The Alps are certainly among the best candidates to play a leadership role in this research.

7.6.2. *Calibration of rifting processes*

Research on rifted margins notoriously lack data, in particular in deep water distal domains. While many high-resolution 3D seismic blocks of distal margins became recently available to researchers and although numerical modeling is able to reproduce some of the key rift structures in 2D and/or 3D, the access to samples from these domains remains the biggest problem. Deep-sea drilling is expensive and remains exceptional. Therefore, a major asset for research on rifted margins in the Alps is the possibility to put specific outcrops into the framework of a present-day rifted margin. The access to analogue rock data from the field (ages, PTt information, chemistry, etc.) enables geologists to test and calibrate assumptions made at present-day rift systems.

Using Alpine Tethys remnants to calibrate present-day rift systems will require us to change our way of working with Alpine datasets. In the last two decades, the Alpine Tethys rift system was restored in a way that is compatible with observations at present-day rifted margins (e.g. Mohn *et al.* 2010; Masini *et al.* 2013; Ribes *et al.* 2019a). Current and future research (will) attempt to better define the 3D spatial framework and temporal evolution of rifting processes. This includes the analysis of vertical and horizontal movements during rifting, the characterization of the different deformation events and the recognition of depositional systems.

Hyperextended rift systems where mantle is exhumed provide first-order sampling of the lithosphere, from the upper to the lower crust and from the upper to the lower lithospheric mantle. Moreover, magmatic additions emplaced during mantle exhumation or early seafloor spreading provide information on the interactions between the asthenosphere and the lithospheric mantle. Last but not least, the hydrothermal fluids interacting with the crust, mantle and sediments during rifting are not only efficient ways to transport elements from one reservoir to the other, but may also provide access to new resources and fundamental insights into the origins of life.

7.6.3. *The benefit of data-based scientific debate*

When Alpine geologists proposed nappe tectonics; when Gustav Steinmann (1905) suggested that the association of serpentinized peridotite, “diabase” and radiolarite, mapped throughout the Alps, was characteristic of a deep ocean floor;

when Emile Argand (1911, 1916, 1924a, 1924b) reconstructed paleogeographic domains; when Decandia and Elter (1969) proposed mantle exhumation in the Alpine Tethys and when Bernoulli (1964) proposed that fault-bounded extensional basins represent the legacy of a former rifted margin, they all created "alternative" views that conflicted with the well-established and generally accepted views. The legacy of all of the debates fought with so much passion by Alpine geologists taught us that research can only advance where debates exist. The success of the Alps, and the Alpine Tethys as a concept, comes from the coexistence of different generations of researchers and schools of thought using one and the same dataset, but not agreeing on one and the same model. In this sense, we do not consider the view of the Alpine Tethys we presented in this chapter as definitive, but as a snapshot of ongoing research.

FURTHER READING.– The above descriptions are abbreviated and often simplified. If interested in reading and learning further, the reader is referred to the following list of publications and references.

– A selection of the most influential papers: (Froitzheim and Eberli 1990; Lemoine *et al.* 1987; Manatschal 2004).

– Historical review papers: (Şengör 1990; Trümpy 2003).

– A selection of papers at the cutting edge of present-day research on rift systems:

- on mantle composition and its evolution: (Picazo *et al.* 2016);

- on necking zones and necking processes: (Chenin *et al.* 2019; Mohn *et al.* 2014);

- on ocean–continent transitions and ophiolites: Manatschal and Müntener 2009);

- on the tectono-stratigraphic evolution: (Ribes *et al.* 2019a).

7.7. References

Amann, M., Ulrich, M., Manatschal, G., Pelt, E., Epin, M.E., Autin, J., Sauter, D. (2020). Geochemical characteristics of basalts related to incipient oceanization: The example from the Alpine-Tethys OCTs. *Terra Nov.*, 32, 75–88. https://doi.org/10.1111/ter.12438.

Ampferer, O. and Hammer, W. (1911). Geologischer Querschnitt durch die Ostalpen vom Allgäu zum Gardasee. *Jahrb. der Kais. Geol. Reichsanstalt*, 61, 532–710.

Argand, E. (1911). Les nappes de recouvrement des Alpes pennines et leurs prolongements structuraux. *Mater. Carte géol. Suisse*, N.S., XXXI.

Argand, E. (1916). *Sur l'arc des Alpes Occidentales.* Bridel, Lausanne.

Argand, E. (1924a). Les Alpes et l'Afrique. *Bull. Société vaudoise des Sci. Nat.*, 55.

Argand, E. (1924b). La tectonique de l'Asie. *13th International Geological Congress*, Brussels, 171–372.

Barboza, S.A., Bergantz, G.W., Brown, M. (1999). Regional granulite facies metamorphism in the Ivrea zone: Is the Mafic Complex the smoking gun or a red herring? *Geology*, 27, 447–450. https://doi.org/10.1130/0091-7613(1999)027<0447:rgfmit>2.3.co;2.

Barnett-Moore, N., Hosseinpour, M., Maus, S. (2016). Assessing discrepancies between previous plate kinematic models of Mesozoic Iberia and their constraints. *Tectonics*, 35, 1843–1862. https://doi.org/10.1002/2015TC004019.

Baumgartner, P.O. (2013). Mesozoic radiolarites – Accumulation as a function of sea surface fertility on Tethyan margins and in ocean basins. *Sedimentology*, 60, 292–318. https://doi.org/10.1111/sed.12022.

Beltrando, M., Stockli, D.F., Decarlis, A., Manatschal, G. (2015). A crustal-scale view at rift localization along the fossil Adriatic margin of the Alpine Tethys preserved in NW Italy. *Tectonics*, 34, 1927–1951. https://doi.org/10.1002/2015TC003973.

Bernoulli, D. (1964). Zur Geologie des Monte Generoso (Lombardische Alpen). *Beitr. Geol. Karte Schweiz*, N.F., 118, 1–134.

Bertotti, G. (1991). Early Mesozoic extension and Alpine shortening in the western Southern Alps: The geology of the area between Lugano and Menaggio (Lombardy, Northern Italy). *Mem. Soc. Geol. Padova*, 43, 17–123.

Boillot, G., Recq, M., Winterer, E.L., Meyer, A.W., Applegate, J., Baltuck, M., Bergen, J.A., Comas, M.C., Davies, T.A., Dunham, K., Evans, C.A., Girardeau, J., Goldberg, G., Haggerty, J., Jansa, L.F., Johnson, J.A., Kasahara, J., Loreau, J.P., Luna-Sierra, E., Moullade, M., Ogg, J., Sarti, M., Thurow, J., Williamson, M. (1987). Tectonic denudation of the upper mantle along passive margins: A model based on drilling results (ODP leg 103, western Galicia margin, Spain). *Tectonophysics*, 132, 335–342. https://doi.org/10.1016/0040-1951(87)90352-0.

Bonin, B., Azzouni-Sekkal, A., Bussy, F., Ferrag, S. (1998). Alkali-calcic and alkaline post-orogenic (PO) granite magmatism: Petrologic constraints and geodynamic settings. *Lithos*, 45, 45–70. https://doi.org/10.1016/S0024-4937(98)00025-5.

Cannat, M., Sauter, D., Lavier, L., Bickert, M., Momoh, E., Leroy, S. (2019). On spreading modes and magma supply at slow and ultraslow mid-ocean ridges. *Earth Planet. Sci. Lett.*, 519, 223–233. https://doi.org/10.1016/j.epsl.2019.05.012.

Chenin, P., Manatschal, G., Picazo, S., Müntener, O., Karner, G.D., Johnson, C., Ulrich, M. (2017). Influence of the architecture of magma-poor hyperextended rifted margins on orogens produced by the closure of narrow versus wide oceans. *Geosphere*, 13, 1–18. https://doi.org/10.1130/GES01363.1.

Chenin, P., Picazo, S., Jammes, S., Manatschal, G., Müntener, O., Karner, G. (2018a). Potential role of lithospheric mantle composition in the Wilson cycle: A North Atlantic perspective. In *Fifty Years of the Wilson Cycle Concept in Plate Tectonics*, Wilson, R.W., Houseman, G.A., Mccaffrey, K.J.W., Doré, A.G., Buiter, S.J.H. (eds). Geological Society, London, Special Publications. https://doi.org/10.1144/SP470.10.

Chenin, P., Schmalholz, S.M., Manatschal, G., Karner, G.D. (2018b). Necking of the lithosphere: A reappraisal of basic concepts with thermo-mechanical numerical modeling. *J. Geophys. Res. Solid Earth*, 123, 5279–5299. https://doi.org/10.1029/2017JB014155.

Chenin, P., Manatschal, G., Decarlis, A., Schmalholz, S.M., Duretz, T., Beltrando, M. (2019). Emersion of distal domains in advanced stages of continental rifting explained by asynchronous crust and mantle necking. *Geochem., Geophys. Geosyst.*, 2019GC008357. https://doi.org/10.1029/2019GC008357.

Chenin, P., Schmalholz, S.M., Manatschal, G., Duretz, T. (2020). Impact of crust–mantle mechanical coupling on the topographic and thermal evolutions during the necking phase of "magma-poor" and "sediment-starved" rift systems: A numerical modeling study. *Tectonophysics*, 228472. https://doi.org/10.1016/j.tecto.2020.228472.

Claudel, M.E. and Dumont, T. (1999). A record of multistage continental break-up on the Briançonnais marginal plateau (Western Alps): Early and Middle-Late Jurassic rifting. *Eclogae Geol. Helv.*, 92, 45–61.

Debelmas, J. (1955). Les zones subbriançonnaise et briançonnaise occidentale entre Vallouise et Guillestre (Hautes-Alpes). *Revue de géographie alpine*, 45(2), 423–425; https://www.persee.fr/doc/rga_0035-1121_1957_num_45_2_1908_t1_0423_0000_2.

Decandia, F.A. and Elter, P. (1969). Riflessioni sul problema delle ofioliti nell'Appennino settentrionale (nota preliminare). *Atti della Soc. Toscana di Sci. Nat.*, Mem. A, 76, 1(1–9).

Decarlis, A., Beltrando, M., Manatschal, G., Ferrando, S., Carosi, R. (2017). Architecture of the distal piedmont-Ligurian rifted margin in NW Italy: Hints for a flip of the rift system polarity. *Tectonics*, 36, 2388–2406. https://doi.org/10.1002/2017TC004561.

Decarlis, A., Gillard, M., Tribuzio, R., Epin, M.E., Manatschal, G. (2018). Breaking up continents at magma-poor rifted margins: A seismic vs. outcrop perspective. *J. Geol. Soc. London*, 175(6), 875–882.

Desmurs, L., Manatschal, G., Bernoulli, D. (2001). The Steinmann trinity revisited: Exhumation and magmatism along an ocean-continent transition: The Platta nappe, eastern Switzerland. In *Non-Volcanic Rifting of Continental Margins: A Comparison of Evidence From Land and Sea*, Wilson, R.C.L., Withmarsh, R.B., Taylor, B., Froitzheim, N. (eds). Geological Society of London Special Publication.

Dietrich, V. (1970). Die Stratigraphie der Platta-Decke: Fazielle Zusammenhänge zwischen Oberpenninikum und Unterostalpin. Geologisches Institut der Eidg, Technischen Hochschule und der Universität Zürich, Zurich.

Dürr, S. (1992). Structural history of the Arosa Zone between Platta and Err Nappes east of Marmorera (Grisons): Multi-phase deformation at the Penninic-Austroalpine Plate boundary. *Eclogae Geol. Helv.*, 85, 361–374.

Eberli, G.P. (1988). The evolution of the southern continental margin of the Jurassic Tethys Ocen as recorded in the Allgäu Formation of the Austroalpine Nappes of Graubünden (Switzerland). *Eclogae Geol. Helv.*, 81, 175–214.

Epin, M.E. and Manatschal, G. (2018). Three-dimensional architecture, structural evolution, and role of inheritance controlling detachment faulting at a hyperextended distal margin: The example of the Err detachment system (SE Switzerland). *Tectonics*, 37, 4494–4514. https://doi.org/10.1029/2018TC005125.

Epin, M.E., Manatschal, G., Amann, M. (2017). Defining diagnostic criteria to describe the role of rift inheritance in collisional orogens: The case of the Err-Platta nappes (Switzerland). *Swiss J. Geosci.*, 110, 419–438. https://doi.org/10.1007/s00015-017-0271-6.

Epin, M.-E., Manatschal, G., Amann, M., Ribes, C., Clausse, A., Guffon, T., Lescanne, M. (2019). Polyphase tectono-magmatic evolution during mantle exhumation in an ultra-distal, magma-poor rift domain: Example of the fossil Platta ophiolite, SE Switzerland. *Int. J. Earth Sci.* https://doi.org/10.1007/s00531-019-01772-0.

Florineth, D. and Froitzheim, N. (1994). Transition from continental to oceanic basement in the Tasna Nappe: Evidence for Early Cretaceous opening of the Valais Ocean. *Schweiz. Miner. Petrogr. Mitt.*, 74, 437–448.

Frisch, W. (1979). Tectonic progradation and plate tectonic evolution of the Alps. *Tectonophysics*, 60, 121–139. https://doi.org/10.1016/0040-1951(79)90155-0.

Frizon de Lamotte, D., Fourdan, B., Leleu, S., Leparmentier, F., Clarens, P. (2015). Style of rifting and the stages of Pangea breakup. *Tectonics*, 34, 1–21. https://doi.org/10.1002/2014TC003760.

Froitzheim, N. and Eberli, G.P. (1990). Extensional detachment faulting in the evolution of a Tethys passive continental margin, Eastern Alps, Switzerland. *Geol. Soc. Am. Bull.*, 102, 1297–1308. https://doi.org/10.1130/0016-7606(1990)102.

Froitzheim, N. and Manatschal, G. (1996). Kinematics of Jurassic rifting, mantle exhumation, and passive-margin formation in the Austroalpine and Penninic nappes (eastern Switzerland). *Bull. Geol. Soc. Am.*, 108, 1120–1133. https://doi.org/10.1130/0016-7606(1996)108<1120:KOJRME>2.3.CO;2.

Froitzheim, N., Schmid, S.M., Conti, P. (1994). Repeated change from crustal shortening to orogen-parallel extension in the Austroalpine units of Graubunden. *Eclogae Geol. Helv.*, 87, 559–612.

Galster, F., Cavargna-Sani, M., Epard, J.L., Masson, H. (2012). New stratigraphic data from the Lower Penninic between the Adula nappe and the Gotthard massif and consequences for the tectonics and the paleogeography of the Central Alps. *Tectonophysics*, 579, 37–55. https://doi.org/10.1016/j.tecto.2012.05.029.

Handy, M.R., Schmid, S.M., Bousquet, R., Kissling, E., Bernoulli, D. (2010). Reconciling plate-tectonic reconstructions of Alpine Tethys with the geological-geophysical record of spreading and subduction in the Alps. *Earth-Science Rev.*, 102, 121–158. https://doi.org/10.1016/j.earscirev.2010.06.002.

Haug, E. (1909). *Les géosynclinaux de la chaîne des Alpes pendant les temps secondaires*. Gauthier-Villars.

Haupert, I., Manatschal, G., Decarlis, A., Unternehr, P. (2016). Upper-plate magma-poor rifted margins: Stratigraphic architecture and structural evolution. *Mar. Pet. Geol.*, 69, 241–261. https://doi.org/10.1016/j.marpetgeo.2015.10.020.

Hermann, J., Müntener, O., Trommsdorff, V., Hansmann, W., Piccardo, G.B. (1997). Fossil crust-to-mantle transition, Val Malenco (Italian Alps). *J. Geophys. Res.*, 102, 20123. https://doi.org/10.1029/97JB01510.

Hsü, K.J. (1995). *The Geology of Switzerland: An Introduction to Tectonic Facies*. Princeton University Press, Princeton.

Hsü, K.J. and Briegel, U. (1991). *Geologie der Schweiz*. Springer, Switzerland.

Incerpi, N., Martire, L., Manatschal, G., Bernasconi, S.M. (2017). Evidence of hydrothermal fluid flow in a hyperextended rifted margin: The case study of the Err nappe (SE Switzerland). *Swiss J. Geosci.*, 110, 439–456. https://doi.org/10.1007/s00015-016-0235-2.

Incerpi, N., Martire, L., Manatschal, G., Bernasconi, S.M., Gerdes, A., Czuppon, G., Palcsu, L., Karner, G.D., Johnson, C.A., Figueredo, P.H. (2020). Hydrothermal fluid flow associated to the extensional evolution of the Adriatic rifted margin: Insights from the pre- to post-rift sedimentary sequence (SE Switzerland, N Italy). *Basin Res.*, 32, 91–115. https://doi.org/10.1111/bre.12370.

Jagoutz, O., Müntener, O., Manatschal, G., Rubatto, D., Peron-Pinvidic, G., Turrin, B.D., Villa, I.M. (2007). The rift-to-drift transition in the North Atlantic: A stuttering start of the MORB machine? *Geology*, 35, 1087–1090. https://doi.org/10.1130/G23613A.1.

Jammes, S., Manatschal, G., Lavier, L., Masini, E. (2009). Tectonosedimentary evolution related to extreme crustal thinning ahead of a propagating ocean: Example of the western Pyrenees. *Tectonics*, 28, TC4012. https://doi.org/10.1029/2008TC002406.

Kaczmarek, M.-A. and Müntener, O. (2008). Juxtaposition of melt impregnation and high-temperature shear zones in the upper mantle: Field and petrological constraints from the Lanzo Peridotite (Northern Italy). *J. Petrol.*, 49, 2187–2220. https://doi.org/10.1093/petrology/egn065.

Kilian, W. (1891). *Sur quelques Céphalopodes nouveaux ou peu connus de la période secondaire. 2 brochures*. F. Allier père et fils, Grenoble.

Lagabrielle, Y. (1994). Ophiolites of the southwestern Alps and the structure of the Tethyan oceanic lithosphere. *Ofioliti*, 19, 413–434.

Lagabrielle Y. and Cannat, M. (1990). Alpine Jurassic ophiolites resemble the modern central Atlantic basement. *Geology*, 18, 319–322. doi: 10.1130/0091-7613.

Lagabrielle, Y., Labaume, P., de Saint Blanquat, M. (2010). Mantle exhumation, crustal denudation, and gravity tectonics during Cretaceous rifting in the Pyrenean realm (SW Europe): Insights from the geological setting of the lherzolite bodies. *Tectonics*, 29. https://doi.org/10.1029/2009TC002588.

Lagabrielle, Y., Vitale Brovarone, A., Ildefonse, B. (2015). Fossil oceanic core complexes recognized in the blueschist metaophiolites of Western Alps and Corsica. *Earth-Science Rev.* https://doi.org/10.1016/j.earscirev.2014.11.004.

Lavier, L.L. and Manatschal, G. (2006). A mechanism to thin the continental lithosphere at magma-poor margins. *Nature*, 440, 324–328. https://doi.org/10.1038/nature04608.

Lemoine, M. (1961). La marge externe de la fosse piémontaise dans les Alpes occidentales. *Rev. Géogr. Phys. Géol. Dyn.*, 2, 163–180.

Lemoine, M., Bas, T., Arnaud-Vanneau, A., Arnaud, H., Dumont, T., Gidon, M., Bourbon, M., de Graciansky, P.C., Rudkiewicz, J.L., Megard-Galli, J., Tricart, P. (1986). The continental margin of the Mesozoic Tethys in the Western Alps. *Mar. Pet. Geol.*, 3, 179–199. https://doi.org/10.1016/0264-8172(86)90044-9.

Lemoine, M., Tricart, P., Boillot, G. (1987). Ultramafic and gabbroic ocean floor of the Ligurian Tethys (Alps, Corsica, Apennines): In search of a genetic imodel. *Geology*, 15, 622–625. https://doi.org/10.1130/0091-7613(1987)15.

Lescoutre, R. and Manatschal, G. (2020). Role of rift-inheritance and segmentation for orogenic evolution: Example from the Pyrenean-Cantabrian system. *Tectonics*. 40(5). doi: 10.1029/2020TC006134.

Li, X.-H.H., Faure, M., Lin, W., Manatschal, G. (2013). New isotopic constraints on age and magma genesis of an embryonic oceanic crust: The Chenaillet Ophiolite in the Western Alps. *Lithos*, 160–161, 283–291. https://doi.org/10.1016/j.lithos.2012.12.016.

Lory C. (1860). Description géologique du Dauphiné pour servir à l'explication de la carte géologique de cette province. *Bull. Soc. Stat. Isère*, 2(7), 79.

Lory, C. (1866). Sur le gisement de la Terebratula diphya dans les calcaires de la Porte-de-France, aux environs de Grenoble et de Chambéry. *Bull. la Soc. Geol. Fr.*, 2, 516–521.

Manatschal, G. (2004). New models for evolution of magma-poor rifted margins based on a review of data and concepts from West Iberia and the Alps. *Int. J. Earth Sci.*, 93, 432–466. https://doi.org/10.1007/s00531-004-0394-7.

Manatschal, G. and Müntener, O. (2009). A type sequence across an ancient magma-poor ocean–continent transition: The example of the western Alpine Tethys ophiolites. *Tectonophysics*, 473, 4–19. https://doi.org/10.1016/j.tecto.2008.07.021.

Manatschal, G. and Nievergelt, P. (1997). A continent-ocean transition recorded in the Err and Platta nappes (Eastern Switzerland). *Eclogae Geol. Helv.*, 90, 3–27.

Manatschal, G., Engström, A., Desmurs, L., Schaltegger, U., Cosca, M., Müntener, O., Bernoulli, D. (2006). What is the tectono-metamorphic evolution of continental break-up: The example of the Tasna Ocean-Continent Transition. *J. Struct. Geol.*, 28, 1849–1869. https://doi.org/10.1016/j.jsg.2006.07.014.

Manatschal, G., Muntener, O., Lavier, L.L., Minshull, T.A., Peron-Pinvidic, G. (2007). Observations from the Alpine Tethys and Iberia-Newfoundland margins pertinent to the interpretation of continental breakup. *Geol. Soc. Spec. Publ.*, 282, 291–324. https://doi.org/10.1144/SP282.14.

Manatschal, G., Sauter, D., Karpoff, A.M., Masini, E., Mohn, G., Lagabrielle, Y. (2011). The Chenaillet Ophiolite in the French/Italian Alps: An ancient analogue for an Oceanic Core Complex? *Lithos*, 124, 169–184. https://doi.org/10.1016/j.lithos.2010.10.017.

Manatschal G., Chenin P., Ghienne J.F., Ribes C., Masini E. (2021). The syn-rift tectono-stratigraphic record of rifted margins (Part 1): Insights from the Alpine Tethys. *Basin Research*. doi: 10.1111/(ISSN)1365-2117.

Marroni, M., Feroni, A.C., Di Biase, D., Ottria, G., Pandolfi, L., Taini, A. (2002). Polyphase folding at upper structural levels in the Borbera Valley (northern Apennines, Italy): Implications for the tectonic evolution of the linkage area between Alps and Apennines. *Comptes Rendus – Geosci.*, 334, 565–572. https://doi.org/10.1016/S1631-0713(02)01784-4.

Masini, E., Manatschal, G., Mohn, G., Unternehr, P. (2012). Anatomy and tectono-sedimentary evolution of a rift-related detachment system: The example of the Err detachment (central Alps, SE Switzerland). *Bull. Geol. Soc. Am.*, 124, 1535–1551. https://doi.org/10.1130/B30557.1.

Masini, E., Manatschal, G., Mohn, G. (2013). The Alpine Tethys rifted margins: Reconciling old and new ideas to understand the stratigraphic architecture of magma-poor rifted margins. *Sedimentology*, 60, 174–196. https://doi.org/10.1111/sed.12017.

McCarthy, A. and Müntener, O. (2015). Ancient depletion and mantle heterogeneity: Revisiting the Permian-Jurassic paradox of Alpine peridotites. *Geology*, 43, 255–258. https://doi.org/10.1130/G36340.1.

McCarthy, A., Chelle-Michou, C., Müntener, O., Arculus, R., Blundy, J. (2018). Subduction initiation without magmatism: The case of the missing Alpine magmatic arc. *Geol. Soc. Am. Geol.*, 46, 1–4. https://doi.org/10.1130/G45366.1.

McCarthy, A., Tugend, J., Mohn, G., Candioti, L., Chelle-Michou, C., Arculus, R., Schmalholz, S.M., Müntener, O. (2020). A case of Ampferer-type subduction and consequences for the Alps and the Pyrenees. *Am. J. Sci.*, 320, 313–372. https://doi.org/10.2475/04.2020.01.

Mohn, G., Manatschal, G., Müntener, O., Beltrando, M., Masini, E. (2010). Unravelling the interaction between tectonic and sedimentary processes during lithospheric thinning in the Alpine Tethys margins. *Int. J. Earth Sci.*, 99, 75–101. https://doi.org/10.1007/s00531-010-0566-6.

Mohn, G., Manatschal, G., Masini, E., Müntener, O. (2011). Rift-related inheritance in orogens: A case study from the Austroalpine nappes in Central Alps (SE-Switzerland and N-Italy). *Int. J. Earth Sci.*, 100, 937–961. https://doi.org/10.1007/s00531-010-0630-2.

Mohn, G., Manatschal, G., Beltrando, M., Masini, E., Kusznir, N. (2012). Necking of continental crust in magma-poor rifted margins: Evidence from the fossil Alpine Tethys margins. *Tectonics*, 31, TC1012. https://doi.org/10.1029/2011TC002961.

Mohn, G., Manatschal, G., Beltrando, M., Haupert, I. (2014). The role of rift-inherited hyper-extension in Alpine-type orogens. *Terra Nov.*, 26, 347–353. https://doi.org/10.1111/ter.12104.

Montadert, L., de Charpal, O., Roberts, D., Guennoc, P., Sibuet, J.-C. (1979). Northeast Atlantic passive continental margins: Rifting and subsidence processes. In *Deep Drilling Results in the Atlantic Ocean: Continental Margins and Paleoenvironment*, Talwani, M., Hay, W., Ryan, W.B.F. (eds). American Geophysical Union (AGU). https://doi.org/10.1029/me003p0154.

Müntener, O. and Piccardo, G.B. (2003). Melt migration in ophiolitic peridotites: The message from Alpine-Apennine peridotites and implications for embryonic ocean basins. *Geol. Soc., London, Spec. Publ.*, 218, 69–89. https://doi.org/10.1144/GSL.SP.2003.218.01.05.

Müntener, O., Pettke, T., Desmurs, L., Meier, M., Schaltegger, U. (2004). Refertilization of mantle peridotite in embryonic ocean basins: Trace element and Nd isotopic evidence and implications for crust-mantle relationships. *Earth Planet. Sci. Lett.*, 221, 293–308. https://doi.org/10.1016/S0012-821X(04)00073-1.

Müntener, O., Manatschal, G., Desmurs, L., Pettke, T. (2010). Plagioclase peridotites in ocean-continent transitions: Refertilized mantle domains generated by melt stagnation in the shallow mantle lithosphere. *J. Petrol.*, 51, 255–294. https://doi.org/10.1093/petrology/egp087.

Nirrengarten, M., Manatschal, G., Tugend, J., Kusznir, N.J., Sauter, D. (2017). Nature and origin of the J-magnetic anomaly offshore Iberia – Newfoundland: Implications for plate reconstructions. *Terra Nov.*, 29, 20–28. https://doi.org/10.1111/ter.12240.

Nirrengarten, M., Manatschal, G., Tugend, J., Kusznir, N., Sauter, D. (2018). Kinematic evolution of the southern north Atlantic: Implications for the formation of hyperextended rift systems. *Tectonics*, 37, 89–118. https://doi.org/10.1002/2017TC004495.

Petri, B., Mohn, G., Štípská, P., Schulmann, K., Manatschal, G. (2016). The Sondalo gabbro contact aureole (Campo unit, Eastern Alps): Implications for mid-crustal mafic magma emplacement. *Contrib. to Mineral. Petrol.*, 171, 52. https://doi.org/10.1007/s00410-016-1263-7.

Petri, B., Skrzypek, E., Mohn, G., Mateeva, T., Galster, F., Manatschal, G. (2017). U–Pb geochronology of the Sondalo gabbroic complex (Central Alps) and its position within the Permian post-Variscan extension. *Int. J. Earth Sci.*, https://doi.org/10.1007/s00531-017-1465-x.

Petri, B., Duretz, T., Mohn, G., Schmalholz, S.M., Karner, G.D., Müntener, O. (2019). Thinning mechanisms of heterogeneous continental lithosphere. *Earth Planet. Sci. Lett.*, 512, 147–162. https://doi.org/10.1016/J.EPSL.2019.02.007.

Picazo, S., Müntener, O., Manatschal, G., Bauville, A., Karner, G.D., Johnson, C. (2016). Mapping the nature of mantle domains in Western and Central Europe based on clinopyroxene and spinel chemistry: Evidence for mantle modification during an extensional cycle. *Lithos*. https://doi.org/10.1016/j.lithos.2016.08.02.

Piccardo, G.B., Rampone, E., Vannucci, R. (1990). Upper mantle evolution during continental rifting and ocean formation: Evidences from peridotite bodies of the western Alpine-northern Apennine system. *Mémoires la Société Géologique Fr.*, 156, 323–333.

Piccardo, G.B., Müntener, O., Zanetti, A., Romairone, A., Bruzzone, S., Poggi, E., Spagnolo, G. (2004). The lanzo south peridotite: Melt/peridotite interaction in the mantle lithosphere of the Jurassic Ligurian Tethys. *Ofioliti*, 29, 37–62. https://doi.org/10.4454/OFIOLITI.V29I1.205.

Piccardo, G.B., Zanetti, A., Müntener, O. (2007). Melt/peridotite interaction in the Southern Lanzo peridotite: Field, textural and geochemical evidence. *Lithos*, 94, 181–209. https://doi.org/10.1016/j.lithos.2006.07.002.

Pinto, V.H.G., Manatschal, G., Karpoff, A.-M., Viana, A. (2015). Tracing mantle-reacted fluids in magma-poor rifted margins: The example of Alpine Tethyan rifted margins. *Geochem., Geophys. Geosyst.*, 16, 3271–3308. https://doi.org/10.1002/2015GC005830.

Popier, A. (1991). A la recherche de l'océan perdu dans les Alpes de Briançon. *Bull. Mens. la Société Linnéenne Lyon*, 60th year, 10, 388–391.

Quick, J.E., Sinigoi, S., Negrini, L., Demarchi, G., Mayer, A. (1992). Synmagmatic deformation in the underplated igneous complex of the Ivrea-Verbano zone. *Geology*, 20, 613–616. https://doi.org/10.1130/0091-7613(1992)020<0613:sditui>2.3.co;2.

Rampone, E., Hofmann, A.W., Piccardo, G.B., Vannucci, R., Bottazzi, P., Ottolini, L. (1995). Petrology, mineral and isotope geochemistry of the external liguride peridotites (Northern Apennines, Italy). *J. Petrol.*, 36, 81–105. https://doi.org/10.1093/petrology/36.1.81.

Rampone, E., Piccardo, G.B., Vannucci, R., Bottazzi, P. (1997). Chemistry and origin of trapped melts in ophiolitic peridotites. *Geochim. Cosmochim. Acta*, 61, 4557–4569. https://doi.org/10.1016/s0016-7037(97)00260-3.

Rampone, E., Hofmann, A.W., Raczek, I. (1998). Isotopic contrasts within the Internal Liguride ophiolite (N. Italy): The lack of a genetic mantle–crust link. *Earth Planet. Sci. Lett.*, 163, 175–189. https://doi.org/10.1016/S0012-821X(98)00185-X.

Rampone, E., Borghini, G., Basch, V. (2020). Melt migration and melt-rock reaction in the Alpine-Apennine peridotites: Insights on mantle dynamics in extending lithosphere. *Geosci. Front.*, 11, 151–166. https://doi.org/10.1016/j.gsf.2018.11.001.

Renna, M.R., Tribuzio, R., Sanfilippo, A., Thirlwall, M. (2018). Role of melting process and melt–rock reaction in the formation of Jurassic MORB-type basalts (Alpine ophiolites). *Contrib. to Mineral. Petrol.*, 173, 1–21. https://doi.org/10.1007/s00410-018-1456-3.

Ribes, C., Manatschal, G., Ghienne, J.-F., Karner, G.D., Johnson, C.A., Figueredo, P.H., Incerpi, N.N., Epin, M. (2019a). The syn-rift stratigraphic record across a fossil hyper-extended rifted margin: The example of the northwestern Adriatic margin exposed in the Central Alps. *Int. J. Earth Sci.*, 108, 2071–2095. https://doi.org/DOI: 10.1007/s00531-019-01750-6.

Ribes, C., Petri, B., Ghienne, J., Manatschal, G., Galster, F., Karner, G.D., Figueredo, P.H., Johnson, C.A., Karpoff, A. (2019b). Tectono-sedimentary evolution of a fossil ocean-continent transition: Tasna nappe, central Alps (SE Switzerland). *Geol. Soc. Am. Bull.*, 1–20. https://doi.org/10.1130/B35310.1/4863878/b35310.pdf.

Ribes, C., Ghienne, J.F., Manatschal, G., Dall'Asta, N., Stockli, D.F., Galster, F., Gillard, M., Karner, G.D. (2020). The Grès Singuliers of the Mont Blanc region (France and Switzerland): Stratigraphic response to rifting and crustal necking in the Alpine Tethys. *Int. J. Earth Sci.*, 109, 2325–2352. https://doi.org/10.1007/s00531-020-01902-z.

Rosenbaum, G., Lister, G.S., Duboz, C. (2002). Relative motions of Africa, Iberia and Europe during Alpine orogeny. *Tectonophysics*, 359, 117–129. https://doi.org/10.1016/S0040-1951(02)00442-0.

Sanfilippo, A., Salters, V., Tribuzio, R., Zanetti, A. (2019). Role of ancient, ultra-depleted mantle in Mid-Ocean-Ridge magmatism. *Earth Planet. Sci. Lett.* 511, 89–98. https://doi.org/10.1016/j.epsl.2019.01.018.

Sauter, D., Cannat, M., Rouméjon, S., Andreani, M., Birot, D., Bronner, A., Brunelli, D., Carlut, J., Delacour, A., Guyader, V., MacLeod, C.J., Manatschal, G., Mendel, V., Ménez, B., Pasini, V., Ruellan, E., Searle, R. (2013). Continuous exhumation of mantle-derived rocks at the Southwest Indian Ridge for 11 million years. *Nat. Geosci.*, 6, 314–320. https://doi.org/10.1038/ngeo1771.

Schmid, S.M., Pfiffner, O.A., Froitzheim, N., Schönborn, G., Kissling, E. (1996). Geophysical-geological transect and tectonic evolution of the Swiss-Italian Alps. *Tectonics*, 15, 1036–1064. https://doi.org/10.1029/96TC00433.

Schmid, S.M., Fgenschuh, B., Kissling, E., Schuster, R., Fügenschuh, B. (2004). Tectonic map and overall architecture of the Alpine orogen. *Eclogae Geol. Helv.*, 97, 93–117. https://doi.org/10.1007/s00015-004-1113-x.

Şengör, A.M.C. (1991). Orogenic architecture as a guide to size of ocean lost in collisional mountain belts. *Bull. Tech. Univ., Istanbul*, 44, 43–74.

Şengör, A.M.C. (2014). Eduard Suess and global tectonics: An illustrated "Short Guide". *Austrian J. Earth Sci.*, 107, 6–82.

Speranza, F., Minelli, L., Pignatelli, A., Chiappini, M. (2012). The Ionian Sea: The oldest in situ ocean fragment of the world? *J. Geophys. Res. B Solid Earth*, 117. https://doi.org/10.1029/2012JB009475.

Stampfli, G.M., Borel, G.D., Marchant, R., Mosar, J. (2002). Western Alps geological constraints on western Tethyan reconstructions. *J. Virtual Explor.*, 8, 77–106.

Steinmann, G. (1905). Geologische Beobachtungen in den Alpen. Wagners Universitäts-Buchdruckerei.

Suess, E. (1883). *Das Antlitz der Erde, Ia (Erste Abtheilung)*. F. Tempsky, Prag G. Freytag, Leipzig, 310.

Tavani, S., Bertok, C., Granado, P., Piana, F., Salas, R., Vigna, B., Muñoz, J.A. (2018). The Iberia-Eurasia plate boundary east of the Pyrenees. *Earth-Science Rev.* https://doi.org/10.1016/j.earscirev.2018.10.008.

Tribuzio, R., Thirlwall, M.F., Vannucci, R. (2004). Origin of the Gabbro–Peridotite Association from the Northern Apennine Ophiolites (Italy). *J. Petrol.*, 45, 1109–1124. https://doi.org/10.1093/petrology/egh006.

Tribuzio, R., Garzetti, F., Corfu, F., Tiepolo, M., Renna, M.R. (2016). U-Pb zircon geochronology of the Ligurian ophiolites (Northern Apennine, Italy): Implications for continental breakup to slow seafloor spreading. *Tectonophysics*, 666, 220–243. https://doi.org/10.1016/j.tecto.2015.10.024.

Trümpy, R. (1975). Penninic-Austroalpine boundary in the Swiss Alps: A presumed former continental margins and its problems. *Am. J. Sci.*, 275, 209–238.

Trümpy, R. (2001). Why plate tectonics was not invented in the Alps. *Int. J. Earth Sci.*, 90, 477–483. https://doi.org/10.1007/s005310000175.

Tugend, J., Manatschal, G., Kusznir, N.J., Masini, E. (2014). Characterizing and identifying structural domains at rifted continental margins: Application to the Bay of Biscay margins and its Western Pyrenean fossil remnants. *Geol. Soc. London Spec. Publ.*, 413, SP413-3. https://doi.org/10.1002/2014TC003529.

Vissers, R.L.M. and Meijer, P.T. (2012). Mesozoic rotation of Iberia: Subduction in the Pyrenees? *Earth-Science Rev.*, 110, 93–110. https://doi.org/10.1016/j.earscirev.2011.11.001.

Wernicke, B. (1985). Uniform-sense normal simple shear of the continental lithosphere. *Can. J. Earth Sci.*, 22, 108–125. https://doi.org/10.1139/e85-009.

Whitmarsh, R.B., Manatschal, G., Minshull, T.A. (2001). Evolution of magma-poor continental margins from rifting to seafloor spreading. *Nature*, 413, 150–154. https://doi.org/10.1038/35093085.

8

A Fossil Magma-rich Rifted Margin in the Scandinavian Caledonides

Johannes JAKOB[1], Hans Jørgen KJØLL[2] and Torgeir Bjørge ANDERSEN[3]

[1] *Geological Survey of Norway (NGU), Trondheim, Norway*

[2] *Section for Geology and Geophysics, Department of Geosciences, University of Oslo, Norway*

[3] *Centre for Earth Evolution and Dynamics, Department of Geosciences, University of Oslo, Norway*

Due to their inaccessibility, the understanding of present-day rifted margins is mostly based on geophysical studies and by deep-sea drill hole data. Important information, however, can be drawn from partially preserved fossil examples of rifted margins that have been incorporated into mountain belts during orogenic events. Although remnants of fossil magma-poor rifted margins have been identified in several mountain belts, examples of well-preserved fossil magma-rich rifted margins are rare. One example of a fossil magma-rich rifted margin that formed during the opening of the Iapetus Ocean in the Ediacaran is preserved in the allochthons of the Scandinavian Caledonides. The magmatism is interpreted to be related to elevated asthenosphere temperatures, possibly caused by a mantle plume that led to the formation and emplacement of the Central Iapetus Magmatic Province (CIMP) and aided the formation of the Iapetus magma-rich rifted margin. During the margin formation, mafic magmatic rocks of the more than 900-km-long

For a color version of all the figures in this chapter, see www.iste.co.uk/peronpinvidic/margins2.zip.

Continental Rifted Margins 2,
coordinated by Gwenn PERON-PINVIDIC.

Scandinavian Dyke Complex were emplaced in mid-crustal levels in the distal domain of the rifted margin, whereas mafic-ultramafic rocks of the Seiland Igneous Province were emplaced in lower crustal rocks and are interpreted to represent a deep-seated magmatic conduit system.

8.1. Introduction

Rifted margins are commonly classified based on the volume of magmatic additions emplaced during rifting and lithospheric breakup, that is, magma-rich and magma-poor rifted margins as well as margins of uncertain classification (e.g. Reston and Manatschal 2011; Franke 2013). Studying the different styles and domains of present-day rifted margins in detail, however, is hampered due to their inaccessibility. In particular, the more distal domains of rifted margins can primarily only be studied by geophysical methods and by deep-sea drill hole data. The different architectural elements of a rifted margin, for example, the proximal, necking, distal and outer domains (Peron-Pinvidic and Manatschal 2009; Peron-Pinvidic *et al.* 2013), may however be partially preserved in mountain belts after the inversion of a rifted margin during orogenic events (Beltrando *et al.* 2014; Mohn *et al.* 2014; Jakob *et al.* 2019). The study of remnants of fossil rifted margins that have been incorporated into accretionary or collisional orogens, where a wide range of these rocks are accessible on-land, is of great value in improving our understanding of rifted margins. These partially preserved fossil rifted margins, however, are almost exclusively remnants of magma-poor rifted margins (Manatschal and Nievergelt 1997; Desmurs *et al.* 2001; Manatschal 2004; Mohn *et al.* 2010, 2014; Andersen *et al.* 2012; Clerc *et al.* 2012; van Staal *et al.* 2013; Epin *et al.* 2017; Lagabrielle *et al.* 2019). Examples of well-preserved fossil magma-rich rifted margins are scarce due to their low buoyancy and proneness to be recycled into the mantle during collisions (e.g. Turino *et al.* 2021).

The study of partially preserved magma-poor rifted margins in mountain belts and reconstructions of their pre-orogenic architecture in combination with comparisons with the data from present-day rifted margins, greatly improved the understanding of magma-poor rifted margins and led to a revision of their architectures and the process that formed them (e.g. Manatschal 2004; Peron-Pinvidic and Manatschal 2009, 2010). Thus, the identification of the remnants of magma-rich rifted margins in mountain belts is much warranted and would allow for the study of otherwise inaccessible parts of this style of rifted margins.

The Scandinavian Caledonides encompass remnants of the Iapetus Ocean with its rifted margins (Corfu *et al.* 2014). The orogen formed during the Silurian–Devonian continental collision between the continents of Laurentia and

Baltica(-Avalonia) after closure of the Iapetus Ocean (see, for example, Torsvik and Cocks (2005) Domeier (2015, 2018); for a paleogeographic overview). During the collision, the Baltican continental margin was buried to mantle depth and overthrust by an intricate stack of nappes (e.g. Hacker *et al.* 2010). The nappe stack from base to top includes units interpreted to have been derived from the Baltican, Iapetan and Laurentian realms (e.g. Stephens and Gee 1989). The remnants of the Iapetus magma-rich rifted margin presented below are preserved within allochthons that were thrust over the Baltican margin during the continental collision.

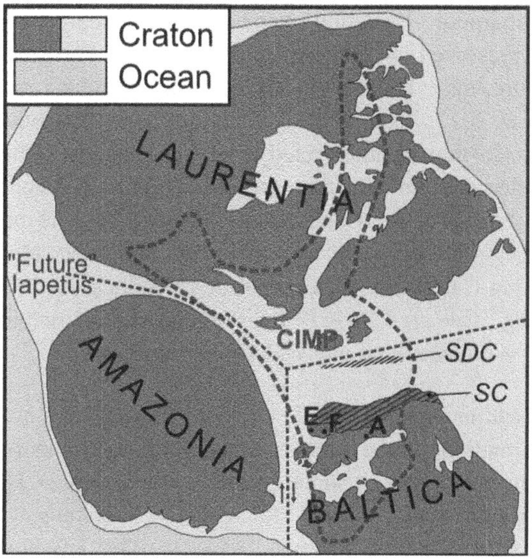

Figure 8.1. *Paleogeographic reconstruction of East Rodinia before opening of the Iapetus Ocean between Laurentia and Baltica. The estimated extent of the Central Iapetus Magmatic Province (CIMP) is indicated by the red dotted line. SDC, Scandinavian Dyke Complex; SC, Scandinavian Caledonides (present-day extends); E, Egersund Dykes; F, Fen Carbonatite Complex; A, Alnö Carbonatite Complex (source: modified after Tegner* et al. *2019)*

8.2. Opening of the Iapetus Ocean and formation of the Central Iapetus Magmatic Province

Available evidence suggests that the Iapetus opened in the Neoproterozoic (Wilson 1966; Merdith *et al.* 2017; Robert *et al.* 2020). The breakup was preceded by a long-lasting phase of crustal extension (Nystuen *et al.* 2008) and is thought to have been aided by the emplacement of a Large Igneous Province, the Central Iapetus Magmatic Province (CIMP) (Figure 8.1) (Puffer 2002; Tegner *et al.* 2019). It is estimated that the CIMP erupted over a vast area of more than 9 million square kilometers, that it formed over a prolonged period that lasted from 615 and 550 Ma and that it included several pulses of magmatic activity (Tegner *et al.* 2019). The magma-rich rifted margin that is preserved within the Scandinavian Caledonides is interpreted to have formed as part of the CIMP, but in a much shorter timeframe of only ca. 12 million years.

The architecture of the pre-Caledonian rifted margin of Baltica was complex and may have included a magma-rich and magma-poor section that linked up through a magma-rich–magma-poor transition zone within Neoproterozoic sedimentary successions now preserved in southern Central Norway. Furthermore, the complex margin architecture may have included continental slivers or microcontinents (e.g. Andersen *et al.* 2012; Jakob *et al.* 2019, 2017). This contribution focuses on the dated magmatic rocks that are interpreted to be the remnants of the distal domain of a magma-rich rifted margin and that are preserved in the central and northern segments of the Scandinavian Caledonides (Figure 8.2). The units in which the magma-rich margin is preserved are commonly but not universally interpreted to be remnants of the pre-Caledonian rifted margin of Baltica (e.g. Corfu *et al.* 2007; Kirkland *et al.* 2007; Gee *et al.* 2016) – a discussion of the possible paleogeographic origins, however, is outside the framework of this contribution.

Within the Scandinavian Caledonides, almost all rocks that are interpreted to be related to the formation of the Iapetus magma-rich rifted margin are preserved within the less deformed cores of nappe complexes (Figure 8.2). However, there is a paucity of magmatic rocks related to breakup magmatism in the Baltican autochthon. The autochthonous magmatic rocks that are interpreted to be directly related to the opening of the Iapetus and the CIMP in western Baltica are the ~ 616 Ma Egersund dykes in southern Norway (Bingen *et al.* 1998) as well as the ~ 584 Ma Fen and Alnö carbonatite complexes in Norway and Sweden, respectively (Figures 8.1 and 8.2) (Meert *et al.* 2007, 1998). In the Scandian allochthons, the magmatic rocks interpreted to be related to the CIMP occur in Tonian and Cryogenian–Ediacaran metasedimentary complexes (Figure 8.2). The Tonian metasedimentary complexes are preserved within the Kalak Nappe Complex and host the igneous rocks of the Seiland Igneous Province (SIP). The Cryogenian–Ediacaran metasedimentary complexes are the country rocks to the Scandinavian Dyke Complex (SDC) and are preserved within the Särv, Seve and Corrovarre Nappes. In the following, we briefly present some of the geological information that has been published about the allochthonous units, in which most of the magmatic products of the CIMP are preserved. The geological history of these magmatic complexes and their metasedimentary host rocks can be used to derive information about the structural positions that these units occupied within the evolving rifted margin at the time of CIMP magmatism.

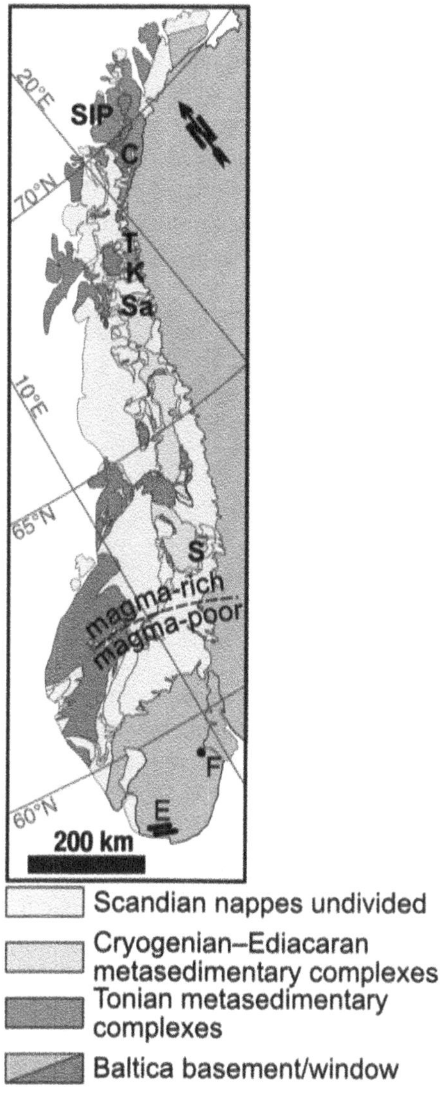

Figure 8.2. *Simplified map of the central and northern Scandinavian Caledonides. E, Egersund dykes; F, Fen Carbonatite Complex; S, Särv area; Sa, Sarek area; K, Kebnekaise area; T, Indre Troms area; C; Corrovarre area; SIP, Seiland Igneous Province*

8.2.1. *The Seiland Igneous Province*

The SIP occurs in northern Norway (Figure 8.2) and is mostly composed of mafic and ultramafic rocks that intruded Tonian metasediments of the Sørøy Succession of the upper Kalak Nappe Complex. The Sørøy Succession was deposited between ~910 and 840 Ma and was deformed, metamorphosed and intruded by granitoids between 850–820 Ma and 710–700 Ma (Kirkland *et al.* 2006, 2007; Corfu *et al.* 2011, 2007; Gasser *et al.* 2015, 2006). Temperatures and pressures during a metamorphic event dated ~700 Ma in paragneisses correlated with the Sørøy Succession have been estimated at ~730–775°C and ~6.3–9.8 kbar, respectively (Gasser *et al.* 2015), indicating that these rocks were at lower crustal levels already at the beginning of the Cryogenian.

Figure 8.3. *Photomosaic from the "Favoritkammen" locality in Sarek National Park. Modified from Kjøll et al. (2019b). Sedimentary layering (white dashed line) and younging direction are indicated. Dykes are color-coded with respect to their orientation. Note that the dykes of the two different orientations mutually crosscut each other and that at several localities the dykes bend and change direction*

The magmatic rocks of the SIP that intrude the Tonian metasedimentary complexes include gabbroic and ultramafic rocks as well as minor amounts of carbonatites and syenites (Larsen *et al.* 2018). The carbonatites and syenites have been emplaced at the beginning and end of the magmatic activity of the SIP, and their emplacement has been dated at 580–570 and 540–520 Ma (Pedersen *et al.* 1989; Cadow 1993; Roberts *et al.* 2010). ε_{Nd} values of around +4 from the alkaline–carbonatite rocks indicate that these earliest and latest melts emplaced in the SIP were derived directly from the mantle (Cadow 1993; Roberts *et al.* 2010; Larsen *et al.* 2018). The mafic–ultramafic rocks, however, constitute most of the magmatic rocks of the SIP and have been emplaced between 580 and 560 Ma. Note that more than 90 % of the mafic–ultramafic igneous rocks have been emplaced within a short time span of just ~ 4 Ma between 570 and 560 Ma (Roberts *et al.* 2010, 2006). The bulk of the mafic magmas is of ocean island basalt (OIB) composition and intruded into the country rock at lower crustal levels creating 1- to 3-km-thick contact aureoles (Tegner *et al.* 1999; Roberts *et al.* 2010; Larsen *et al.* 2018). Grant *et al.* (2016, 2020) and Larsen *et al.* (2018) estimated that the ambient pressures during emplacement of the SIP at 570–560 Ma were 7.9 ± 1.1 kbar. These estimates are corroborated by pressure estimates derived from the metamorphic mineral assemblages from the contact aureoles (Elvevold *et al.* 1994; Tegner *et al.* 1999). The SIP is interpreted to preserve remnants of a deep-seated magmatic conduit system that intruded the Tonian metasedimentary complexes at lower crustal levels during the formation of the CIMP (Larsen *et al.* 2018).

8.2.2. *The Scandinavian Dyke Complex*

The Scandinavian Dyke Complex is preserved within Cryogenian–Ediacaran metasedimentary successions that are now preserved over more than 900 km along strike of the Scandinavian Caledonides (Figures 8.2 and 8.3). In the Kebnekaise and Corrovarre areas, the units containing the SDC are structurally and directly overlying the Tonian metasedimentary successions and orthogneisses (Andréasson *et al.* 2018; Kjøll *et al.* 2019a). However, the sedimentary successions are commonly truncated by thrust faults at the top and at the base (e.g. Kjøll 2020), and their depositional substratum is essentially unknown. The stratigraphic thicknesses of the sedimentary successions varies along strike of the orogen but in some areas reaches up to 3.5–4 km (e.g. Kumpulainen 1980; Svenningsen 1994a, 1994b, 1995; Kjøll 2020). The sedimentary successions include carbonaceous and arkosic meta-sandstones, meta-siltstones, meta-evaporites, carbonates and conglomeratic layers. These sedimentary successions are interpreted to have been deposited in shallow-marine and sabkha-type environments (*op.cit.*). Bedding and way-up criteria are locally well preserved within the sedimentary successions. In the Särv and Sarek area, conglomeratic layers have been described as diamictites and are interpreted to be of glaciogenic origin (Kumpulainen 1980, 2011; Kjøll 2020). A depositional age of these successions between 700 and 608 Ma has been suggested (Kjøll 2020), and a Marinoan age (~ 635 Ma; Hoffmann *et al.* 2004) for the glaciogenic units is commonly accepted (e.g. Kumpulainen *et al.* 2016; Nystuen *et al.* 2016; Gee *et al.* 2020).

The entire sedimentary succession in all areas described above is cut by mafic dykes belonging to the SDC (Figure 8.3). The percentage of mafic dykes relative to the metasedimentary host rocks can vary from 50 to 100 vol.% (Svenningsen 1994a;

Kjøll *et al.* 2019a). Radiometric ages from the mafic rocks of the SDC and associated granitoid bodies have been presented from the Särv (Kumpulainen *et al.* 2016), Sarek (Svenningsen 2001), Kebnekaise (Baird *et al.* 2014; Kirsch and Svenningsen 2015) and Corrovarre areas (Gee *et al.* 2016; Kjøll *et al.* 2019a) (Figure 8.1b). These crystallization ages from the mafic rocks indicate that the majority of the SDC was emplaced over a distance of ~ 1,000 km during only a short phase of intense magmatism between 608 and 596 Ma, that is, ~ 30 Ma after deposition of the sediments. Granitoid bodies in the Corrovarre area that are locally cut by the SDC have been dated at 612, 608 and 602 Ma (Corfu *et al.* 2007; Gee *et al.* 2016; Kjøll *et al.* 2019a). The formation of these granites has been attributed to the heating of the crust and partial melting of the surrounding metasediments during, as well as already before, the final emplacement of the SDC (Kjøll *et al.* 2019a, 2019b). The elevation of the geotherm prior to the emplacement of the SDC is suggested to be caused by the buildup of large mafic magma reservoirs at the base of the lithosphere (Kjøll *et al.* 2019a, 2019b).

Ambient pressures during the emplacement of the SDC of 2.5–4.5 kbar have been estimated in the Särv, Sarek, Kebnekaise, Troms and Corrovarre areas (Kjøll *et al.* 2019a). These pressures indicate that the SDC intruded the metasediments at mid-crustal levels (10–15 km depth). The dykes were emplaced at a high angle to the bedding of the sediments, and two dyke populations that mutually crosscut each other at an acute angle of ~ 30° can be distinguished (Figure 8.3). Other dykes can be observed changing orientation over short distances. The mutual cross-cutting relationships indicate that the dyke populations have been emplaced contemporaneously and are not the result of successive phases of diking before and after tilting of the sedimentary successions. It is estimated that the emplacement of the dykes due to their angular relationship resulted in an overall ~ 94 % horizontal stretching and ~ 27 % vertical thickening of the crust (Kjøll *et al.* 2019b). In addition, shortening structures were observed in the host rock next to the dykes, suggesting that even in this rift-scenario, the dykes were emplaced in a forceful manner and did not accommodate passively for the tectonic stretching (Kjøll *et al.* 2019b). Kjøll *et al.* (2019b) also demonstrated that the diking was associated with a shallowing of the brittle–ductile transition, which is interpreted to have been caused by the additional heat added to the crust during dyking, which may have played an important role for the weakening of the crust, leading up to the final breakup (Kjøll *et al.* 2019b).

Tegner *et al.* (2019) showed that the SDC exhibits a lateral geochemical zonation from enriched to depleted basaltic compositions from south to north. The same authors concluded that this geochemical zoning indicates melting related to a broad mantle plume at elevated ambient mantle potential temperatures (ΔTP of

75–250°C). The lateral variation shows that the plume was enriched (primitive mantle) in the south and depleted (depleted MORB mantle) in the north. Furthermore, the formation of the CIMP may have involved several mantle plumes sourced from the plume generation zones along the edges of a Large Low Shear-wave Velocity Province (LLSVP) at the core–mantle boundary (Torsvik *et al.* 2006; Tegner *et al.* 2019).

8.3. Discussion

Although the original relationships between the Tonian and Cryogenian–Ediacaran metasedimentary complexes and the magmatic phases during CIMP magmatism are uncertain, the respective structural position of the SIP and SDC in the crust at the time of CIMP magmatism can be derived from the emplacement history of the 608–560 Ma magmatic products. At the time of SDC magmatism (608–596 Ma), the Cryogenian–Ediacaran units were in depths corresponding to 2.5–4.5 kbar, that is, in the middle crust. The depositional age of the sediments with respect to the magmatism and inferred margin formation suggests that these sediments have been deposited during an early stretching phase. Moreover, the volume of the magmatism and the mid-crustal position suggest that there might have been a rapid accumulation of thick sequences of volcanic rocks above the Cryogenian–Ediacaran successions (SDRs, Figure 8.4) during SDC-magmatism (Kjøll *et al.* 2019a). These volcanic sequences may have been comparable to seaward dipping reflectors that reach thicknesses of 4–7 km in present-day rifted margins (Menzies *et al.* 2002; Abdelmalak *et al.* 2016, 2015; Kjøll *et al.* 2019a). However, rocks that may preserve these upper crustal levels of the distal domain of the magma-rich rifted margin are only sparsely preserved in the Scandinavian Caledonides.

The Tonian metasedimentary complexes were at depths corresponding to ~ 8 kbar at the time of mafic–ultramafic magmatism (580–560 Ma) (Roberts *et al.* 2010; Larsen *et al.* 2018), that is, at lower crustal levels. The Tonian metasedimentary complexes apparently occupied this structural level since at least ~ 700 Ma (Gasser *et al.* 2015), indicating that the Tonian metasedimentary complexes were already in lower crustal levels even before the Cryogenian–Ediacaran successions were deposited. A common paleogeographic origin of the Tonian and Cryogenian–Ediacaran sedimentary successions, however, has been suggested from detrital zircon spectra as well as the lack of evidence of sutures interrupting the Tonian and Cryogenian–Ediacaran tectonostratigraphic successions (e.g. Kirkland *et al.* 2011; Gee *et al.* 2016). Therefore, (1) the lower crustal position during CIMP magmatism; (2) the older depositional ages; (3) the differences in the pre-Ediacaran metamorphic and plutonic history of the metasedimentary complexes and (4) the structural position of the Tonian units within the Scandian nappe stack below the Corrovarre Nappe indicate a position stratigraphically below or paleogeographically inboard (with respect to Baltica) of the Tonian units relative to the Cryogenian–Ediacaran complexes before onset of Scandian nappe thrusting (Figure 8.4). The Tonian metasedimentary complexes, therefore, may represent a piece of lower continental crust that was invaded by CIMP-related magmas penecontemporaneous to the formation of the Iapetus magma-rich rifted margin.

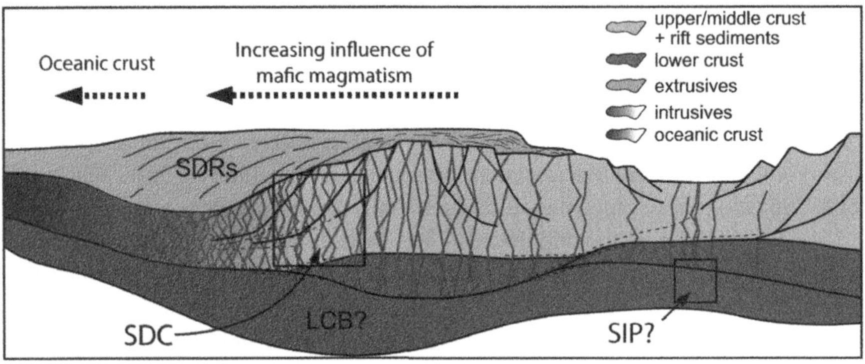

Figure 8.4. *Diagram of the outer domain of the Iapetus magma-rich rifted margin that is now preserved within the nappe complexes of the Scandinavian Caledonides. SDC, Scandinavian Dyke Complex; SIP, Seiland Igneous Province; SDRs, seaward dipping reflectors; LCB, lower crustal body (source: modified from Kjøll* et al. *2019a)*

8.4. Conclusion

Remnants of the CIMP are preserved within the allochthons of the Scandinavian Caledonides (Tegner *et al.* 2019). These include the SIP as well as the SDC. The emplacement of these magmatic rocks into their host rocks is interpreted to have coincided with and to have aided continental breakup and opening of the Iapetus Ocean between Baltica and Laurentia in the Ediacaran (Kjøll *et al.* 2019b, 2019a; Tegner *et al.* 2019). The SIP and SDC are interpreted to represent the remnants of a magma-rich rifted margin. The Cryogenian–Ediacaran successions and the SDC are interpreted to represent the intermediate to lower levels of a distal magma-rich rifted margin (Svenningsen 1994a; Kjøll *et al.* 2019a, 2019b), whereas the SIP is interpreted to represent a deep-seated magmatic conduit system that intruded their host rock at lower crustal levels (Larsen *et al.* 2018).

8.5. Acknowledgments

The Centre for Earth Evolution and Dynamics is funded by CoE-Grant 223272 from the Research Council of Norway. The project "Hyperextension in magma-poor and magma-rich domains along the pre-Caledonian passive margin of Baltica" was funded by Grant 250327/F20 from the Research Council of Norway.

FURTHER READING.– The above descriptions are abbreviated and often simplified. If interested in reading and learning further, the reader is referred to the following list of publications and references.

– General: (Abdelmalak *et al.* 2015; Domeier 2015; Epin *et al.* 2017; Jakob *et al.* 2019a; Kjøll *et al.* 2019b; Larsen *et al.* 2018; Robert *et al.* 2020; Tegner *et al.* 2019; Tugend *et al.* 2018).

8.6. References

Abdelmalak, M.M., Andersen, T.B., Planke, S., Faleide, J.I., Corfu, F., Tegner, C., Shephard, G.E., Zastrozhnov, D., Myklebust, R. (2015). The ocean-continent transition in the mid-Norwegian margin: Insight from seismic data and an onshore caledonian field analogue. *Geology*, 43, 1011–1014. https://doi.org/10.1130/G37086.1.

Abdelmalak, M.M., Planke, S., Faleide, J.I., Jerram, D.A., Zastrozhnov, D., Eide, S., Myklebust, R. (2016). The development of volcanic sequences at rifted margins: New insights from the structure and morphology of the Vøring Escarpment, mid-Norwegian Margin. *J. Geophys. Res., Solid Earth*, 121, 1–25.

Andersen, T.B., Corfu, F., Labrousse, L., Osmundsen, P.T. (2012). Evidence for hyperextension along the pre-Caledonian margin of Baltica. *J. Geol. Soc., London*, 169, 601–612. https://doi.org/10.1144/0016-76492012-011.

Andréasson, P.G., Allen, A., Aurell, O., Boman, D., Ekestubbe, J., Goerke, U., Lundgren, A., Nilsson, P., Sandelin, S. (2018). Seve terranes of the Kebnekaise Mts., Swedish Caledonides, and their amalgamation, accretion and affinity. *GFF*, 140, 264–291. https://doi.org/10.1080/11035897.2018.1470200.

Baird, G.B., Figg, S.A., Chamberlain, K.R. (2014). Intrusive age and geochemistry of the kebne dyke complex in the seve nappe complex, Kebnekaise Massif, arctic Sweden Caledonides. *GFF*, 136, 556–570. https://doi.org/10.1080/11035897.2014.924553.

Beltrando, M., Manatschal, G., Mohn, G., Dal Piaz, G.V., Vitale Brovarone, A., Masini, E. (2014). Recognizing remnants of magma-poor rifted margins in high-pressure orogenic belts: The alpine case study. *Earth-Science Rev.*, 131, 88–115. https://doi.org/10.1016/j.earscirev.2014.01.001.

Bingen, B., Demaiffe, D., van Breemen, O. (1998). The 616 ma old Egersund basaltic dike swarm, SW Norway, and late neoproterozoic opening of the Iapetus Ocean. *J. Geol.*, 106, 565–574. https://doi.org/10.1086/516042.

Cadow, R. (1993). Sm-Nd and Rb-Sr ages of hornblende clinopyroxenite and metagabbro from the Lillebukt Alkaline Complex, Seiland Igneous Province. *Nor. Geol. Tidsskr.*, 73, 243–249.

Clerc, C., Lagabrielle, Y., Neumaier, M., Reynaud, J.-Y., De Saint Blanquat, M. (2012). Exhumation of subcontinental mantle rocks: Evidence from ultramafic-bearing clastic deposits nearby the Lherz peridotite body, French Pyrenees. *Bull. la Soc. Géologique Fr.*, 183, 443–459.

Corfu, F., Roberts, R.J., Torsvik, T.H., Ashwal, L.D., Ramsay, D.M. (2007). Peri-Gondwanan elements in the Caledonian Nappes of Finnmark, Northern norway: Implications for the paleogeographic framework of the Scandinavian Caledonides. *Am. J. Sci.*, 307, 434–458. https://doi.org/10.2475/02.2007.05.

Corfu, F., Gerber, M., Andersen, T.B., Torsvik, T.H., Ashwal, L.D. (2011). Age and significance of Grenvillian and Silurian orogenic events in the Finnmarkian Caledonides, northern Norway. *Can. J. Earth Sci.*, 48, 419–440. https://doi.org/Doi10.1139/E10-043.

Corfu, F., Andersen, T.B., Gasser, D. (2014). The Scandinavian Caledonides: Main features, conceptual advances and critical questions. *Geol. Soc. Spec. Publ.*, 390, 9–43. https://doi.org/10.1144/SP390.25.

Desmurs, L., Manatschal, G., Bernoulli, D. (2001). The Steinmann Trinity revisited: Mantle exhumation and magmatism along an ocean-continent transition: The Platta nappe, Eastern Switzerland. *Geol. Soc. Spec. Publ.*, 187, 235–266. https://doi.org/10.1144/GSL.SP.2001.187.01.12.

Domeier, M. (2015). A plate tectonic scenario for the Iapetus and Rheic Oceans. *Gondwana Res.*, 36, 257–295.

Domeier, M. (2018). Early Paleozoic tectonics of Asia: Towards a full-plate model. *Geosci. Front.*, 9, 789–862. https://doi.org/10.1016/j.gsf.2017.11.012.

Elvevold, S., Reginiussen, H., Krogh, E.J., Bjørklund, F. (1994). Reworking of deep-seated gabbros and associated contact metamorphosed paragneisses in the southeastern part of the Seiland Igneous Province, Northern Norway. *Metamorph. Geol.*, 12, 539–556.

Epin, M.E., Manatschal, G., Amann, M. (2017). Defining diagnostic criteria to describe the role of rift inheritance in collisional orogens: The case of the Err-Platta nappes (Switzerland). *Swiss J. Geosci.*, 110, 419–438. https://doi.org/10.1007/s00015-017-0271-6.

Franke, D. (2013). Rifting, lithosphere breakup and volcanism: Comparison of magma-poor and volcanic rifted margins. *Mar. Pet. Geol.*, 43, 63–87. https://doi.org/10.1016/j.marpetgeo.2012.11.003.

Gasser, D., Jeřábek, P., Faber, C., Stünitz, H., Menegon, L., Corfu, F., Erambert, M., Whitehouse, M.J. (2015). Behaviour of geochronometers and timing of metamorphic reactions during deformation at lower crustal conditions: Phase equilibrium modelling and U-Pb dating of zircon, monazite, rutile and titanite from the Kalak Nappe Complex, northern Norway. *J. Metamorph. Geol.*, 33, 513–534. https://doi.org/10.1111/jmg.12131.

Gee, D.G., Andréasson, P.G., Li, Y., Krill, A. (2016). Baltoscandian margin, Sveconorwegian crust lost by subduction during Caledonian collisional orogeny. *GFF*, 139, 36–51. https://doi.org/10.1080/11035897.2016.1200667.

Gee, D.G., Klonowska, I., Andréasson, P.G., Stephens, M.B. (2020). Chapter 21: Middle thrust sheets in the Caledonide orogen, Sweden: The outer margin of Baltica, the continent – ocean transition zone and late Cambrian – Ordovician subduction – accretion, in: Sweden: Lithotectonic Framework. *Tect. Evol. Miner. Resour.*, 517–548.

Grant, T.B., Larsen, R.B., Anker-Rasch, L., Grannes, K.R., Iljina, M., McEnroe, S., Nikolaisen, E., Schanche, M., Øen, E. (2016). Anatomy of a deep crustal volcanic conduit system; The Reinfjord Ultramafic Complex, Seiland Igneous Province, Northern Norway. *Lithos*, 252–253, 200–215. https://doi.org/10.1016/j.lithos.2016.02.020.

Grant, T.B., Larsen, R.B., Brown, E.L., Müller, A.B., McEnroe, S. (2020). Mixing of heterogeneous, high-MgO, plume-derived magmas at the base of the crust in the Central Iapetus Magmatic Province (Ma 610–550): Origin of parental magmas to a global LIP event. *Lithos*, 364–365. https://doi.org/10.1016/j.lithos.2020.105535.

Hacker, B.R., Andersen, T.B., Johnston, S.M., Kylander-Clark, A.R.C., Peterman, E.M., Walsh, E.O., Young, D. (2010). High-temperature deformation during continental-margin subduction & exhumation: The ultrahigh-pressure Western Gneiss Region of Norway. *Tectonophysics*, 480, 149–171.

Hoffmann, K.H., Condon, D.J., Bowring, S.A., Crowley, J.L. (2004). U-Pb zircon date from the Neoproterozoic Ghaub Formation Namibia: Constraints on Marinoan glaciation. *Geology*, 32, 817–820. https://doi.org/10.1130/G20519.1.

Jakob, J., Alsaif, M., Corfu, F., Andersen, T.B. (2017). Age and origin of thin discontinuous gneiss sheets in the distal domain of the magma-poor hyperextended pre-Caledonian margin of Baltica, Southern Norway. *J. Geol. Soc., London*, 174, 557–571. https://doi.org/10.1144/jgs2016-049.

Jakob, J., Andersen, T.B., Kjøll, H.J. (2019). A review and reinterpretation of the architecture of the South and South-Central Scandinavian Caledonides – A magma-poor to magma-rich transition and the significance of the reactivation of rift inherited structures. *Earth-Science Rev.*, 192, 513–528. https://doi.org/10.1016/j.earscirev.2019.01.004.

Kirkland, C.L., Daly, J.S., Whitehouse, M.J. (2006). Granitic magmatism of grenvillian and late neoproterozoic age in Finnmark, Arctic Norway – Constraining pre-Scandian deformation in the Kalak Nappe Complex. *Precambrian Res.*, 145, 24–52. https://doi.org/10.1016/j.precamres.2005.11.012.

Kirkland, C.L., Daly, J.S., Whitehouse, M.J. (2007). Provenance and terrane evolution of the Kalak Nappe Complex, Norwegian Caledonides: Implications for neoproterozoic paleogeography and tectonics. *J. Geol.*, 115, 21–41. https://doi.org/10.1086/509247.

Kirkland, C.L., Bingen, B., Whitehouse, M.J., Beyer, E., Griffin, W.L. (2011). Neoproterozoic palaeogeography in the North Atlantic Region: Inferences from the Akkajaure and Seve Nappes of the Scandinavian Caledonides. *Precambrian Res.*, 186, 127–146. https://doi.org/10.1016/j.precamres.2011.01.010.

Kirsch, M. and Svenningsen, O.M. (2015). Root zone of a continental rift: The Neoproterozoic Kebnekaise Intrusive Complex, Northern Swedish Caledonides. *GFF*, 138, 31–53.

Kjøll, H.J. (2020). Late Neoproterozoic basin evolution of the magma rich Iapetus margin of Baltica. *Nor. J. Geol.*, 1–32. https://doi.org/10.17850/njg100-1-6.

Kjøll, H.J., Andersen, T.B., Corfu, F., Labrousse, L., Tegner, C., Abdelmalak, M.M., Planke, S. (2019a). Timing of breakup and thermal evolution of a pre-Caledonian Neoproterozoic exhumed magma-rich rifted margin. *Tectonics*, 38, 1843–1862. https://doi.org/10.1029/2018TC005375.

Kjøll, H.J., Galland, O., Labrousse, L., Andersen, T.B. (2019b). Emplacement mechanisms of a dyke swarm across the brittle-ductile transition and the geodynamic implications for magma-rich margins. *Earth Planet. Sci. Lett.*, 518, 223–235. https://doi.org/10.1016/j.epsl.2019.04.016.

Kumpulainen, R.A. (1980). Upper proterozoic stratigraphy and depositional environments of the Tossåsfjället group, Särv nappe, Southern Swedish Caledonides. *GFF*, 102, 531–550. https://doi.org/10.1080/11035898009454506.

Kumpulainen, R.A. (2011). The neoproterozoic glaciogenic Lillfjället formation, southern Swedish Caledonides. *Geol. Soc. Mem.*, 36, 629–634. https://doi.org/10.1144/m36.61.

Kumpulainen, R.A., Hamilton, M.A., Söderlund, U., Nystuen, J.P. (2016). A new U-Pb baddeleyite age for the Ottfjället dolerite dyke swarm in the Scandinavian Caledonides – A minimum age for Late Neoproterozoic glaciation in Baltica. *Nordic Geolgical Wintermeeting*, Helsinki, Finland. https://doi.org/10.1007/s13398-014-0173-7.2.

Lagabrielle, Y., Asti, R., Fourcade, S., Corre, B., Labaume, P., Uzel, J., Clerc, C., Lafay, R., Picazo, S. (2019). Mantle exhumation at magma-poor passive continental margins. Part II: Tectonic and metasomatic evolution of large-displacement detachment faults preserved in a fossil distal margin domain (Saraillé lherzolites, northwestern Pyrenees, France). *BSGF – Earth Sci. Bull.*, 190, 14. https://doi.org/10.1051/bsgf/2019013.

Larsen, R.B., Grant, T., Sørensen, B.E., Tegner, C., McEnroe, S., Pastore, Z., Fichler, C., Nikolaisen, E., Grannes, K.R., Church, N., ter Maat, G.W., Michels, A., ter Maat, G.W. (2018). Portrait of a giant deep-seated magmatic conduit system: The Seiland Igneous Province. *Lithos*, 296–299, 600–622. https://doi.org/10.1016/j.lithos.2017.11.013.

Manatschal, G. (2004). New models for evolution of magma-poor rifted margins based on a review of data and concepts from West Iberia and the Alps. *Int. J. Earth Sci.*, 93, 432–466.

Manatschal, G. and Nievergelt, P. (1997). A continent-ocean transition recorded in the Err and Platta nappes (Eastern Switzerland). *Eclogae Geol. Helv.*, 90, 3–27.

Meert, J.G., Torsvik, T.H., Eide, E.A., Dahlgren, S. (1998). Tectonic significance of the Fen Province, S. Norway: Constraints from geochronology and paleomagnetism. *J. Geol.*, 106, 553–564. https://doi.org/10.1086/516041.

Meert, J.G., Walderhaug, H.J., Torsvik, T.H., Hendriks, B.W.H.H. (2007). Age and paleomagnetic signature of the Alnø carbonatite complex (NE Sweden): Additional controversy for the Neoproterozoic paleoposition of Baltica. *Precambrian Res.*, 154, 159–174. https://doi.org/10.1016/j.precamres.2006.12.008.

Menzies, M.A., Klemperer, S.L., Ebinger, C.J., Baker, J. (2002). Characteristics of volcanic rifted margins. *Spec. Pap. Geol. Soc. Am.*, 362, 1–14. https://doi.org/10.1130/0-8137-2362-0.1.

Merdith, A.S., Collins, A.S., Williams, S.E., Pisarevsky, S., Foden, J.D., Archibald, D.B., Blades, M.L., Alessio, B.L., Armistead, S., Plavsa, D., Clark, C., Müller, R.D. (2017). A full-plate global reconstruction of the neoproterozoic a full-plate global reconstruction of the Neoproterozoic. *Gondwana Res.* https://doi.org/10.1016/j.gr.2017.04.001.

Mohn, G., Manatschal, G., Müntener, O., Beltrando, M., Masini, E. (2010). Unravelling the interaction between tectonic and sedimentary processes during lithospheric thinning in the Alpine Tethys margins. *Int. J. Earth Sci.*, 99, 75–101. https://doi.org/10.1007/s00531-010-0566-6.

Mohn, G., Manatschal, G., Beltrando, M., Haupert, I. (2014). The role of rift-inherited hyper-extension in Alpine-type orogens. *Terra Nov.*, 26, 347–353. https://doi.org/10.1111/ter.12104.

Nystuen, J.P., Andresen, A., Kumpulainen, R.A., Siedlecka, A.W.G. (2008). Neoproterozoic basin evolution in Fennoscandia, East Greenland and Svalbard. *Episodes*, 31, 35–43.

Nystuen, J.P., Kumpulainen, R.A., Söderlund, U., Hamilton, M.A. (2016). The Varangerian/Marinoan glaciation in Scandinavia – New age constraints. *Nord. Geol.*, Wintermeeting.

Pedersen, R.B., Dunning, G.R., Robins, B., Dunningr, G.R., Robins, B. (1989). U–Pb ages of nepheline syenite pegmatites from the Seiland Magmatic Province, N Norway. *The Caledonide Geology of Scandinavia*, 3–8. https://doi.org/10.1007/978-94-009-2549-6_1.

Peron-Pinvidic, G. and Manatschal, G. (2009). The final rifting evolution at deep magma-poor passive margins from Iberia-Newfoundland: A new point of view. *Int. J. Earth Sci.*, 98, 1581–1597. https://doi.org/10.1007/s00531-008-0337-9.

Peron-Pinvidic, G. and Manatschal, G. (2010). From microcontinents to extensional allochthons: Witnesses of how continents rift and break apart? *Pet. Geosci.*, 16, 189–197.

Peron-Pinvidic, G., Manatschal, G., Osmundsen, P.T. (2013). Structural comparison of archetypal Atlantic rifted margins (Angola-Esperito Santo, Iberia-Newfoundland, mid. Norway-East Greenland). *Mar. Pet. Geolgoy*, 43, 21–47.

Puffer, J.H. (2002). A late Neoproterozoic eastern Laurentian superplume: Location, size, chemical composition, and environmental impact. *Am. J. Sci.*, 302, 1–27. https://doi.org/10.2475/ajs.302.1.1.

Reston, T. and Manatschal, G. (2011). Rifted margins: Building blocks of later collision. In *Arc-Continent Collision, Frontiers in Earth Sciences*, Brown, D. and Ryan, P.D. (eds). https://doi.org/10.1007/978-3-540-88558-0.

Robert, B., Domeier, M., Jakob, J. (2020). Iapetan Oceans: An analog of Tethys? *Geology*, 48, 929–933. https://doi.org/https://doi.org/10.1130/G47513.1.

Roberts, R.J., Corfu, F., Torsvik, T.H., Ashwal, L.D., Ramsay, D.M. (2006). Short-lived mafic magmatism at 560–570 Ma in the northern Norwegian Caledonides: U-Pb zircon ages from the Seiland Igneous Province. *Geol. Mag.*, 143, 887–903. https://doi.org/10.1017/S0016756806002512.

Roberts, R.J., Corfu, F., Torsvik, T.H., Hetherington, C.J., Ashwal, L.D. (2010). Age of alkaline rocks in the Seiland Igneous Province, northern Norway. *J. Geol. Soc., London*, 167, 71–81. https://doi.org/10.1144/0016-76492009-014.

van Staal, C.R., Chew, D.M., Zagorevski, A., Hibbard, J., Castonguay, S., Escayola, M.P., Paul, J. (2013). Evidence of Late Ediacaran Hyperextension of the Lau – Rentian Iapetan Margin in the Birchy Complex, Baie Verte Peninsula, Northwest Newfoundland: Implications for the Opening of Iapetus, Formation of Peri-Laurentian Microcontinents and Taconic – Grampi. *Geosci., Canada*, 40, 1–25.

Stephens, M.B. and Gee, D.G. (1989). Terranes and polyphase accretionary history in the Scandinavian Caledonides. *Geol. Soc. Am. Spec. Pap.*, 230, 17–30.

Svenningsen, O.M. (1994a). The Baltica–Iapetus passive margin dyke complex in the Sarektjåkkå Nappe, northern Swedish Caledonides. *Geol. J.*, 29, 323–354. https://doi.org/10.1002/gj.3350290403.

Svenningsen, O.M. (1994b). Tectonic significance of the meta-evaporitic magnesite and scapolite deposits in the Seve Nappes, Sarek Mts., Swedish Caledonides. *Tectonophysics*, 231, 33–44. https://doi.org/10.1016/0040-1951(94)90119-8.

Svenningsen, O.M. (1995). Extensional deformation along the late Precambrian-Cambrian Baltoscandian passive margin: the Sarektjåkkå Nappe, Swedish Caledonides. *Geol. Rundschau*, 84, 649–664. https://doi.org/10.1007/BF00284527.

Svenningsen, O.M. (2001). Onset of seafloor spreading in the Iapetus ocean at 608 Ma: Precise age of the Sarek Dyke Swarm, northern Swedish Caledonides. *Precambrian Res.*, 110, 241–254. https://doi.org/10.1016/S0301-9268(01)00189-9.

Tegner, C., Robins, B., Reginiussen, H., Grundvig, S. (1999). Assimilation of crustal xenoliths in a basaltic magma chamber: Sr and Nd isotopic constraints from the Hasvik layered intrusion, Norway. *J. Petrol.*, 40, 363–380. https://doi.org/10.1093/petroj/40.3.363.

Tegner, C., Andersen, T.B., Kjøll, H.J., Brown, E.L., Hagen-Peter, G., Corfu, F., Planke, S., Torsvik, T.H. (2019). A mantle plume origin for the Scandinavian dyke complex: A "Piercing Point" for 615 Ma Plate Reconstruction of Baltica? *Geochem., Geophys. Geosyst.*, 20, 1075–1094. https://doi.org/10.1029/2018GC007941.

Torsvik, T.H. and Cocks, L.R.M. (2005). Norway in space and time: A centennial cavalcade. *Nor. J. Geol.*, 85, 73–86.

Torsvik, T.H., Smethurst, M.A., Burke, K., Steinberger, B. (2006). Large igneous provinces generated from the margins of the large low-velocity provinces in the deep mantle. *Geophys. J. Int.*, 167, 1447–1460. https://doi.org/10.1111/j.1365-246X.2006.03158.x.

Turino, V., Magni, V., Kjøll, H.J., Jakob, J. (2021). The effect of magma poor and magma rich passive margins on continental collision dynamics. Report, EGU General Assembly.

Wilson, J.T. (1966). Did the Atlantic close and then re-open? *Nature*, 211, 676–681.

9

Cretaceous Rifting at the Pyrenean Iberia–Eurasia Border: Crustal Thinning under a Pre-rift Cover – Application to Passive Margins

Yves LAGABRIELLE
CNRS-Géosciences Rennes, University of Rennes 1, France

9.1. The North Pyrenean area within the Pyrenees

The evolution of the Pyrenean range during the Mesozoic and Cenozoic can be summarized in a succession of three main stages:

1) the establishment of a post-Triassic carbonate platform on the eroded continental basement of an old Variscan chain;

2) the dislocation of this platform in the Lower and Middle Cretaceous during the opening of the North Pyrenean basins. These basins were filled with fine syn-rift turbidites whose base composed the "black flysch group" (Upper Albian–Lower Cenomanian); and finally

---

For a color version of all the figures in this chapter, see www.iste.co.uk/peronpinvidic/margins2.zip.

Continental Rifted Margins 2,
coordinated by Gwenn PERON-PINVIDIC. © ISTE Ltd 2022.

3) the inversion of all the basins and their margins from the Upper Cretaceous (Santonian), ending in the Upper Eocene (Pyrenean–Provençal phase) with the complete uplift of the chain involving Paleozoic basement nappes, mainly tilted towards the south and forming the current culmination of the axial zone (AZ) (Figure 9.1).

The North Pyrenean Zone (NPZ) belongs to the northern flank of the Pyrenean chain and continuously outcrops to the north of the AZ from Perpignan to the Basque Country (Figure 9.1). It is essentially composed of Mesozoic sediments associated with Paleozoic basement massifs (North Pyrenean massifs) and results from the inversion and northward thrusting of the basins formed during the Cretaceous rifting phase (Souquet *et al.* 1977; Canérot *et al.* 1978; Debroas 1978, 1990; Vielzeuf and Kornprobst 1984; Canérot and Delavaux 1986).

The evolution of the Pyrenees is linked to the relative mobility of the Iberian and Eurasian plates. The interpretation of the Atlantic oceanic magnetic anomalies indicates that the opening of the Bay of Biscay is the result of the activity of an E–W-oriented spreading ridge, active from the Albian to the Cenomanian–Turonian. This opening was accompanied by the relative displacement of Iberia towards the SE (Le Pichon *et al.* 1970; Pitman and Talwani 1972; Biju-Duval *et al.* 1977; Boillot and Capdevila 1977; Choukroune and Mattauer 1978; Olivet 1996). The opening of the Cretaceous basins of the future NPZ occurred during this trans-tensional movement between Iberia and Eurasia. The subsequent tectonic inversion of these basins is related to the northward movement of Iberia starting in the Santonian (Choukroune and ECORS team 1989; Roure *et al.* 1989; Muñoz 1992; Deramond *et al.* 1993; Roure and Choukroune 1998; Teixell 1998; Beaumont *et al.* 2000; Vergés and Garcia-Senz 2001; Vergés *et al.* 2002; Mouthereau *et al.* 2014; Ford *et al.* 2016; Pedrera *et al.* 2017; Canérot 2017a; Grool *et al.* 2018; Teixell *et al.* 2018).

Several mechanisms for the opening of the Bay of Biscay have been proposed: scissor, sliding or orthogonal displacements, with poles of rotation located in various places (see the summary in Lescoutre (2019)). Recent published models differ in their timing and in the amplitude of the opened and closed areas (Srivastava *et al.* 1990, 2000; Sibuet *et al.* 2004; Gong *et al.* 2008; Jammes *et al.* 2009; Vissers and Meijer 2012). Kinematic models generally consider that inter-plate deformation was concentrated along the Eurasia–Iberia boundary with non-deformable Iberia, which is certainly not the case, as pointed out by Canérot (2017b). The entire northern half of Iberia indeed constitutes a large unstable domain during the Meso-Cenozoic. This unstable domain comprises a set of basins (Cameros Basin, Iberian Range Basins, Columbrets Basin) separated by continental micro-blocks on land or at sea (Ebro Block, Le Danois Bank, Landes Shoal).

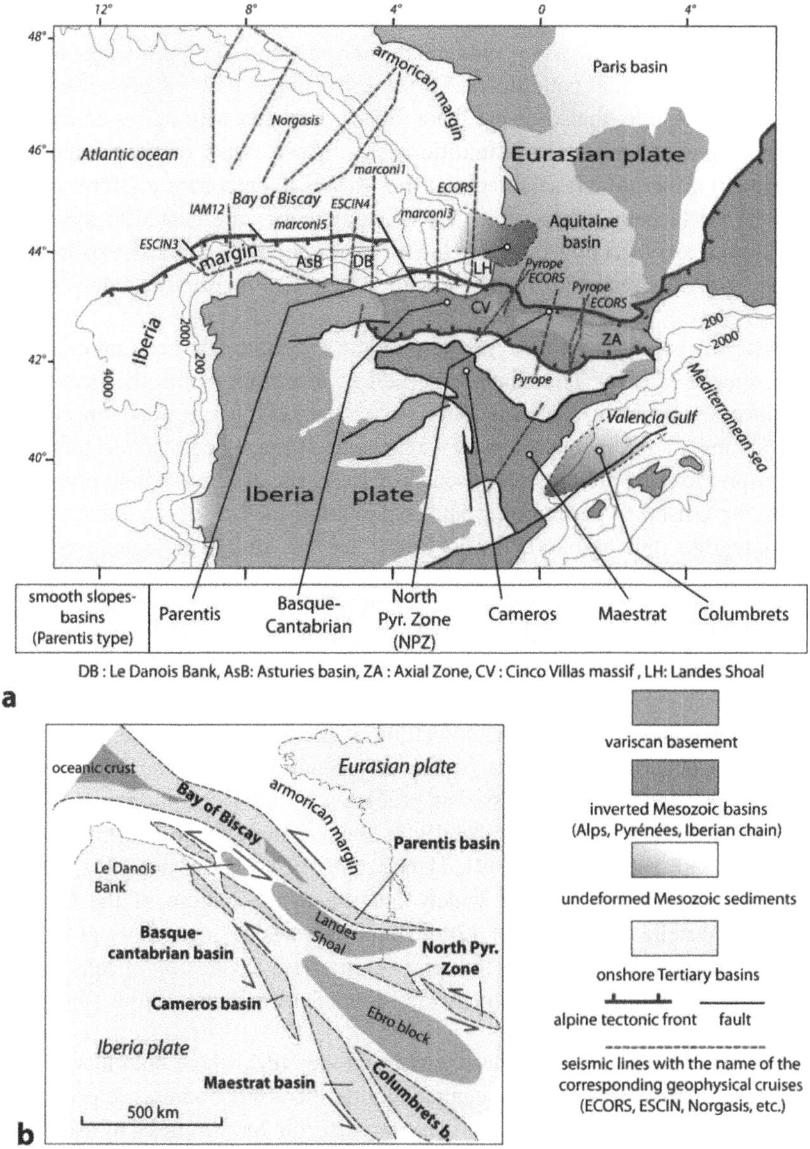

Figure 9.1. *The Pyrenean–Cantabrian chain. a) Simplified map and position of the peri-Pyrenean basins mentioned in this chapter. b) Reconstruction of the position of the Cretaceous basins related to the Pyrenean rifting at the beginning of the Iberian drift (b. for basin). Note the width of the deformed domain forming the Iberian–Eurasian boundary and the position of micro-continents within it (Ebro block, Landes Shoal, Le Danois Bank)*

The geology of the NPZ is unique in the world and of major interest for the study of the formation of passive continental margins and, in particular, for better constraining the thermal conditions of crustal thinning. The NPZ contains remnants of the Earth's mantle, the so-called lherzolites, associated with lenses of continental material, some of which are granulitic rocks. These rocks of deep-seated origin outcrop within the pre-rift sediments of the Mesozoic carbonate platform deformed during the Cretaceous extension. The NPZ sediments were heated to temperatures ranging from 250°C to 600°C, under low-pressure conditions, consistent with burial under the syn-rift Cretaceous flysch (so-called Pyrenean thermo-metamorphism).

A renewal of interest in the geology of the Pyrenees has been taking place in France since the 2010s. The numerous studies carried out within the framework of governmental and industrial programs have, among other results, improved our understanding of the mechanisms of Cretaceous rifting. Geodynamic processes at the lithospheric scale are now proposed to describe the different rifting phases. They account for crustal hyper-thinning and exhumation of the sub-continental mantle. They integrate the remarkably coherent set of sedimentological, structural, geochronological and metamorphic constraints acquired over 150 years in the NPZ.

9.2. Mechanisms of crustal thinning deduced from the study of the NPZ

All authors now agree that the upwelling of the lherzolites resulted from crustal thinning that occurred at the axis of the Cretaceous basins of the NPZ during the translation of Iberia. This process was predicted by many geologists in the 1980s and 1990s (Vielzeuf and Kornprobst 1984; Dauteuil and Ricou 1989; Golberg and Leyreloup 1990; Fabriès *et al.* 1991; Henry *et al.* 1998) but at that time, it was not envisaged that peridotites could widely outcrop at the bottom of the Cretaceous basins. Lagabrielle and Bodinier (2008) and Jammes *et al.* (2009) proposed for the first time models of mantle exposure in these basins by means of large, shallow-dipping normal faults named extensional detachments.

Geological studies in the western massifs of the NPZ, where both lherzolites and small variscan basement units lie against mantle lherzolite bodies, are prime targets for studying the Pyrenean rifting. In these massifs, the temperatures of the Pyrenean thermo-metamorphism did not exceed 350–400°C. At the Urdach-les Pernes and Saraillé sites (Chaînons Béarnais area), preserved original contacts between the mantle rocks and the continental crust are accessible to observation. There, it is possible to decipher rheological behaviors during Cretaceous rifting and to specify the deformation modes of the pre-rift cover during extension, as shown by the works of Corre *et al.* (2016), Asti *et al.* (2019) and Lagabrielle *et al.* (2019a, 2019b).

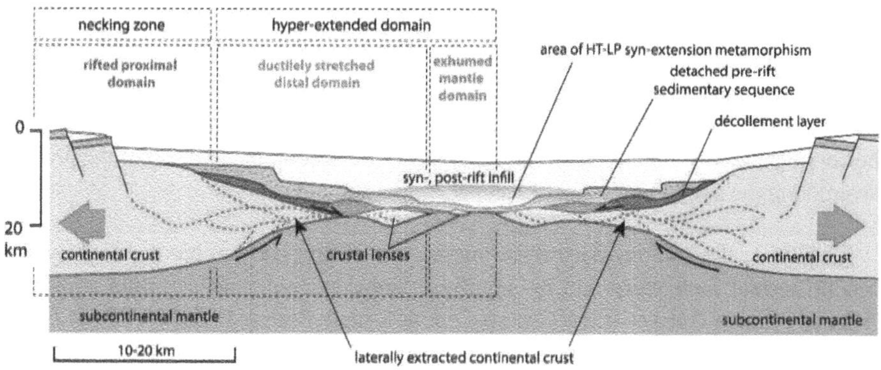

Figure 9.2. *The concept of smooth-slope basins: a model of the Cretaceous basin architecture in the NPZ. Field data allow us to propose a reconstruction of the distal domain of the northern Iberian margin in the Late Cretaceous. Note a highly symmetrical rift domain, dominated by ductile processes in its center and the occurrence of continental bedrock boudins (or tectonic lenses) lying over the exhumed mantle (from Lagabrielle* et al. *2020). The outer domain displays a more classical brittle behavior*

All of these works report the individualization of thin hecto-metric Palaeozoic crustal units remaining adherent to the exhumed mantle in most of the lherzolitic massifs of the NPZ. Recent studies show that at the rift axis, under temperature conditions in the range of 350–400°C, the continental crust ductilely and homogeneously deforms as it is progressively extracted laterally from the center of the basins (Asti *et al.* 2019). Crustal deformation occurs through the combined interplay of anastomosing symmetric mylonitic shear zones and ductile flattening within each of the tectonic lenses. These data led Lagabrielle *et al.* (2020) to define a type of smooth-slope basin characterized by the ductile behavior of its central part (Figure 9.2)[1]. This model completes the models proposed by Clerc *et al.* (2012, 2016), Clerc and Lagabrielle (2014), and de Saint Blanquat et *al.* (2016). It is also close to those proposed by Ducoux (2017), Ducoux *et al.* (2019) and Saspiturry *et al.* (2020). Note that at the maximum temperatures considered in the central domain, the crust never reaches partial melting conditions. Therefore, there is no granitization related to the Pyrenean "hot" rifting.

Based on the description of geological sections and the mineralogical study of the Pyrenean thermo-metamorphism, Lagabrielle *et al.* (2019a, 2019b) identify two major

1 The absence of normal faults only concerns the central part of the system, the hottest, corresponding to the distal margin. The proximal margin is characterized by a classical brittle evolution.

extensive fault zones active in the Middle Cretaceous. These two fault zones located at the crust–mantle boundary and at the base of the detached pre-rift series have been named crust–mantle detachment and cover detachment, respectively (Figure 9.3). They led to (i) the uplift of lherzolites below the basin axis, (ii) the lateral extraction of the continental crust below the rift edges and (iii) the decoupling and slip of the pre-rift cover along the 2 km-thick level of Keuper evaporites and clays.

Finally, in the smooth-slope basin model, the central part is characterized by the lack of normal fault scarp and by significant subsidence devoid of brittle structure. This evolution is consistent with the concept of a sag basin (i.e. a basin without faults cutting the basement and the edges of the system) (Figure 9.3). The notion of relative allochthony of the pre-rift series, mobile on the Upper Triassic layer, is fundamental and represents a major contribution to understanding the evolution of the NPZ. Basement–sediment decoupling allows the pre-rift cover to remain at the bottom of the widening and deepening basin. Heat related to mantle uplift is thus maintained in the distal part of the margins, and the thermal anomaly is sustained there for at least 20 Myr (Saspiturry *et al.* 2020). The pre-rift cover is locally torn to reveal the mantle through tectonic windows, as deduced from the geology of the Urdach Massif (Jammes *et al.* 2009; Debroas *et al.* 2010; Lagabrielle *et al.* 2019a, 2019b).

In the western part of the NPZ, temperatures of the fluids responsible for metamorphic crystallization within the extensive detachments did not exceed 350°C but reached 600°C east of the Pyrenees (Aulus and Boucheville basins: Clerc *et al.* 2015b). Thermicity increased, on the other hand, by burial effect under the depositing flysch. For this reason, the thinned continental crust was not cooled below 350°C at the basin axis.

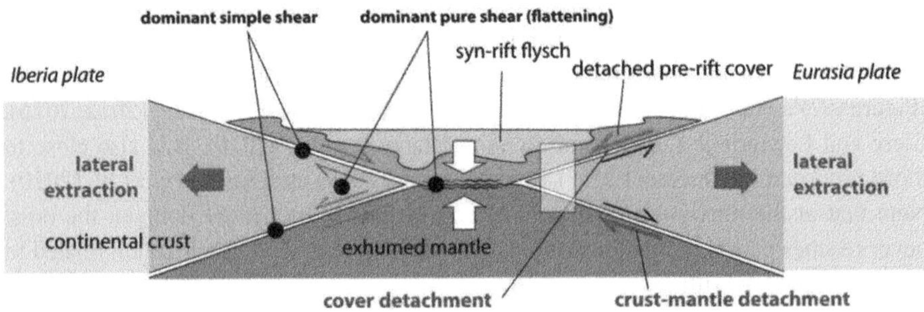

Figure 9.3. *Theoretical diagram showing the geometry of major faults allowing lateral extraction of the hyper-thinned crust from the center of the Cretaceous basins of the future NPZ. The general deformation regime in the different regions of the model and the behavior of the pre-rift and syn-rift sedimentary series during extension are shown (modified from Lagabrielle* et al. *2019a, 2019b)*

9.3. Crustal thinning at the Iberian–Eurasian plate boundary: smooth-slope basins and the legacy of Triassic paleogeography

9.3.1. *Cretaceous basins along the diffuse Iberian–Eurasian border*

As we have seen, the NPZ is the result of the inversion of extensional and transcurrent structures distributed along the Iberian–Eurasian plate boundary. These structures started functioning in the early Aptian–Albian times, possibly earlier, when Iberia began its south-eastward migration away from Armorica (Figure 9.1). The concept of pull-apart basins in which segments of transform faults are linked by rectangular or rhombic relays was used as early as the 1970s (Choukroune and Mattauer 1978; Choukroune 1992). It was later adopted by Jammes *et al.* (2009).

In fact, there are very few constraints to reconstruct the distribution of the basins and major faults that accommodated the displacements between the Iberian and the Eurasian plates. Any attempt to restore these structures depends first on determining the initial position of Iberia, but this is poorly known. Next, it is necessary to quantify the surface of the opened domain that was lost by subduction during the convergence. Shortening rates across the Pyrenees do not exceed 200 km (Teixell *et al.* 2018), so it is likely that the width of the closed domain did not exceed about 100 km. However, for some authors, based on extreme interpretations of oceanic magnetic anomalies, the newly formed domain with oceanic affinity was much larger, thus leading us to envision that large portions of subducted lithosphere disappeared beneath the European margin or within Iberia itself (Sibuet *et al.* 2004; Vissers and Meijer 2012). None of the magmatic markers of subduction zones have ever been found in the Pyrenean domain, and these extreme interpretations have not been confirmed by field studies.

In Figure 9.4, we reproduce an evolutionary model for the paleogeography of the numerous basins that developed along the Iberia–Europa plate boundary during the Cretaceous (R. Asti, pers. comm.,). This model is based on the plate reconstructions by Angrand *et al.* (2020). It favors a lenticular shape and an echelon setting of the basins, and implies numerous and diffuse relay zones between fault segments with various horizontal offsets. Along the Iberia–Eurasia boundary itself, the Basque–Cantabrian basin was connected to the North Pyrenean basins (future NPZ). These two main basins probably displayed an *en échelon* setting with a complex relay zone including the Landes Shoal and the Cinco Villas–Alduides massif (Lescoutre 2019; Lescoutre and Manatschal 2020) (Figure 9.1).

The termination of the ocean-bottom domain propagating eastward from the Bay of Biscay is not found in the Basque–Cantabrian Basin nor in the NPZ, which were located further south, but in the Parentis Basin which terminates in a V-shape within the Aquitaine Platform. Cretaceous basins near the edge of each plate, but further from the boundary, were weakly inverted or completely unaffected by Cenozoic compression. On the Iberian plate, these weakly inverted basins are found north of the Iberian Meseta (e.g. Cameros and Maestrat basins) and in the Iberian chain (see the summary in Rat *et al.* (2019)). The opening of the basins south of the Pyrenees often started in the Lower Cretaceous or even in the very late Jurassic. These basins are therefore younger than those in the NPZ. This shows that the trans-tensional deformation is progressively focused on the core of a narrower and narrower corridor. The mantle will eventually be exhumed along this narrow corridor as is the case in the Basque–Cantabrian Basin, the Mauléon Basin and all the basins of the future NPZ (Figure 9.4).

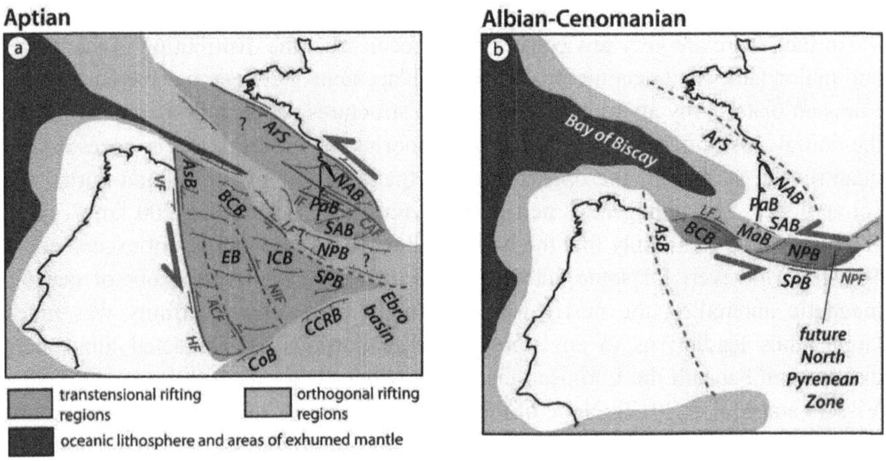

Figure 9.4. *Basins involved in the evolution of the diffuse Eurasian/Iberian boundary during the Aptian and Cenomanian times on a kinematic model by Angrand* et al. *(2020) (R. Asti, pers. comm.). These figures show the highly distributed character of the initial transtensional deformation and its progressive focusing along a narrower corridor. The subcontinental mantle may be exhumed in the Basque–Cantabrian, Mauléon and future NPZ basins, located in the central part of this corridor*

CONTINUATION OF CAPTION FOR FIGURE 9.4. *ACF, Ateca–Castellon Fault; AsB, Asturias Basin; ArS, Armorican margin; BCB, Basque–Cantabrian Basin; CAF, Celtic–Aquitaine flexure; CCRB, Catalan Range Basin; CoB, Columbrets Basin; EB, Ebro Basin; HF, Hesperian Fault; ICB, Iberian Range Basins; IF, Ibis Fault; LF, Leiza Fault; MaB, Mauléon Basin; NAB, North Aquitaine Basins; NIF, North Iberian Fault; NPB, North Pyrenean Basins; PaB, Parentis Basin; SAB, South Aquitaine Basins; SPB, South Pyrenean Basins; VF, Ventaniella Fault.*

9.3.2. *General characteristics of "smooth-slope" basins – consideration of continental crust thinning modes*

Smooth-slope basins are distinguished from step-edge basins (which evolve under the control of steeply dipping faults) by the absence of tilted blocks in the center of the hyper-extended zone (Figures 9.2 and 9.3). The predominantly ductile deformation of the continental crust during mantle exhumation is enabled by the slip of the pre-rift cover that remains in the center of the basin during its evolution (Ethève *et al.* 2018; Asti *et al.* 2019).

Smooth-slope basins in the Pyrenean–Iberian domain can be classified according to the degree of total extension they exhibit (Saspiturry *et al.* 2021). An increase in extension rate is associated with a larger volume of uplifted mantle, a higher heat flow, a stronger crustal thinning and greater subsidence and burial. Figure 9.5 shows how reconstructed sections in different regions of the peri-Pyrenean domain and in the NPZ allow for reconstructing a theoretical evolutionary outline of the smooth-slope basins of the diffuse Iberian/Eurasian boundary.

In order to place the thinning mode of the NPZ basins into a more theoretical context, it is important to recall that two major exclusive or complementary deformation regimes control the structure of extended domains: simple shear and pure shear. Most of the concepts attached to these contrasted tectonic regimes were shown and discussed by the pioneers McKenzie (1978), Wernicke (1981, 1985), Lister and Davis (1989) and Lister *et al.* (1991), and many debates on the symmetry or lack of symmetry of lithospheric and crustal thinning processes have thus flourished (Buck *et al.* 1988; Allemand *et al.* 1989; Buck 1991, 1999) (Questions about the symmetry of hyper-extending systems are also discussed when considering the structure of current passive continental margins; Reston *et al.* 1995; Michon and Merle 2003; Huismans and Beaumont 2007; Reston and Pérez-Gussinyé 2007; Sutra *et al.* 2013; Brune *et al.* 2014.)

The fundamental questions regarding the relative importance of ductile versus brittle regimes of deformation during continental extension also apply to the evolution of smooth-slope basins. As discussed by Brun and Beslier (1996) and Nagel and Buck (2004), an apparently symmetrical structure does not imply a dominant pure shear mode, but may result from a combination of simple shears discretely distributed in the crust. In their analogical experiments using sand and silicon to simulate mantle exhumation at the continent–ocean transition, Brun and Beslier (1996) show that the lower crust can ductilely deform while the upper mantle layers remain brittle (Figure 9.11a).

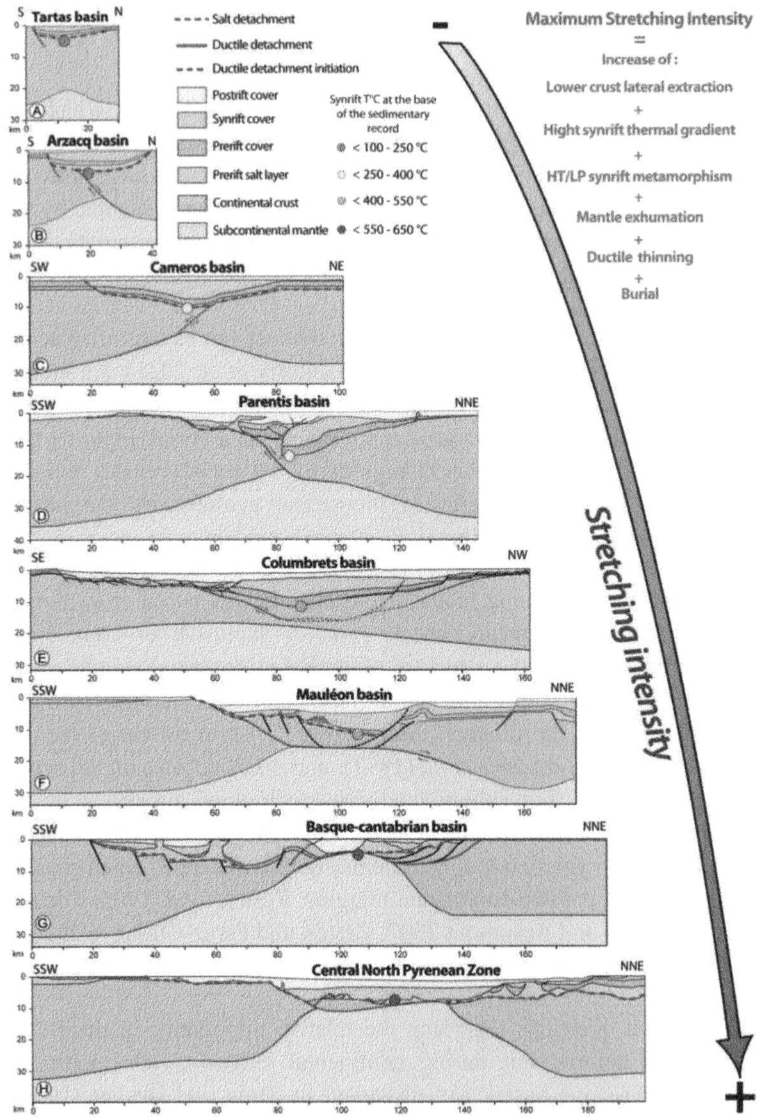

Figure 9.5. *Theoretical ranking of the smooth-sloping basins of the peri-Pyrenean domain and the NPZ according to their extension rate. Increased extension rate is associated with a greater mantle uplift, a higher syn-rift geothermal gradient, a stronger crustal thinning, greater subsidence and burial, etc. These reconstructed sections in different regions of the diffuse Iberian/Eurasian boundary allow us to propose a theoretical evolutionary diagram for smooth-slope basins (modified from Saspiturry* et al. *(2021))*

The model by Brun and Beslier (1996) shows analogies with the behavior of the lithosphere in the case of the Parentis Basin or the NPZ basins (Figure 9.6a). It implies a homogeneous thinning of the crust during its lateral extraction off the rift axis, as suggested by Corre *et al.* (2016) and Teixell *et al.* (2016, 2018) for the western NPZ. It is also consistent with the symmetric ductile thinning model proposed by Asti *et al.* (2019) for the crust in this region. This extension is similar to large-scale homogeneous pure shear deformation but is achieved through the interplay of mylonitic shear zones that separate tectonic lenses that are themselves flattened. Such a deformation mode occurs in the green schist facies around 350–450°C.

Symmetrical modes of deformation have been proposed by geologists who deciphered the thinning modes of the Basin and Range Province basement, as shown in Figure 9.6b. Hamilton (1987) describes tectonic lenses separated by undulating shear contacts. Here, crustal thinning is accompanied by the uplift of deep lenses that come into contact with shallower tectonic lenses. In his analogical models of hyper-thinned zones, such as passive margins, Gartrell (1997) (Figure 9.6c) also proposes large-scale boudinage and describes a succession of necking zones in which the ductile crust is sheared to the extreme. This results in a structure composed of a series of boudins similar to geometries accounting for the origin of the North Pyrenean massifs within the central NPZ (lower profile in Figure 9.5; Espurt *et al.* (2019)).

As discussed above, in their study of the Urdach and Saraillé massifs, Lagabrielle *et al.* (2019a, 2019b) describe two types of detachment faults that accommodated the extension of the rift system: the crust–mantle detachment and the cover detachment (Figures 9.2 and 9.3).

The crust–mantle detachment consists of a 50 m-thick basal part comprising metric tectonic lenses of serpentinites and a 10 m-thick deformed zone of talc-rich, serpentine schist lenses. The cover detachment separates the detached pre-rift rift cover from the Paleozoic basement or the subcontinental mantle. It is a 10 m-to 30 m-thick zone of strong deformation that was the site of hydrothermal circulation and metamorphic crystallization involving Triassic fluids. At the outcrop scale, it is difficult to discriminate between pure versus simple shear deformation processes. Indeed, in the mantle, a major shear zone, such as a detachment fault, can be formed by a succession of relatively symmetrical lenses suggesting, at least locally, a dominant flattening mode.

Finally, in the studied cases, the dominant pure shear mechanisms are concentrated within tectonic lenses of continental crustal material, whereas pure shear processes are concentrated in detachments, if considered at their regional scale. In addition, at the regional scale, pure shear flattening mechanisms develop within the pre-rift cover which ductilely thins during extension, as shown in Figures 9.3 and 9.5.

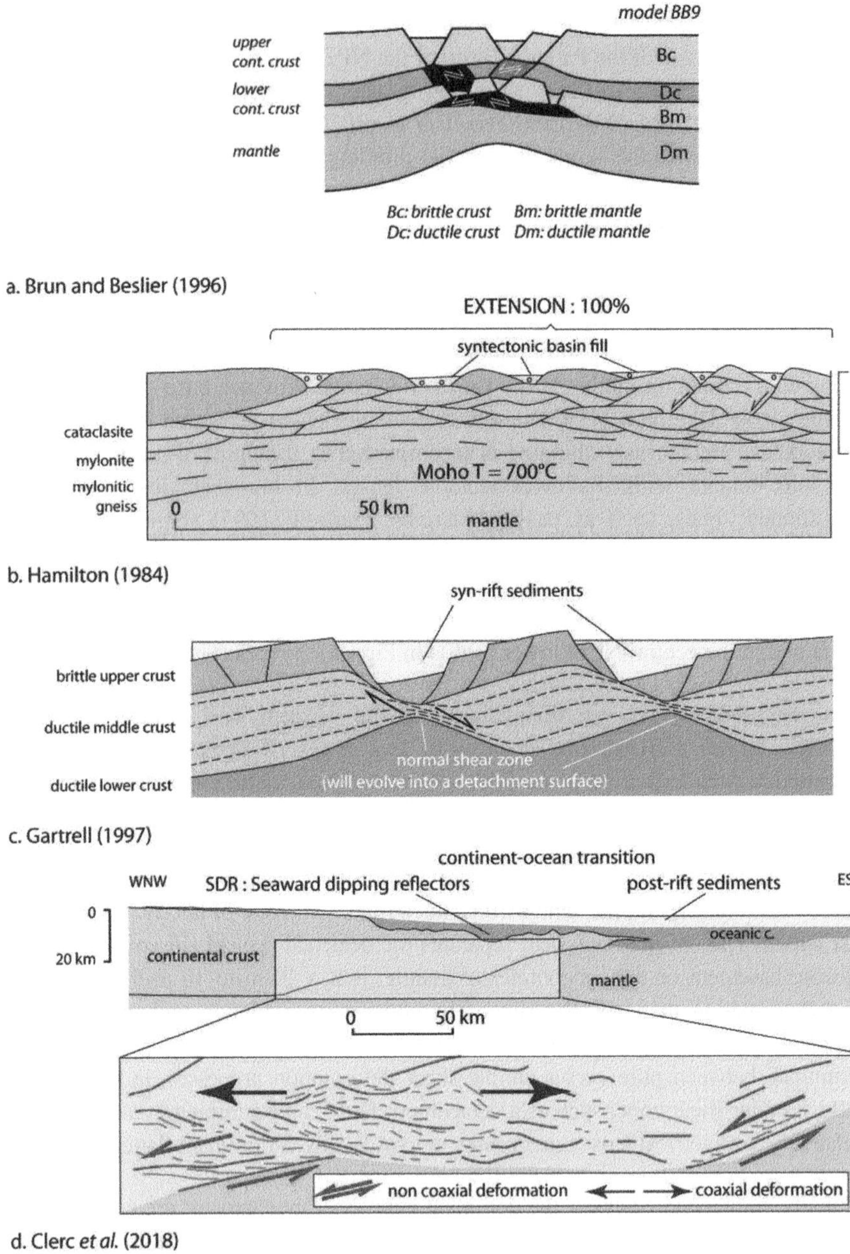

Figure 9.6. *Mechanisms of continental crustal thinning and mantle exhumation. Compilation of figures related to the examples discussed in the text*

9.3.3. *The opposition between Pyrenean- and Alpine-type rifts: role of Triassic palaeogeography on the modes of crustal thinning*

As we have seen, most recent models based on geological studies around the Paleozoic massifs, lherzolites and metamorphic units of the NPZ account for the absence of Cretaceous extensive brittle structures in the pre-rift sediments and basement in the vicinity of the exhumed mantle. In this sense, the evolution of the Pyrenean smooth-slope basins contrasts with brittle-dominated models of crustal thinnings, as exemplified along the Iberian–Newfoundland passive margins. In these margins, crustal thinning involves crustal disruption into discrete rigid blocks separated by listric faults and/or by detachments (Reston *et al.* 1995; Manatschal *et al.* 2001, 2006; Peron-Pinvidic and Manatschal 2009; Jammes *et al.* 2010c; Mohn *et al.* 2010, 2012, 2015; Osmundsen and Peron-Pinvidic 2018). In the necking zone, tilted blocks slide against each other and occupy the entire slope. In the distal margin zone, subcontinental mantle is exposed between rigid blocks of continental crust representing extensional allochthons. The pre-rift cover remains stuck to these blocks and slides with them along the slope (Figure 9.7). Because they are basically asymmetrical, these models use "lower- and upper-plate" terminology. Each conjugate margin evolves according to its original position, above or below a major detachment splitting the system into two parts with a different structure (Manatschal 2004; Sutra *et al.* 2013).

Using the example of the Alpine fossil margins, Mohn *et al.* (2012) propose a three-layer continental crust model where the brittle lower and upper crust are separated by a highly ductile middle crust. Crustal thinning occurs at the necking zone and results from interactions between detachments in the brittle layers and decoupling in the quartzo-feldspathic middle crust. Removal of the ductile middle crust leads to coupling of the upper and lower crusts and progressive hardening of the entire system. This allows detachment faults to intersect the mantle, triggering its exposure at the foot of the margin. This model differs from the Pyrenean one by a fully brittle behavior during mantle exhumation due to a general cooling of the system during margin evolution. It contrasts with the Pyrenean case by the absence of detachment of the pre-rift cover, which here remains welded to the tilted blocks.

Attempts to apply classical models of hyper-extended margins based on the study of conjugate Iberian–Newfoundland-type systems to Pyrenean rifting are not numerous (Masini *et al.* 2014; Tugend *et al.* 2014; Lescoutre 2019). Clerc and Lagabrielle (2014) showed the limitations of the Iberian–Newfoundland-type rifting model for the Pyrenean case. The main differences are schematically shown in Figure 9.7.

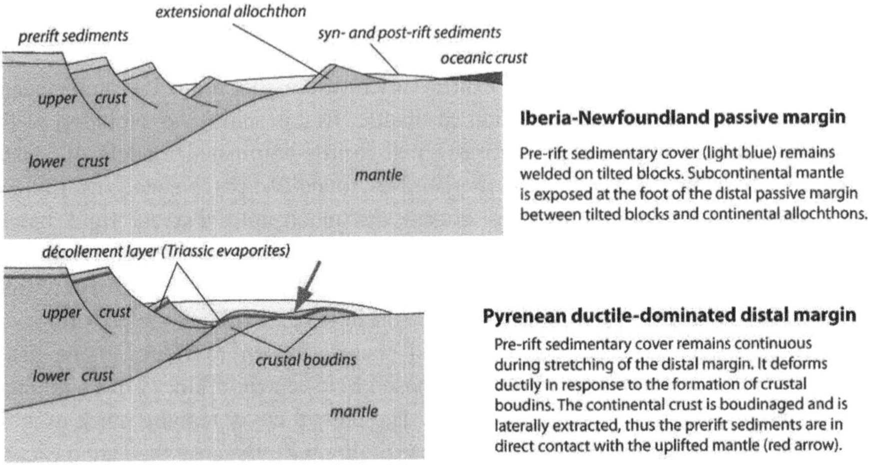

Figure 9.7. *Differences between the Iberian–Newfoundland and the Pyrenean-type margin models (modified from Clerc and Lagabrielle (2014))*

Common to all peri-Pyrenean basins is the presence of a very low-strength detachment layer represented by the evaporite clays and Keuper salt. As we have seen, efficient decoupling along this horizon allows the pre-rift cover to remain trapped within the basin while the crust is extracted laterally. The lithospheric mantle rises rapidly below the basin axis causing thermo-metamorphism and, at greater depths, local partial melting of peridotites yielding the alkaline magmatic products typical of the Pyrenean province. The deposition of the syn-rift flysch sustains the initial thermal anomaly which inhibits brittle deformation of the bedrock and promotes ductile deformation of the crust and the axial part of the pre- and syn-rift series (Clerc and Lagabrielle 2014; Clerc *et al.* 2015b; Asti *et al.* 2019; Duretz *et al.* 2019; Saspiturry *et al.* 2021) (Figure 9.5).

The question of the paleogeographic distribution of the Keuper group on the Iberian–Eurasian plate scale must now be addressed in order to test a possible link between rifting mechanisms and the presence or absence of a thick evaporite sequence at the base of the pre-rift series.

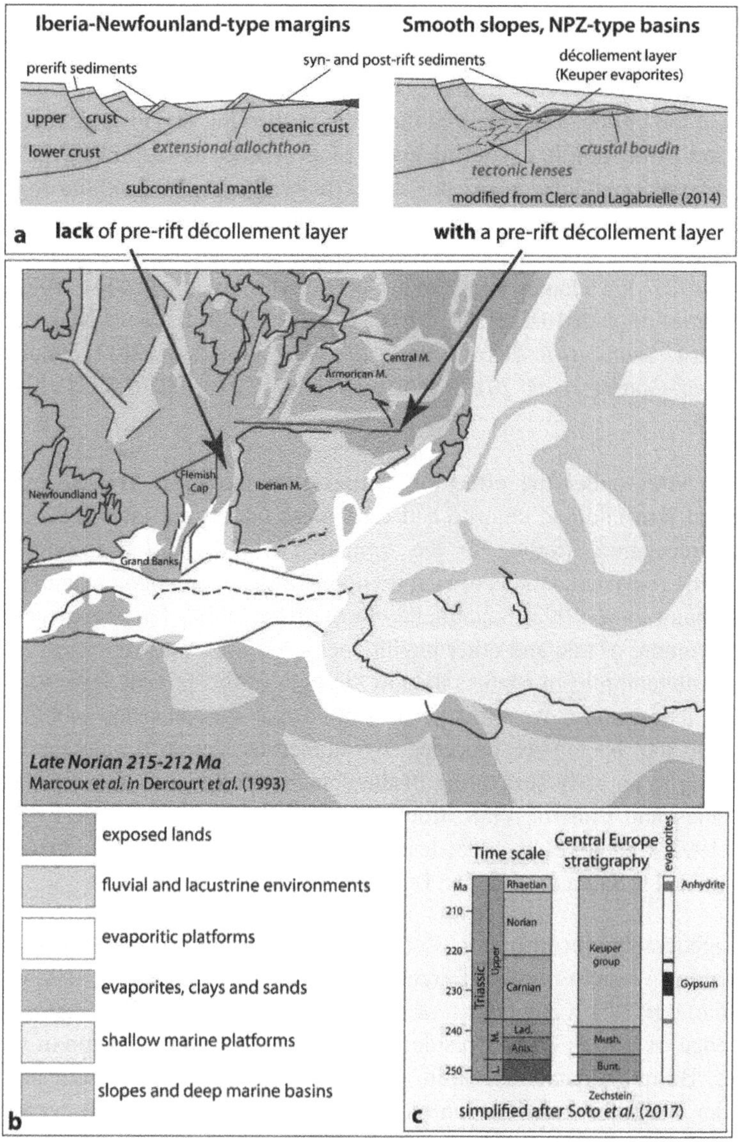

Figure 9.8. *One of several paleogeographic reconstructions available for the Triassic of Pangea centered on Iberia (from Dercourt* et al. *(1993) and Soto* et al. *(2017), modified from Lagabrielle* et al. *(2020)). Note the correspondence between the distribution of Keuper evaporites and the structure of the Iberian–Newfoundland or Pyrenean-type margins and smooth-slope basins. This map suggests a link between the occurrence of a thick evaporite layer and a ductile-dominated mode of crustal thinning. Comments in the text*

Following the Permian times, the Iberian platform was the site of a Triassic extension giving rise to a series of basins in the regions of the future Gulf of Valencia and the future Iberian, Basque–Cantabrian and Pyrenean chains, prefiguring the basins individualized during the following periods (see the synthesis in Angrand *et al.* (2020)). The thickness of evaporites is significant in all these regions. Based on field data in the southern Pyrenees and borehole logs in the Mauléon and Aquitaine basins, the maximum thickness reached by the Keuper salt and salt clays in the whole Pyrenean domain (including the future NPZ) is calculated to be about 2.7 km (James and Canérot 1988; McClay *et al.* 2004; Biteau *et al.* 2006; Jammes *et al.* 2010a, 2010b, 2010c; Roca *et al.* 2011; Lopez-Mir *et al.* 2014; Rowan 2014; Saura *et al.* 2016; Soto *et al.* 2017; Orti *et al.* 2017; Zamora *et al.* 2017; López-Gómez *et al.* 2019; Saspiturry *et al.* 2019, 2020, 2021; Lagabrielle *et al.* 2020).

In the NPZ, under the effect of thermo-metamorphism, Triassic clays were transformed into talc and chlorite and carbonates underwent intense fluid-assisted tectonic brecciation, leading to the formation of *cargneules* accompanied by metamorphic recrystallizations of talc, tremolite dolomite, phlogopite and muscovite (e.g. Thiébault *et al.* 1992; Lagabrielle *et al.* 2019a, 2019b for references). Due to the large volume of talc and other phyllite neo-crystallizations replacing evaporites and clays, metamorphism enables the low strength of the Triassic layer to persist at increasing temperature. We must keep in mind that syn-extension mobility of the Triassic in the NPZ was preceded by numerous ante-rift halokinetic events participating in an early structuring of the domain subjected to slip (Canérot 1988, 1989; James and Canérot 1988; Canérot and Lenoble 1989, 1993; Lenoble and Canérot 1992; Canérot *et al.* 2005; Jammes *et al.* 2009; Jammes *et al.* 2010a, 2010b; Roca *et al.* 2011; Saura *et al.* 2016; Teixell *et al.* 2016).

Paleogeographic reconstructions are available for the period under consideration at the scale of Western Europe (Dercourt *et al.* 1986, 1993; Ziegler 1988; Ortí *et al.* 2017; Soto *et al.* 2017; Angrand *et al.* 2020). Evaporites are well represented on the eastern edge of Iberia (Tethysian side) and in open rifts at the location of the future Pyrenees, Basque–Cantabrian Basin, Asturias Basin and the southern part of the Armorican margin. At the location of the future North Atlantic Rift, evaporites are restricted to the Peniche, Lusitania, Alentejo and Algarve basins along the southern part of Portugal only. Evaporites are therefore absent along the northern part of the Iberian margin at the level of the Iberian–Newfoundland transect. This domain has been intensively studied since the discovery of the peridotites at the foot of the margin by Boillot *et al.* (1980). On the American part of the transect, evaporites are very narrowly distributed and are only known at the base of the Jeanne d'Arc Basin (Figure 9.8).

Finally, Keuper evaporites are only known in abundance in Germanic facies on the European side of the Atlantic rift system (Figure 9.8). This highly contrasting paleogeographic distribution is consistent with the distribution of rift systems corresponding to the "smooth-slope" basins on one side and the Alps–Iberia–Newfoundland-type margins on the other side. Therefore, we propose a link between the presence of a thick pre-rift salt layer and the rheological parameters that control the evolution of crustal thinning processes at depth. Due to effective detachment at the base of the Mesozoic sediments, the evaporites remain in the center of the opening basins, together with the overlying pre-rift cover. The presence of this allochthonous assemblage thus maintains the high thermal gradient in the AZ of the rift, which leads to ductile deformation of the thinned continental bedrock, the pre-rift series and the base of the syn-rift flysch. The absence of a thick detachment level at the base of the pre-rift series in the Alpine paleo-margins or in the present-day Iberian–Newfoundland margins may explain why the sediments remained stuck and coupled with the basement during extension. In these cases, the system cooled efficiently, and only cold syn-rift sediments may overlie the exhumed deep levels, as illustrated in the models by Peron-Pinvidic *et al.* (2007), Peron-Pinvidic and Manatschal (2009) and Mohn *et al.* (2012).

9.4. Conclusion: from the Pyrenean basins to the formation mechanisms of passive continental margins

We have shown that the dominant type of deformation in the continental bedrock during crustal thinning related to Pyrenean rifting consists of tectonic lenses flattened by pure shear separated by shear zones where displacement is concentrated. Similar or related patterns are found at all scales in many representations of passive continental margin structures (generally non-volcanic). In Figure 9.9, we have referred to these tectonic lenses as "*hyper-thinned continental crustal units*". They are observed, for example, beneath the margins of the Barents Sea in the interpretation by Gernigon *et al.* (2014) (Figure 9.9a). These authors represent continental blocks with a strongly ductile thinning base, reminiscent of the shape proposed for the North Pyrenean massifs in the NPZ, including the Mauléon basin (Wang *et al.* 2016). This same elongated lens shape is proposed by Roca *et al.* (2011) for the North Iberian margin north of the Cantabrian Range. These patterns are also found at the base of the lower continental crust beneath the British Isles (Reston 1988), or as discussed above, in the models by Gartrell (1997) (Figure 9.6c).

Structures consisting of an assemblage of flattened tectonic lenses are found in some interpretations of the structure of the Norwegian margins (Osmundsen and Ebbing 2008) (Figure 9.9b). In this case, the hyper-extended continental crust occurs as very thin lenses isolated by anastomosing shear zones. In the case of the Labrador and West Greenland margins, these crustal lenses, only 10 km thick and 100 km long, are now separated by extensive regions of exhumed mantle (Chalmers and Pulvertaft 2001; Reston and Pérez-Gussinyé 2007) (Figure 9.9c,d). They are also known along the Angola margin (Brune *et al.* 2014) and in the very distal domain of the Gulf of Lion margin (Jolivet *et al.* 2015) (Figure 9.9e,f). These hyper-thinned crustal units are probably themselves composed of an assemblage of heterogeneous lenses formed by a large-scale shearing process. They are remarkably reproduced by thermo-mechanical modeling (Huismans and Beaumont 2011, 2014; Brune *et al.* 2014), especially when using the geological parameters of the Pyrenean rifting (Duretz *et al.* 2019). Finally, in a compilation of profiles from numerous passive margins around the world (Uruguay, Namibia, Gabon, South China Sea and Barents Sea), Clerc *et al.* (2015a, 2018) suggest that the lower crust is less resistant than conventionally accepted and also undergoes ductile shearing. This again results in the boudinage of the lower part of the system and the individualization of continental dipping flat normal faults cutting vast tectonic lenses (Figure 9.6d).

Finally, the thin continental units of the Labrador–Greenland or Angola and Gulf of Lion margins shown in Figures 9.6 and 9.9 do not correspond to the so-called "extensional allochthons" known from the Iberian–Newfoundland margins and formed by processes dominated by simple shearing along "cold" detachments. Although much larger in volume, the tectonic lenses found on the various margins examined in this chapter show affinities in shape with the crustal boudins that were extracted from the Variscan crust during Cretaceous extension into the future NPZ. Examples include the Barousse and Agly boudins as shown in the sections of Clerc *et al.* (2016) and Espurt *et al.* (2019). Thus, the deformation mode of the continental crust during its hyper-thinning under the Pyrenean basins of the NPZ can provide a fruitful picture for understanding how passive margins behave during rifting under a mobile sedimentary cover. Deformation occurs under greenschist facies conditions, at temperatures around 350–450°C and mylonites are concentrated in a dense network of anastomosing shear zones. Boudinage at all scales is a dominant pattern, leading to the juxtaposition on the same horizontal level of lenses from various structural levels, as highlighted by Hamilton (1987).

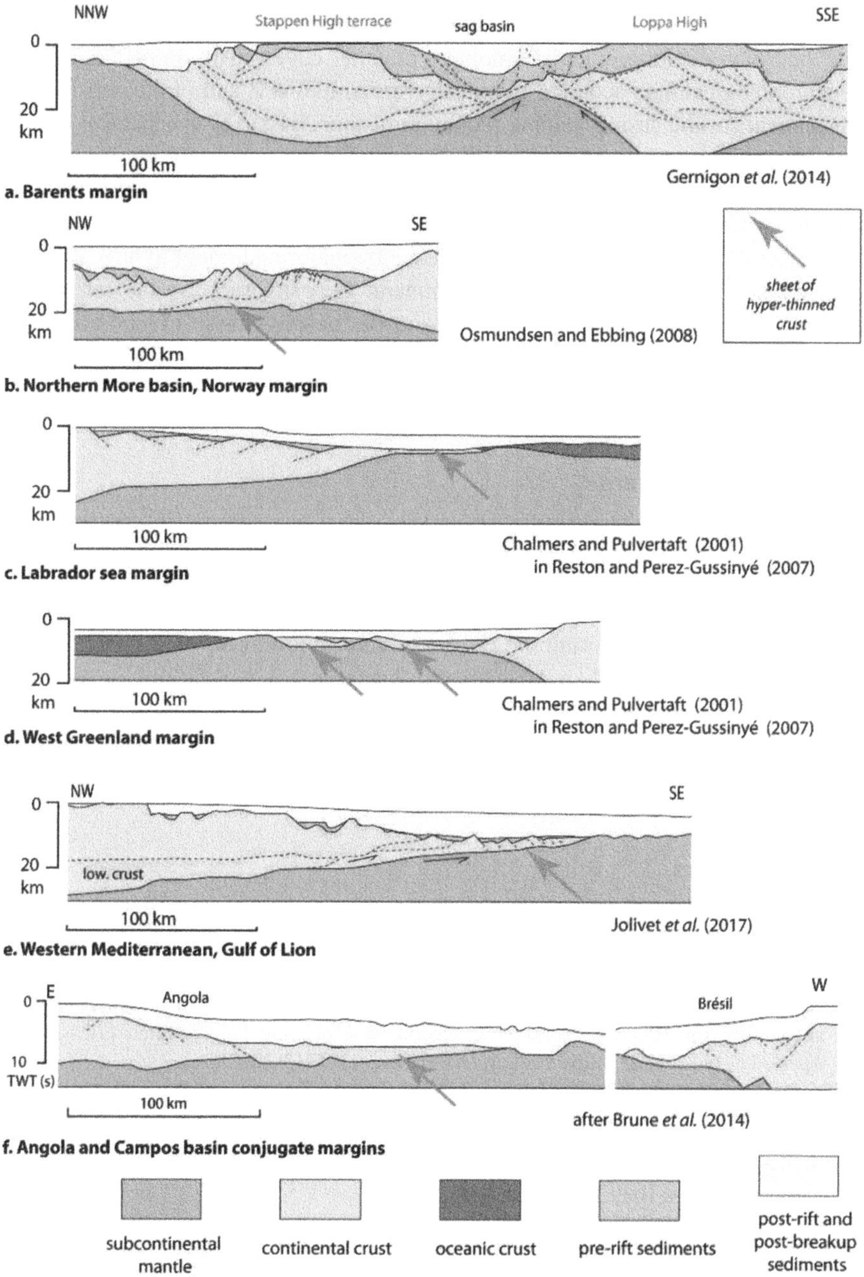

Figure 9.9. *Structure of non-volcanic passive hyper-thinned continental margins. Compilation of sections referred to in the text*

Let us conclude by recalling the major fluid/rock interactions induced by this margin evolution under mobile pre-rift cover. All mineralogical and geochronological data acquired in recent years in the NPZ and in the peri-Pyrenean basins confirm the strong volume reduction of the Triassic by mechanical crushing and dissolution of evaporites, whose elements participate in the massive metasomatosis of the continental cover and crust up to the major faults of the outer margin where temperatures sometimes reach 550°C. These interactions are accompanied by transfers of chemical elements (major and trace) in huge quantities and are at the origin of world-class mineral deposits such as the talc–chlorite associations of the NPZ bedrock in the Saint Barthélémy massif (Trimouns deposit) (Schärer *et al.* 1999; Boulvais *et al.* 2007; Fallourd *et al.* 2014; Boutin *et al.* 2015; Quesnel *et al.* 2019; Odlum and Stockli 2020).

9.5. Acknowledgments

The concepts gathered in this chapter are the result of discussions and reflections between many geologists for more than 14 years. I would like to thank all the colleagues, post-PhD and PhD students and master's students with whom we have exchanged since 2006, during the rich years that followed my first missions in the Pyrenees. I would also like to thank the managers of the various governmental and industrial programs (ANR's Pyramid, BRGM's RGF and Total-BRGM-CNRS/INSU's Orogen programs) who have placed their trust in us and funded our research.

9.6. References

Allemand, P., Brun J.-P., Davy P., Van Den Driessche, J. (1989). Symétrie et asymétrie des rifts et mécanismes d'amincissement de la lithosphère. *Bull. Soc. Géol. Fr.*, 5, 445–451.

Angrand, P., Mouthereau, F., Masini, E., Asti, R. (2020). A reconstruction of Iberia accounting for W-Tethys/N-Atlantic kinematics since the late Permian-Triassic. *Solid Earth*, 11, 1313–1332. https://doi.org/10.5194/se-11-1313-2020.

Asti, R., Lagabrielle, Y., Fourcade, S., Corre, B., Monié, P. (2019). How do continents deform during mantle exhumation? Insights from the northern Iberia inverted paleopassive margin, western Pyrenees (France). *Tectonics*, 38, 1666–1693. https://doi.org/10.1029/2018TC005428.

Beaumont, C., Muñoz, J.A., Hamilton, J., Fullsack, P. (2000). Factors controlling the Alpine evolution of the central Pyrenees inferred from a comparison of observations and geodynamical models. *J. Geophys. Res. Solid Earth*, 105, 8121–8145. https://doi:10.1029/1999JB900390.

Biju-Duval, B., Dercourt, J., Le Pichon, X. (1977). From the Thetys ocean to the Mediterranean Basin: A plate tectonic model of the evolution of the western alpine system. In *Internat. Symp. on the Structural History of the Mediterranean Basins*, Biju-Duval, B., Montadert, L. (eds). Editions Technip, Paris.

Biteau, J.-J., Le Marrec, A., Le Vot, M., Masset, J.-M. (2006). The aquitaine basin. *Pet. Geosci.*, 12, 247–273.

Boillot G. and Capdevila R. (1977). The Pyrenees: Subduction and collision. *Earth Planet. Sci. Lett.*, 35, 151–160.

Boillot, G., Grimaud, S., Mauffret, A., Mougenot, D., Kornprobst, J., Mergoil-Daniel, J., Torrent, G. (1980). Ocean-continent boundary off the Iberian margin: A serpentinite diapir west of the Galicia bank. *Earth Planet. Sci. Lett.*, 48, 23–34.

Boulvais, P., Ruffet, G., Cornichet, J., Mermet, M. (2007). Cretaceous albitization and dequartzification of Hercynian peraluminous granite in the Salvezines Massif (French Pyrénées). *Lithos*, 93(1–2), 89–106. https://doi:10.1016/j.lithos.2006.05.001.

Boutin, A., de Saint Blanquat, M., Poujol, M., Boulvais, P., de Parseval, P., Rouleau, C., Robert, J.F. (2015). Succession of Permian and Mesozoic metasomatic events in the eastern Pyrenees with emphasis on the Trimouns talc–chlorite deposit. *Int. J. Earth Sci.*, 105(3), 747–770.

Brun, J.P. (1999). Narrow rifts versus wide rifts: Inferences for the mechanics of rifting from laboratory experiments. *Philos. Trans. R. Soc. London*, A, 357, 695–712.

Brun, J.P. and Beslier M.O. (1996). Mantle exhumation at passive margin. *Earth Planet. Sci. Lett.*, 142, 161–173.

Brune, S., Heine, C., Pérez-Gussinyé, M., Sobolev, S.V. (2014). Rift migration explains continental margin asymmetry and crustal hyper-extension. *Nat. Commun.*, 5. https://doi.org/10.1038/ncomms5014.

Buck, W.R. (1991). Modes of continental lithospheric extension. *J. Geophys. Res.*, 96, 20161–20178.

Buck, W.R., Martinez, F., Steckler, M.S., Cochran, J.R. (1988). Thermal consequences of lithospheric extension: Pure and simple. *Tectonics*, 7, 213–234.

Canérot, J. (1988). Manifestations de l'halocinèse dans les Chaînons Béarnais (Zone Nord-Pyrénéenne) au Crétacé inférieur. *C. R. Acad. Sci. Paris*, II, 306, 1099–1102.

Canérot, J. (1989). Rifting éocrétacé et halocinèse sur la marge ibérique des Pyrénées Occidentales (France). Conséquences structurales. *Bull. Cent. Rech. Explor. Prod. Elf-Aquitaine*, 13, 87–99.

Canérot, J. (2017a). The pull apart-type Tardets-Mauléon Basin, a key to understand the formation of the Pyrenees. *Bull. Soc. Géol. Fr.*, 188(6), 35. https://doi.org/10.1051/bsgf/2017198.

Canérot, J. (2017b). Origine de la chaîne des Pyrénées : collision entre les plaques ibérique et européenne ou inversion d'un ancien rift continental avorté ? *Bull. Soc. Hist. Nat. Toulouse*, 153, 95–110.

Canérot, J. and Delavaux, F. (1986). Tectonic and sedimentation on the north Iberian margin, Chainons Béarnais south Pyrenean zone (Pyrenees basco-béarnaises): New data about the signification of the lherzolites in the Saraillé area. *C. R. Acad. Sci. Paris*, II, 302(15), 951–956.

Canérot, J. and Lenoble, J.-L. (1989). Le diapir du Lichançumendy (Pyrénées-Atlantiques), nouvel élément de la marge ibérique des Pyrénées occidentales. *C. R. Acad. Sci. Paris*, II, 308, 1467–1472.

Canérot, J. and Lenoble, J.L. (1993). Diapirisme crétacé sur la marge ibérique des Pyrénées occidentales. Exemple du pic de Lauriolle ; comparaisons avec l'Aquitaine, les Pyrénées centrales et orientales. *Bull. Soc. Géol. Fr.*, 164, 719–726.

Canérot, J., Peybernès, B., Ciszak, R. (1978). Présence d'une marge méridionale à l'emplacement des Chaînons Béarnais (Pyrénées basco-béarnaises). *Bull. Soc. Géol. Fr.*, 7(20), 673–676.

Canérot, J., Hudec, M.R., Rockenbauch, K. (2005). Mesozoic diapirism in the Pyrenean orogen: Salt tectonics on a transform plate boundary. *AAPG Bull.*, 89, 211–229. https://doi.org/10.1306/09170404007.

Chalmers, J.A. and Pulvertaft, T.C.R. (2001). Development of the continental margins of the Labrador Sea: A review. In *Non-Volcanic Rifting of Continental Margins: A Comparison of Evidence from Land and Sea*, Wilson, R.C.C.L., Whitmarsh, R.B., Taylor, B., Froitzheim, N. (eds). Geological Society of London, Special Publication.

Choukroune P. (1992). Tectonic evolution of the Pyrenees. *Annu. Rev. of Earth Planet. Sci.*, 20, 143–158.

Choukroune, P. and ECORS Team (1989). The ECORS Pyrenean deep seismic profile reflection data and the overall structure of an orogenic belt. *Tectonics*, 8, 23–39.

Choukroune, P. and Mattauer, M. (1978). Tectonique des plaques et Pyrénées. Sur le fonctionnement de la faille transformante nord-pyrénéenne : comparaisons avec des modèles actuels. *Bull. Soc. Géol. Fr.*, 7, 689–700.

Clerc, C. and Lagabrielle, Y. (2014). Thermal control on the modes of crustal thinning leading to mantle exhumation: Insights from the Cretaceous Pyrenean hot paleomargins. *Tectonics*, 33, 1340–1359. https://doi.org/10.1002/2013TC003471.

Clerc, C., Lagabrielle, Y., Neumaier, M,. Reynaud, J.-Y., de Saint-Blanquat. M. (2012). Exhumation of subcontinental mantle rocks: Evidence from ultramafic-bearing clastic deposits nearby the Lherz Peridotite Body, French Pyrenees. *Bull. Soc. Géol. Fr.*, 183(5), 443–459. https://doi.org/10.2113/gssgfbull.183.5.443.

Clerc, C., Boulvais, P., Lagabrielle, Y., de Saint Blanquat, M. (2014). Ophicalcites from the Northern Pyrenean Belt: A field, petrographic and stable isotope study. *International Journal of Earth Sciences*, 1–23. https://doi.org/10.1007/s00531-013-0927-z.

Clerc, C., Jolivet, L., Ringenbach, J.-C. (2015a). Ductile extensional shear zones in the lower crust of a passive margin. *Earth Planet. Sci. Lett.*, 431, 1–7. https://doi.org/10.1016/j.epsl.2015.08.038.

Clerc, C., Lahfid, A., Monié, P., Lagabrielle, Y., Chopin, C., Poujol, M., Boulvais, P., Ringenbach, J.C., Masini, E., de St Blanquat, M. (2015b). High-temperature metamorphism during extreme thinning of the continental crust: A reappraisal of the North Pyrenean passive paleomargin. *Solid Earth*, 6, 643–668.

Clerc, C., Lagabrielle, Y., Labaume, P., Ringenbach, J.-C., Vauchez, A., Nalpas, T., Bousquet, R., Ballard, J.-F., Lahfid, A., Fourcade, S. (2016). Basement – Cover decoupling and progressive exhumation of metamorphic sediments at hot rifted margin. Insights from the Northeastern Pyrenean analog. *Tectonophysics*, 686, 82–97. https://doi.org/10.1016/j.tecto.2016.07.022.

Clerc, C., Ringenbach, J.-C., Jolivet, L., Ballard, J.-F. (2018). Rifted margins: Ductile deformation, boudinage, continentward-dipping normal faults and the role of the weak lower crust. *Gondwana Res.*, 53, 20–40. https://doi.org/10.1016/j.gr.2017.04.030

Corre, B., Lagabrielle, Y., Labaume, P., Fourcade, S., Clerc, C., Ballèvre, M. (2016). Deformation associated with mantle exhumation in a distal, hot passive margin environment: New constraints from the Saraillé Massif (Chaînons Béarnais, North-Pyrenean Zone). *C. R. Geosci.*, 348, 279–289. https://doi.org/10.1016/j.crte.2015.11.007.

Dauteuil, O. and Ricou, L. (1989). Une circulation de fluides de haute température à l'origine du métamorphisme Crétacé nord-Pyrénéen. *Geodin. Acta*, 3, 237–250.

Debroas E.-J. (1978). Evolution de la fosse du flysch ardoisier de l'Albien supérieur au Sénonien inférieur (zone interne métamorphique des Pyrénées navarro-languedociennes), *Bull. Soc. Géol. Fr.*, 7, XX, 639–648.

Debroas, E.-J. (1990). Le flysch noir albo-cénomanien témoin de la structuration albienne à sénonienne de la Zone nord-pyrénéenne en Bigorre (Hautes-Pyrénées, France). *Bull. Soc. Géol. Fr.*, VI, 273–285. https://doi.org/10.2113/gssgfbull.VI.2.273.

Debroas, E.-J., Canérot, J., Billotte. M. (2010). Les brèches d'Urdach, témoins de l'exhumation du manteau pyrénéen dans un escarpement de faille vraconnien-cénomanien inférieur (zone nord-Pyrénéenne, Pyrénées-Atlantiques, France). *Géologie de la France*, 2, 53–63.

Deramond, J., Souquet, P., Fondecave-Wallez, M.-J., Specht, M. (1993). Relationships between thrust tectonics and sequence stratigraphy surfaces in foredeeps: Model and examples from the Pyrenees (Cretaceous-Eocene, France, Spain). *Geol. Soc. Lond. Spec. Publ.*, 71, 193–219.

Dercourt, J., Zonenshain, L.P., Ricou, L.E., Kazmin, V.G., Le Pichon, X., Knipper, A.L., Grandjacquet, C., Sbortshikov, I.M., Geyssant, J., Lepvrier, C., Pechersky, D.H., Boulin, J., Sibuet, J.C., Savostin, L.A., Sorokhtin, O., Westphal, M., Bazhenov, M.L., Lauer, J.P., Biju-Duval, B. (1986). Geological evolution of the Tethys belt from the Atlantic to the Pamirs since the Lias. *Tectonophysics*, 123(1), 241–315.

Dercourt, J., Ricou, L.E., Vrielynck, B. (1993). *Atlas Tethys: Palaeoenvironmental Maps Atlas and Explanatory Notes*. Gauthier Villars Editions, CGMW, Paris.

Ducoux, M. (2017). Structure, thermicité et évolution géodynamique de la Zone Interme Métamorphique des Pyrénées. Unpublished PhD thesis, Université d'Orléans, France.

Ducoux, M., Jolivet, L., Cagnard, F., Gumiaux, C., Baudin, T., Masini, E., Callot, J.P., Aubourg, C., Lahfid, A., Homonnay, E. (2019). The Nappe des Marbres unit of the Basque-Cantabrian basin: The tectono-thermal evolution of a fossil hyperextended rift basin. *Tectonics*. https://doi.org/10.1029/2018TC005348.

Duretz, T., Asti, R., Lagabrielle, Y., Brun, J.P., Jourdon, A., Clerc, C., Corre, B., (2019). Numerical modelling of syn-rift salt tectonics during Cretaceous Pyrenean Rifting. *Basin Res.*, 32(4), 652–667. https://doi.org/10.1111/bre.12389.

Espurt, N., Angrand, P., Teixell, A., Labaume, P., Ford, M., de Saint Blanquat, M., Chevrot, S. (2019). Crustal-scale balanced cross-section and restorations of the Central Pyrenean belt (Nestes-Cinca transect): Highlighting the structural control of Variscan belt and Permian-Mesozoic rift systems on mountain building. *Tectonophysics*, 764, 25–45. https://doi.org/10.1016/j.tecto.2019.1004.1026.

Ethève, N., Mohn, G., Frizon de Lamotte, D., Roca, E., Tugend, J., Gómez-Romeu, J. (2018). Extreme Mesozoic crustal thinning in the Eastern Iberia Margin: The example of the Columbrets Basin (Valencia Trough). *Tectonics*, 37, 636–662.

Fabriès, J., Lorand, J.-P., Bodinier, J.L., Dupuy, C. (1991). Evolution of the upper mantle beneath the Pyrenees during the Mesozoïc: Evidences from spinel lherzolite orogenic massifs. In *Orogenic Lherzolites and Mantle Processes*, Menzies, M.A., Dupuy, C., Nicolas, A.J. (eds.). J. Petrol., Special Lherzolites Issue.

Fallourd, S., Poujol, M., Boulvais, P., Paquette, J.L., de Saint Blanquat, M., Rémy. P. (2014). In situ LA-ICP-MS U–Pb titanite dating of Na–Ca metasomatism in orogenic belts: The North Pyrenean example. *Int. J. Earth Sci.*, 1–16. https://doi.org/10.1007/s00531-013-0978-1.

Ford, M., Hemmer, L., Vacherat, A., Gallagher, K., Christophoul, F. (2016). Retro-wedge foreland basin evolution along the ECORS line, eastern Pyrenees, France. *J. Geol. Soc.*, 173, 419–437. https://doi.org/10.1144/jgs2015-129.

Gartrell, A.P. (1997). Evolution of rift basins and low-angle detachments in multilayer analog models. *Geology*, 25, 615–618. https://doi.org/10.1130/0091.7613.

Gernigon, L., Brönner, M., Roberts, D., Olesen, O., Nasuti, A., Yamasaki, T. (2014). Crustal and basin evolution of the southwestern Barents Sea: From Caledonian orogeny to continental breakup: Evolution of the Barents Sea. *Tectonics*, 33, 347–373. https://doi.org/10.1002/2013TC003439.

Golberg, J.M. and Leyreloup, A.F. (1990). High temperature-low pressure Cretaceous metamorphism related to crustal thinning (Eastern North Pyrenean Zone, France). *Contrib. Mineral. Petrol*, 104, 194–207. https://doi.org/10.1007/BF00306443.

Gong, Z., Langereis, C.G., Mullender, T.A.T. (2008). The rotation of Iberia during the Aptian and the opening of the Bay of Biscay. *Earth Planet. Sci. Lett.*, 273, 80–93. https://doi.org/10.1016/j.epsl.2008.06.016.

Grool, A.R., Ford, M., Vergés, J., Huismans, R.S., Christophoul, F., Dielforder, A. (2018). Insights into the crustal-scale dynamics of a doubly vergent orogen from a quantitative analysis of its forelands: A case study of the Eastern Pyrenees. *Tectonics*, 37. https://doi.org/10.1002/2017TC004731.

Hamilton, W. (1987). Crustal extension in the Basin and Range Province, southwestern United States. *Geol. Soc. Lond. Spec. Publ.*, 28, 155–176. https://doi.org/10.1144/GSL.SP.1987.028.01.12.

Henry, P., Azambre, B., Montigny, R., Rossy, M., Stevenson, R.K. (1998). Late mantle evolution of the Pyrenean sub-continental lithospheric mantle in the light of new ^{40}Ar-^{39}Ar and Sm-Nd ages on pyroxenites and peridotites (Pyrenees, France). *Tectonophysics*, 296, 103–123.

Huismans, R.S. and Beaumont, C. (2007). Roles of lithospheric strain softening and heterogeneity in determining the geometry of rifts and continental margins. *Geological Society, London, Special Publications*, 282, 111–138. http://dx.doi.org/10.1144/SP282.6.

Huismans, R.S. and Beaumont, C. (2011). Depth-dependent extension, two-stage breakup and cratonic underplating at rifted margins. *Nature*, 473, 74–78. https://doi.org/10.1038/nature09988.

Huismans, R.S. and Beaumont, C. (2014). Rifted continental margins: The case for depth-dependent extension. *Earth Planet. Sci. Lett.*, 407, 148–162. https://doi.10.1016/j.epsl.2014.09.032.

James, V. and Canérot, J. (1988). Diapirisme et structuration post-triasique des Pyrénées occidentales et de l'Aquitaine méridionale (France). *Eclogae Geol. Helveticae*, 92, 63–72.

Jammes, S., Manatschal, G., Lavier, L., Masini, E. (2009). Tectonosedimentary evolution related to extreme crustal thinning ahead of a propagating ocean: Example of the western Pyrenees. *Tectonics*, 28. https://doi.10.1029/2008TC002406.

Jammes, S., Lavier, L., Manatschal, G., Persaud, N. (2010a). Extreme crustal thinning in the Bay of Biscay and the Western Pyrenees: From observations to modeling. *Geochem. Geophys. Geosyst.*, 11. https://doi.org/10.1029/2010GC003218.

Jammes, S., Tiberi, C., Manatschal, G. (2010b). 3D architecture of a complex transcurrent rift system: The example of the Bay of Biscay–Western Pyrenees. *Tectonophysics*, 489, 210–226. https://doi.org/10.1016/j.tecto.2010.04.023.

Jammes, S., Manatschal, G., Lavier, L. (2010c). Interaction between prerift salt and detachment faulting in hyperextended rift systems: The example of the Parentis and Mauléon basins (Bay of Biscay and western Pyrenees). *AAPG Bull.*, 94, 957–975. https://doi.org/10.1306/12090909116.

Jolivet, L., Gorini, C., Smit, J., Leroy, S. (2015). Continental breakup and the dynamics of rifting in back-arc basins: The Gulf of Lion margin: Backarc rift and lower crust extraction. *Tectonics*, 34, 662–679. https://doi.org/10.1002/2014TC003570.

Lagabrielle, Y. and Bodinier, J.-L. (2008). Submarine reworking of exhumed subcontinental mantle rocks: Field evidence from the Lherz peridotites, French Pyrenees. *Terra Nova*, 20, 11–21. https://doi.org/10.1111/j.1365-3121.2007.00781.x.

Lagabrielle, Y., Clerc, C., Vauchez, A., Lahfid, A., Labaume, P., Azambre, B., Fourcade, S., Dautria, J.-M. (2016). Very high geothermal gradient during mantle exhumation recorded in mylonitic marbles and carbonate breccias from a Mesozoic Pyrenean palaeomargin (Lherz area, North Pyrenean Zone, France). *C. R. Geosci.*, 348, 290–300. https://doi.org/10.1016/j.crte.2015.11.004.

Lagabrielle, Y., Asti, R., Fourcade, S., Corre, B., Poujol, M., Uzel, J., Labaume, P., Clerc, C., Lafay, R., Picazo, S., Maury, R. (2019a). Mantle exhumation at magma-poor passive continental margins. Part I. 3D architecture and metasomatic evolution of a fossil exhumed mantle domain (Urdach lherzolite, north-western Pyrenees, France), *Bull. Soc. Géol. Fr.*, 190, 8. https://doi.org/10.1051/bsgf/2019007.

Lagabrielle, Y., Asti, R., Fourcade, S., Corre, B., Uzel, J., Labaume, P., Clerc, C., Lafay, R., Picazo, S. (2019b). The mechanisms of mantle exhumation at magma-poor passive continental margins. Part II. Insights from high-displacement, low-angle faults preserved in a fossil distal margin domain (Saraillé lherzolites, north-western Pyrenees, France). *Bull. Soc. Géol. Fr.*, 190, 14. https://doi.org/10.1051/bsgf/2019013.

Lagabrielle, Y., Asti, R., Duretz, T., Clerc, C., Fourcade, S., Teixell, A., Labaume, P., Corre, B., Saspiturry, N. (2020). A review of Cretaceous smooth-slope extensional basins along the Iberia-Eurasia plate boundary: How pre-rift salt controls the modes of continental rifting and mantle exhumation. *Earth-Sci. Rev.*, 201, 103071. https://doi.org/10.1016/j.earscirev.2019.103071.

Le Pichon, X., Bonnin, J., Sibuet, J.-C. (1970). La faille nord-pyrénéenne : faille transformante liée à l'ouverture du golfe de gascogne. *C. R. Acad. Sci. Paris*, 271, 1941–1944.

Lenoble, J.-L. and Canérot, J. (1992). La lame extrusive de Pont Suzon (Zone Nord-Pyrénéenne en Vallée d'Aspe) : reprise pyrénéenne d'une ride diapirique transverse d'âge crétacé. *C. R. Acad. Sci. Paris*, 2, 314, 387–391.

Lescoutre, R. (2019). Formation et réactivation du système de rift pyrénéo-cantabrique : héritage, segmentation et évolution thermique. Unpublished PhD thesis, Université de Strasbourg, France.

Lescoutre, R. and Manatschal, G. (2020). Role of rift-inheritance and segmentation for orogenic evolution: Example from the Pyrenean-Cantabrian system. *Bull. Soc. Géol. Fr.*, 191, 18. https://doi.org/10.1051/bsgf/2020021.

Lescoutre, R., Tugend, J., Brune, S., Masini, E., Manatschal, G. (2019). Thermal evolution of asymmetric hyperextended magma-poor rift systems: Results from numerical modeling and Pyrenean field observations. *Geochem., Geophys., Geosyst.*, 20. https://doi.org/10.1029/2019GC008600.

Lister, G.S. and Davis, G.A. (1989). The origin of metamorphic core complexes and detachment faults formed during Tertiary continental extension in the northern Colorado River region, U.S.A. *J. Struct. Geol.*, 11, 112, 65–94.

Lister, G.S., Etheridge, M.A., Symonds, P.A. (1991). Detachment models for the formation of passive continental margins. *Tectonics*, 10, 1038–1064.

López-Gómez, J., Alonso-Azcárate, J., Arche, A., Arribas, J., Fernández Barrenechea, J., Borruel-Abadía, V., Bourquin, S., Cadenas, P., Cuevas, J., De la Horra, R., Bienvenido Díez, J., Escudero-Mozo, M.J., Fernández-Viejo, G., Galán-Abellán, B., Galé, C., Gaspar-Escribano, J., Gisbert Aguilar, J., Gómez-Gras, D., Goy, A., Gretter, N., Heredia Carballo, N., Lago, M., Lloret, J., Luque, J., Márquez, L., Márquez-Aliaga, A., Martín-Algarra, A., Martín-Chivelet, J., Martín-González, F., Marzo, M., Mercedes-Martín, R., Ortí, F., Pérez-López, A., Pérez-Valera, F., Pérez-Valera, J.A., Plasencia, P., Ramos, E., Rodríguez-Méndez, L., Ronchi, A., Salas, R., Sánchez-Fernández, D., Sánchez-Moya, Y., Sopeña, A., Suárez-Rodríguez, A., María Tubía, J., Ubide, T., Garcés, B.V., Vargas, H., Viseras, C. (2019). Permian-Triassic rifting stage. In *The Geology of Iberia: A Geodynamic Approach*, Quesada, C. and Oliveira, J. (eds). Regional Geology Reviews. Springer. https://doi.org/10.1007/978-3-030-11295-0_3.

Lopez-Mir, B., Muñoz, J.A., García Senz, J. (2014). Restoration of basins driven by extension and salt tectonics: Example from the Cotiella Basin in the central Pyrenees. *J. Struct. Geol.*, 69(Part A), 147–162. https://doi.org/10.1016/j.jsg.2014.09.022.

Manatschal, G. (2004). New models for evolution of magma-poor rifted margins based on a review of data and concepts from West Iberia and the Alps. *Int. J. Earth Sci.*, 93. https://doi.org/10.1007/s00531-004-0394-7.

Manatschal, G., Froitzheim, N., Rubenach, M., Turrin, B.D. (2001). The role of detachment faulting in the formation of an ocean-continent transition: Insights from the Iberia Abyssal Plain. *Geol. Soc. Lond. Spec. Publ.*, 187, 405–428. https://doi.org/10.1144/GSL.SP.2001.187.01.20.

Manatschal, G., Engström, A., Desmurs, L., Schaltegger, U., Cosca, M., Müntener, O., Bernoulli, D. (2006). What is the tectono-metamorphic evolution of continental break-up: The example of the Tasna Ocean–Continent Transition. *J. Struct. Geol.*, 28, 1849–1869. https://doi.org/10.1016/j.jsg.2006.07.014

Masini, E., Manatschal, G., Tugend, J., Mohn, G., Flament, J.-M. (2014). The tectono-sedimentary evolution of a hyper-extended rift basin: The example of the Arzacq–Mauléon rift system (Western Pyrenees, SW France). *Int. J. Earth Sci.*, 103, 1569–1596. https://doi.org/10.1007/s00531-014-1023-8.

McClay, K., Muñoz, J.-A., García-Senz, J. (2004). Extensional salt tectonics in a contractional orogen: A newly identified tectonic event in the Spanish Pyrenees. *Geology*, 32, 737–740. https://doi.org/10.1130/G20565.1.

McKenzie, D. (1978). Some remarks on the development of sedimentary basins. *Earth Planet. Sci. Lett.*, 40, 25–32. https://doi.org/10.1016/0012-821X(78)90071-7.

Michon, L. and Merle, O. (2003). Mode of lithospheric extension: Conceptual models from analogue modeling. *Tectonics*, 22(4). https://doi.org/10.1029/2002TC001435.

Mohn, G., Manatschal, G., Müntener, O., Beltrando, M., Masini, E. (2010). Unravelling the interaction between tectonic and sedimentary processes during lithospheric thinning in the Alpine Tethys margins. *Int. J. Earth Sci.*, 99, 75–101. https://doi.org/10.1007/s00531-010-0566-6.

Mohn, G., Manatschal, G., Beltrando, M., Masini, E., Kusznir, N. (2012). Necking of continental crust in magma-poor rifted margins: Evidence from the fossil Alpine Tethys margins. *Tectonics*, 31, 1012–1040. https://doi.org/10.1029/2011TC002961.

Mohn, G., Karner, G.D., Manatschal, G., Johnson, C.A. (2015). Structural and stratigraphic evolution of the Iberia–Newfoundland hyper-extended rifted margin: A quantitative modelling approach. *Geol. Soc. Lond. Spec. Publ.*, 413, 53–89. https://doi.org/10.1144/SP413.9.

Mouthereau, F., Filleaudeau, P.Y., Vacherat, A., Pik, R., Lacombe, O., Fellin, M.G., Castelltort, S., Christophoul, F., Masini, E. (2014). Placing limits to shortening evolution in the Pyrenees: Role of margin architecture and implications for the Iberia/Europe convergence. *Tectonics*, 33, 2283–2314. https://doi.org/10.1002/2014TC003663.

Muñoz, J.A. (1992). Evolution of a continental collision belt: ECORS-Pyrenees crustal balanced cross-section. In *Thrust Tectonics*, McClay, K.R., Chapman & Hall, London.

Nagel, T.J. and Buck, W.R. (2004). Symmetric alternative to asymmetric rifting models. *Geology*, 32(11), 937–940. https://doi.org/10.1130/G20785.1.

Odlum, M. and Stockli, D.F. (2020). Geochronologic constraints on deformation and metasomatism along an exhumed mylonitic shear zone using apatite U-Pb, geochemistry, and microtextural analysis. *Earth and Planetary Science Letters*, 536, 116–177. https://doi.org/10.1016/j.epsl.2020.116177.

Olivet, J.-L. (1996). La cinématique de la plaque Ibérique. *Bull. Cent. Rech. Explor.-Prod. Elf- Aquitaine*. 20, 131–195.

Ortí, F., Pérez-López, A., Salvany, J.M. (2017). Triassic evaporites of Iberia: Sedimentological and palaeogeographical implications for the western Neotethys evolution during the Middle Triassic–Earliest Jurassic. *Palaeogeogr. Palaeoclimatol. Palaeoecol.*, 471, 157–180. https://doi.org/10.1016/j.palaeo.2017.01.025.

Osmundsen, P.T. and Ebbing, J. (2008). Styles of extension offshore mid-Norway and implications for mechanisms of crustal thinning at passive margins. *Tectonics*, 27, https://doi.org/10.1029/2007TC002242.

Osmundsen, P.T. and Peron-Pinvidic, G. (2018). Crustal-scale fault interaction at rifted margins and the formation of domain-bounding breakaway complexes: Insights from offshore Norway. *Tectonics* 37, 935–964. https://doi.org/10.1002/2017TC004792.

Pedrera, A., García-Senz, J., Ayala, C., Ruiz-Constán, A., Rodríguez-Fernández, L.R., Robador, A., González Menéndez, L. (2017). Reconstruction of the exhumed mantle across the North Iberian Margin by crustal-scale 3-D gravity inversion and geological cross section. *Tectonics*, 36, 3155–3177. https://doi.org/10.1002/2017TC004716.

Peron-Pinvidic, G. and Manatschal, G. (2009). The final rifting evolution at deep magma-poor passive margins from Iberia-Newfoundland: A new point of view. *Int. J. Earth Sci.*, 98, 1581–1597. https://doi.org/10.1007/s00531-008-0337-9.

Peron-Pinvidic, G., Manatschal, G., Minshull, T.A., Sawyer, D.S. (2007). Tectonosedimentary evolution of the deep Iberia-Newfoundland margins: Evidence for a complex breakup history. *Tectonics*, 26, 1–19. https://doi.org/10.1029/2006TC001970.

Pitman III, C. and Talwani, M. (1972). Sea-floor spreading in the North Atlantic. *Geol. Soc. Am. Bull.*, 83, 619–646.

Quesnel, B., Boiron, M.C., Cathelineau, M., Truche, L., Rigaudier, T., Bardoux, G., Agrinier, P., de Saint Blanquat, M., Masini, E., Gaucher, E. (2019). Nature and origin of mineralizing fluids in hyperextensional systems: The case of Cretaceous Mg metasomatism in the Pyrenees. *Geofluids*, 18. https://doi.org/10.1155/2019/7213050.

Rat, J., Mouthereau, F., Brichau, S., Crémades, A., Bernet, M., Balvay, M., Ganne, J., Lahfid, A., Gautheron, C. (2019). Tectonothermal evolution of the Cameros Basin: Implications for Tectonics of North Iberia. *Tectonics*, 38, 440–469. https://doi.org/10.1029/2018TC005294.

Reston, T.J. (1988). Evidence for shear zones in the lower crust offshore Britain. *Tectonics*, 7(5), 929–945.

Reston, T.J. and Pérez-Gussinyé, M. (2007). Lithospheric extension from rifting to continental breakup at magma-poor margins: Rheology, serpentinisation and symmetry. *Int. J. Earth Sci.*, 96, 1033–1046. https://doi.org/10.1007/s00531-006-0161-z.

Reston, T.J., Krawczyk, C.M., Hoffmann, H.J. (1995). Detachment tectonics during Atlantic rifting: Analysis and interpretation of the S reflection, the west Galicia margin. *Geol. Soc. Lond. Spec. Publ.*, 90, 93–109. https://doi.org/10.1144/GSL.SP.1995.090.01.05.

Roca, E., Muñoz, J.A., Ferrer, O., Ellouz, N. (2011). The role of the Bay of Biscay Mesozoic extensional structure in the configuration of the Pyrenean orogen: Constraints from the MARCONI deep seismic reflection survey. *Tectonics*, 30. https://doi.org/10.1029/2010TC002735.

Roure, F. and Choukroune, P. (1998). Contribution of the ECORS seismic data to the Pyrenean geology: Crustal architecture and geodynamic evolution of the Pyrenees. *Mém. Soc. Géol. Fr.*, 173, 37–52.

Roure, F., Choukroune, P., Berastegui, X., Muñoz, J.A., Villien, A., Matheron, P., Bareyt, M., Seguret, M., Camara, P., Deramond, J. (1989). ECORS deep seismic data and balanced cross sections; geometric constraints on the evolution of the Pyrenees. *Tectonics*, 8, 41–50. https://doi.org.10.1029/TC008i001p00041.

Rowan, M.G. (2014). Passive-margin salt basins: Hyperextension, evaporite deposition, and salt tectonics. *Basin Res.*, 26, 154–182. https://doi.org/10.1111/bre.12043.

de Saint Blanquat, M., Bajolet, F., Grand'Homme, A., Proietti, A., Zanti, M., Boutin, A., Clerc, C., Lagabrielle, Y., Labaume, P. (2016). Cretaceous mantle exhumation in the central Pyrenees: New constraints from the peridotites in eastern Ariège (North Pyrenean zone, France). *C. R. Geosci.*, 348, 268–278. https://doi.org/10.1016/j.crte.2015.12.003.

Saspiturry, N., Razin, P., Baudin, T., Serrano, O., Issautier, B., Lasseur, E., Allanic, C., Thinon, I., Leleu, S. (2019). Symmetry vs. asymmetry of a hyper-thinned rift: Example of the Mauléon Basin (Western Pyrenees, France). *Mar. Pet. Geol.*, 104, 86–105. https://doi.org/10.1016/j.marpetgeo.2019.03.031.

Saspiturry, N., Lahfid, A., Baudin, T., Guillou–Frottier, L., Razin, P., Issautier, B., Le Bayon, B., Serrano, O., Lagabrielle, Y., Corre, B. (2020). Paleogeothermal gradients across an inverted hyperextended rift system: Example of the Mauléon fossil rift (Western Pyrenees). *Tectonics*, 39. https://doi.org/10.1029/2020TC006206.

Saspiturry, N., Issautier, B., Razin, P., Baudin, T., Asti, R., Lagabrielle, Y., Duretz, T., Allanic, C., Serrano, O. (2021). Review of Iberia-Eurasia plate-boundary basins: Role of sedimentary burial on depth-dependent continental crust ductile thinning during rifting and continental breakup. *Basin Res.*, 00, 1–36. https://doi.org/10.1111/bre.12529.

Saura, E., Ardèvol i Oró, L., Teixell, A., Vergés, J. (2016). Rising and falling diapirs, shifting depocenters, and flap overturning in the Cretaceous Sopeira and Sant Gervàs subbasins (Ribagorça Basin, southern Pyrenees): Southern Pyrenees Cretaceous diapirism. *Tectonics*, 35, 638–662. https://doi.org/10.1002/2015TC004001.

Schärer, U., de Parseval, P., Polvé, M., de St Blanquat, M. (1999). Formation of the Trimouns talc-chlorite deposit (Pyrenees) from persistent hydrothermal activity between 112 and 97 Ma. *Terra Nova*, 11, 1, 30–37. https://doi.org/10.1046/j.1365-3121.1999.00224.x.

Sibuet, J.-C., Srivastava, S.P., Spakman, W. (2004). Pyrenean orogeny and plate kinematics. *J. Geophys. Res. Solid Earth*, 109, B08104. https://doi.org/10.1029/2003JB002514.

Soto, J.I., Flinch, J.F., Tari, G. (2017). Permo-Triassic salt provinces of Europe, North Africa and the Atlantic margins: A synthesis. In *Permo-Triassic Salt Provinces of Europe, North Africa and the Atlantic Margins: Tectonics and Hydrocarbon Potential*, Soto, J.I., Flinch, J., Tari, G. (eds.). Elsevier, Amsterdam.

Souquet, P., Peybèrnes, B., Bilotte, M., Debroas E.-J. (1977). La chaine alpine des Pyrénées. *Géologie alpine*, 53(2), 193–216.

Srivastava, S.P., Roest, W.R., Kovacs, L.C., Oakey, G., Levesque, S., Verhoef, J., Macnab, R., (1990). Motion of Iberia since the Late Jurassic: Results from detailed aeromagnetic measurements in the Newfoundland Basin. *Tectonophysics*, 184, 229–260.

Srivastava, S.P., Sibuet, J.C., Cande, S., Roest, W.R., Reid, I.D. (2000). Magnetic evidence for slow seafloor spreading during the formation of the Newfoundland and Iberian margins. *Earth Planet. Sci. Lett.*, 182, 61–76.

Sutra, E., Manatschal, G., Mohn, G., Unternehr, P. (2013). Quantification and restoration of extensional deformation along the Western Iberia and Newfoundland rifted margins. *Geoch. Geoph. Geosys.*, 14(8), 2575–2597. https://doi.org/10.1002/ggge.20135.

Teixell, A. (1998). Crustal structure and orogenic material budget in the west central Pyrenees. *Tectonics*, 17, 395–406. https://doi.org/10.1029/98TC00561.

Teixell, A., Labaume, P., Lagabrielle, Y. (2016). The crustal evolution of the west-central Pyrenees revisited: Inferences from a new kinematic scenario. *C. R. Geosci.*, 348, 257–267. https://doi.org/10.1016/j.crte.2015.10.010.

Teixell, A., Labaume, P., Ayarza, P., Espurt, N., de Saint Blanquat, M., Lagabrielle, Y. (2018). Crustal structure and evolution of the Pyrenean-Cantabrian belt: A review and new interpretations from recent concepts and data. *Tectonophysics*, 724, 146–170. https://doi.org/10.1016/j.tecto.2018.01.009.

Thiébault, J., Durand-Wackenheim, C., Debeaux, M., Souquet, P. (1992). Métamorphisme des évaporites triasiques du versant nord des Pyrénées centrales et occidentales. *Bull. Soc. Hist. Nat. Toulouse*, 128, 77–84.

Tugend, J., Manatschal, G., Kusznir, N.J., Masini, E., Mohn, G., Thinon, I. (2014). Formation and deformation of hyperextended rift systems: Insights from rift domain mapping in the Bay of Biscay-Pyrenees. *Tectonics*, 33, 1239–1276. https://doi.org/10.1002/2014TC003529.

Vergés, J. and García-Senz, J. (2001). Mesozoic evolution and Cainozoic inversion of the Pyrenean rift. *Mém. Muséum Natl. Hist. Nat.*, 186, 187–212.

Vergés, J., Fernàndez, M., Martínez, A. (2002), The Pyrenean orogen: Pre-, syn-, and post-collisional evolution. *J. Virtual Explor.*, 8, 55–74. https://doi.org/10.3809/jvirtex.2002.00058.

Vielzeuf, D. and Kornprobst, J. (1984). Crustal splitting and the emplacement of Pyrenean lherzolites and granulites. *Earth Planet. Sci. Lett.*, 67, 87–96. https://doi.org/10.1016/0012-821X(84)90041-4.

Vissers, R.L.M. and Meijer, P.T. (2012). Mesozoic rotation of Iberia: Subduction in the Pyrenees? *Earth Science Reviews*, 110, 93–110. https://doi.org/10.1016/j.earscirev.2011.11.001.

Wang, Y., Chevrot, S., Monteiller, V., Komatitsch, D., Mouthereau, F., Manatschal, G., Sylvander, M., Diaz, J., Ruiz, M., Grimaud, F., Benahmed, S., Pauchet, H., Martin, R. (2016). The deep roots of the western Pyrenees revealed by full waveform inversion of teleseismic P waves. *Geology*, 44, 475–478. https://doi.org/10.1130/G37812.1.

Wernicke, B. (1981). Low-angle normal faults in the Basin and Range Province: Nappe tectonics in an extending orogen. *Nature*, 291, 645–648. https://doi.org/10.1038/291645a0.

Wernicke, B. (1985). Uniform-sense normal simple shear of the continental lithosphere. *Can. J. Earth Sci.*, 22, 108–125. https://doi.org/10.1139/e85-009.

Zamora, G., Fleming, M., Gallastegui, J. (2017). Salt tectonics within the offshore Asturian Basin: North Iberian Margin. In *Permo-Triassic Salt Provinces of Europe, North Africa and the Atlantic Margins: Tectonics and Hydrocarbon Potential*, Soto, J.I, Flinch, J.F., Tari, G. (eds). Elsevier, Amsterdam.

Ziegler, P.A. (1988). Evolution of Arctic-North Atlantic and Western Tethys, Tulsa. *AAPG Memoir*, 43, 198.

List of Authors

Torgeir Bjørge ANDERSEN
Centre for Earth Evolution and Dynamics
Department of Geosciences
University of Oslo
Norway

Daniel ASLANIAN
Ifremer
REM/GM/LGS
Plouzané
France

Sung-Ping CHANG
Laboratoire de Géologie de l'ENS
CNRS UMR 8538
PSL Research University
Paris
France

Pauline CHENIN
Université de Strasbourg
CNRS
ITES UMR 7063
Strasbourg
France

Matthias DELESCLUSE
Laboratoire de Géologie de l'ENS
CNRS UMR 8538
PSL Research University
Paris
France

Carmen GAINA
Centre for Earth Evolution and Dynamics
Department of Geosciences
University of Oslo
Norway

Johannes JAKOB
Geological Survey of Norway (NGU)
Trondheim
Norway

Hans Jørgen KJØLL
Section for Geology and Geophysics
Department of Geosciences
University of Oslo
Norway

Yves LAGABRIELLE
CNRS-Géosciences Rennes
University of Rennes 1
France

Sylvie LEROY
Sorbonne Université
CNRS-INSU
Institut des Sciences de la Terre
ISTeP
Paris
France

Gaël LYMER
Fault Analysis Group/iCRAG
(Irish Centre for Research in
Applied Geosciences)
School of Earth Sciences
University College Dublin
Ireland

Gianreto MANATSCHAL
Université de Strasbourg
CNRS
ITES UMR 7063
Strasbourg
France

Maryline MOULIN
Ifremer
REM/GM/LGS
Plouzané
France

Chloé NONN
Sorbonne Université
CNRS-INSU
Institut des Sciences de la Terre
ISTeP
Paris
France

Per Terje OSMUNDSEN
Department of Geoscience and Petroleum
Norwegian University of Science
and Technology
Trondheim
and
Department of Geosciences
University of Oslo
Norway

Gwenn PERON-PINVIDIC
Geological Survey of Norway (NGU)
and
Department of Geoscience and Petroleum
Norwegian University of Science
and Technology
Trondheim
Norway

Manuel PUBELLIER
Laboratoire de Géologie de l'ENS
CNRS UMR 8538
PSL Research University
Paris
France

Tim J. RESTON
School of Geography, Earth and
Environmental Science
University of Birmingham
UK

Index

A, B, C

Afar mantle plume, 93, 94, 96, 103
Angola, 73, 86, 220
Arctic Ocean, 133, 143
asymmetry, 13, 20, 58, 98, 102
basin(s), 36, 40, 45, 47–49, 51–54, 56, 57, 59, 133, 134, 136, 139, 141–144, 203–213, 215–220, 222, 223
  Amundsen, 136, 139
  Canada, 134, 136–138, 141, 142, 144
  Eurasia, 133, 134, 136, 138–142, 144
  Makarov, 134, 136, 141, 142
  Nansen, 136, 139
  Podvodnikov, 136, 142
  smooth-slope, 207–209, 211, 212, 215, 217, 219
Brazil, 73, 83
breakaway complexes, 36, 39, 40, 42, 44–50, 54
Central Iapetus Magmatic Province (CIMP), 185, 187, 188, 191, 193, 194
continental
  breakup, 141, 143, 144
  domain, 98, 102
corrugations, 10, 17
Cretaceous, 203–210, 215, 220, 223
crustal reactivation, 116

D, E, F

data
  aerogeophysical, 139
  magnetic, 139
  seismic, 9, 24, 25
detachment, 1, 3, 5, 9, 12, 14, 15, 18, 21–24, 26
  faults, 102
  systems, 155, 159, 161
ductile stretching, 207, 211, 213, 215–217, 219, 220
dyke emplacement, 185, 187, 190, 192–194
Ediacaran, 185, 188, 191, 193, 194
Europa, 209
evaporites, 208, 217–219, 222
faults, 35, 39, 40, 43, 45–48, 50, 52–59

G, H, I

Galicia, 2, 5, 6, 9, 12–17, 21, 22, 25–32
hyperextension, 24, 35, 39, 42, 48, 49, 53, 54
Iapetus, 185–188, 194
Iberia, 1, 2, 4–6, 9, 12–15, 18, 21, 22, 26–34, 203, 204, 206, 209, 217–219, 222

M, N

magma
-poor, 2, 5, 6, 18, 19, 21, 22, 24
-rich, 185–188, 193, 194
magmatism, 93, 98, 102, 103
mantle exhumation, 15, 25, 137, 141, 142, 157, 159, 160, 165, 166, 168, 169, 172, 173
margin domains, 35, 45
marginal basin, 107, 109, 110, 113, 118, 122, 123
necking domain, 98
Neoproterozoic, 187, 188
North Slope, 134, 141
Norway, 185, 188, 190
Norwegian margin, 35–40, 44, 49–51, 54, 57–59

O, P, R

ocean-continent transition (OCT), 96, 99–102
oceanic domain, 100, 102, 103
Pacific slab, 141
plate boundary, 209
reconstruction, 70, 74, 76–79, 82, 84, 144
ridge, 136, 139
Alpha-Mendeleev, 136, 142
Gakkel, 139, 143
mid-ocean, 134, 136, 137, 139, 143
Northwind, 136, 137, 141
rifting, 107, 108, 110, 111, 113, 115, 117–119, 121, 122, 203–207, 215, 216, 219, 220, 223

S, T, V

Scandinavian
Caledonides, 185–189, 191, 193, 194
Dyke Complex (SDC), 186–188, 191, 194
scientific drilling, 13, 18, 25
segmentation, 94
Seiland Igneous Province (SIP), 186, 188–190, 193, 194
serpentinization, 3, 5, 9, 12, 15, 17, 18, 20, 23–25, 30–32
South
Atlantic, 69, 72–74, 76–79, 82–84
China Sea, 107–113, 115, 116, 118, 119, 121–124
tectonics, 35, 52–54, 56
V-shape geometry, 107, 108, 113, 115, 119

Summary of Volume 1

Preface

Gwenn PERON-PINVIDIC

Part 1. Rifting and Rifted Margins: Definitions

Chapter 1. What is Rifting? Introduction and Basic Definitions

Gwenn PERON-PINVIDIC

1.1. Introduction
1.2. Rift classifications
  1.2.1. Rift mechanism classification: active versus passive rifting
  1.2.2. Plate tectonic setting classification
1.3. Structural features associated with continental rifting
  1.3.1. Extensional mechanisms
  1.3.2. Main structural geometries
  1.3.3. Main basin types
1.4. Subsidence
1.5. Rift-related materials
  1.5.1. Sediments
  1.5.2. Salt
  1.5.3. Magma
1.6. References

Chapter 2. What is a Rifted Margin? From the Early Models to Modern Views and Future Challenges

Gwenn PERON-PINVIDIC

2.1. What is interesting about studying rifted margins?
2.2. Morphological definition
2.3. Classification schemes for rifted margins

2.3.1. Classifications based on the extensional context: the divergent, sheared and transtensional rifted margins
2.3.2. Classifications based on the amount of magma involved: the volcanic and non-volcanic rifted margins
2.3.3. Classifications based on sediment volume: sediment-rich and sediment-starved rifted margins
2.4. Historical review of how our understanding of rifted margins evolved
2.4.1. The pioneering models
2.4.2. The evolution of knowledge
2.5. Continental rifted margins
2.5.1. Structural domains
2.5.2. The 3D perspective
2.5.3. The parameters influencing rift evolution
2.6. Future research
2.6.1. Question 1: What is early rifting?
2.6.2. Question 2: What is final rifting?
2.6.3. Question 3: What is breakup?
2.6.4. Question 4: What is the COB?
2.6.5. Question 5: Why are rifted margins segmented?
2.7. References

Part 2. How Do We Study Rifted Margins?

Summary of Contributions

Chapter 3. How Can We Identify and Study Remnants of Rifted Margins in Orogens?

Geoffroy MOHN, Julie TUGEND, Benoît PETRI, Anders MCCARTHY and Dominique FRIZON DE LAMOTTE

3.1. Introduction
3.2. Brief considerations on orogens
3.2.1. Structure of an orogen
3.2.2. The significance of Ophiolitic assemblages
3.2.3. Classification and types of orogens
3.3. Identifying fossil-rifted margins in orogenic systems: principles and criteria
3.3.1. Identification of coherent tectonic units
3.3.2. Sedimentary/stratigraphic record and depositional environments
3.3.3. Basement lithology and evolution
3.3.4. Rift-related deformation style: classical rift geometries versus supra-detachment basins
3.4. Application to case studies

3.4.1. Recognition of a proximal domain: the case study of the Dauphinois Unit in the Western Alps
3.4.2. Recognition of necking zones: the case studies of the Campo-Grosina Units (Alps) and Southern part of Mauléon Basin (Pyrenees)
3.4.3. Recognition of hyper-extended domains: the case studies of the Senadhja Nappe (Rif belt)
3.5. Conclusion
3.6. Acknowledgments
3.7. References

Chapter 4. What Can We Learn from Marine Geophysics to Study Rifted Margins?

Julia AUTIN and Louise WATREMEZ

4.1. Geophysical methods
4.1.1. Seismic imaging
4.1.2. Potential field methods
4.2. Understanding continental rifted margins using geophysics
4.2.1. Geological objects interpretation
4.2.2. Data complementarity
4.3. Conclusion
4.4. Acknowledgment
4.5. References

Chapter 5. Numerical Modeling of Rifting: An Overview

Marta PÉREZ-GUSSINYÉ and Zhonglan LIU

5.1. Introduction
5.2. A brief historical introduction to numerical modeling of rifting
5.3. Dynamic modeling: description and formulation
5.3.1. Eulerian and Lagrangian frameworks
5.3.2. Governing equations
5.3.3. Rheological models
5.3.4. Strain softening
5.3.5. Initial and boundary conditions
5.3.6. Initial weaknesses
5.4. Dynamic modeling: applications
5.4.1 Margin width and extension mode
5.4.2. Normal fault geometry
5.4.3. Symmetric and asymmetric margins
5.4.4. Melting and serpentinization
5.4.5. Surface processes

5.5. Summary and future perspectives
5.6. References

Chapter 6. Analogue Modeling of Continental Rifting: An Overview

Frank ZWAAN and Guido SCHREURS

6.1. Introduction
6.2. Methodology
6.2.1. Scaling principles
6.2.2. Materials
6.2.3. General experimental setups and boundary conditions
6.2.4. Model analysis techniques
6.3. Model application
6.3.1. A 2D perspective on rifting
6.3.2. Exploring 3D rift processes
6.4. Summary, challenges and future opportunities
6.5. Acknowledgments
6.6. References

Printed and bound by CPI Group (UK) Ltd, Croydon, CR0 4YY

13/09/2026

14960994-0003